Human–Centric AI in Digital Transformation and Entrepreneurship

Sanjay Misra
Institute for Energy Technology, Halden, Norway

Manju Kaushik
Amity University, Jaipur, India

Amit Jain
Amity University, Jaipur, India

Chitresh Banerjee
Amity University, Jaipur, India

IGI Global
Scientific Publishing
Publishing Tomorrow's Research Today

Vice President of Editorial	Melissa Wagner
Managing Editor of Acquisitions	Mikaela Felty
Managing Editor of Book Development	Jocelynn Hessler
Production Manager	Mike Brehm
Cover Design	Phillip Shickler

Published in the United States of America by
IGI Global Scientific Publishing
701 East Chocolate Avenue
Hershey, PA, 17033, USA
Tel: 717-533-8845
Fax: 717-533-8661
Website: https://www.igi-global.com E-mail: cust@igi-global.com

Library of Congress Cataloging-in-Publication Data
Names: Misra, Sanjay, 1971- editor. | Kaushik, Manju, 1979- editor. | Jain,
  Amit, 1977- editor. | Banerjee, Chitresh, editor.
Title: Human-centric AI in digital transformation and entrepreneurship /
  edited by Sanjay Misra, Manju Kaushik, Amit Jain, Chitresh Banerjee.
Description: Hershey, PA : IGI Global, [2025] | Includes bibliographical
  references and index. | Summary: "The increasing integration of IT into
  society and various industries, coupled with the economic convergence of
  entrepreneurship and innovation, necessitates a comprehensive and
  authoritative research book"-- Provided by publisher.
Identifiers: LCCN 2024051856 (print) | LCCN 2024051857 (ebook) | ISBN
  9798369380093 (hardcover) | ISBN 9798369380109 (paperback) | ISBN
  9798369380116 (ebook)
Subjects: LCSH: Entrepreneurship--Technological innovations. | Sustainable
  development.
Classification: LCC HB615 .H855 2025 (print) | LCC HB615 (ebook) | DDC
  338/.04028563--dc23/eng/20241121
LC record available at https://lccn.loc.gov/2024051856
LC ebook record available at https://lccn.loc.gov/2024051857

British Cataloguing in Publication Data
A Cataloguing in Publication record for this book is available from the British Library.

Table of Contents

Detailed Table of Contents

Pallavi Mishra, Amity University, Jaipur, India
Tanushri Mukherjee, Amity University, Jaipur, India
Neharshi Srivastava, Amity University, Jaipur, India

Digital entrepreneurship plays a crucial role in the economic development and innovation landscape of India, propelled by advancements in digital technologies and the widespread availability of the Internet. The chapter addresses a significant research gap by performing a bibliometric analysis of digital entrepreneurship in India, utilizing the Scopus database for the period from 2012 to 2024. By focusing on keywords such as 'digital entrepreneurship' and 'India,' it gathers metadata related to publication titles, authors, dates, journals, and keywords to discern scholarly trends. The analysis delves into academic discussions, revealing patterns in publication, citation networks, authorship, and thematic interests. By employing analytical tools such as R Studio and VOSviewer, it charts the developing research landscape, emphasizing key contributors and emerging themes. This offers a thorough overview of the scholarship surrounding digital entrepreneurship in India and identifies critical areas for future inquiry, providing essential insights into the evolving domain.

Garima Pancholi, Amity University, Jaipur, India
Monika Sharma, Amity University, Jaipur, India
Abhineet Saxena, Amity University, Jaipur, India
Mukesh Kumar Verma, University of Rajasthan, India

Metaverse has gone the extra mile from its origin as a gaming platform to a dynamic digital economy where the users can create, purchase and sell products and services through their avatars. The retailers are increasing brand engagement and awareness by enhancing product visualizations, enabling highly personalized marketing strategies such as virtual try-ons in the virtual world. Metaverse has led to a paradigm shift that can leave behind those who lag. Hence, it is important to consider progress and foresee future growth. The bibliometric analysis was conducted to get a comprehensive awareness of consumer behavior and consumer engagement in Metaverse. This study examines the research landscape of 274 articles retrieved from Scopus database from 1997 to 2024 to achieve our research objectives. The findings of the study can assist the future researchers in getting in-depth understanding of current trends and themes while identifying the research gap, providing valuable insights for future study directions.

Sumit Verma, Chandigarh University, India
Amit Jain, Amity University, Jaipur, India
Garima Pancholi, Amity University, Jaipur, India
Abhineet Saxena, Amity University, Jaipur, India

The widespread adoption of E-commerce and digital platforms has led to cultural shift towards sharing experiences and feedback as a form of social proof to increase sales. Electronic word of mouth (eWOM) is a significant factor in shaping customer perception, reducing perceived risk and enhancing customer satisfaction while making purchase decisions. The purpose of conducting the bibliometric analysis on eWOM and customer satisfaction is to capture knowledge from published articles to identify trends, themes and gaps. Using Scopus database, a comprehensive analysis was conducted on 265 documents which satisfied the inclusion/exclusion criteria for the period between 2002 to 2024. The analysis focused on the contribution of academic, country, sources and top cited articles and Network analysis for cluster analysis to identify different themes. This research is valuable for assessing the significance, influence, and evolution of research across different fields, providing a quantitative basis for evaluating scientific contributions and guiding future research agendas.

Amit Chaurasia, Amity University, Jaipur, India
Umesh kumar Dwivedi, Amity University, Jaipur, India
Amita Chaurasiya, ICFAI University, Jaipur, India

Tools have a major impact on the precision, effectiveness, and cost of manufacturing composite structures. In order to provide lightweight, affordable, long-lasting, and useful tools for producing high-performance composite structures, industry and academia must work closely together to leverage and investigate new developments in materials, structures, and tool manufacturing technologies. This is because the performance of composites components is becoming more and more desired. It is frequently problematic to provide insulation at higher voltages (kV range). In this work, we created a variety of epoxy resin samples utilising nickel and silicon carbide, two conducting and non-conducting fillers. Using the centrifugal casting procedure, samples were created. We examined the composite's breakdown voltage, which will eventually offer insulation at greater voltages. The sample with a higher filler concentration had a good breakdown voltage, according to the data.

Chapter 5
    Neharshi Srivastava, Amity University, Jaipur, India
    Mani Sachdev, Amity University, Jaipur, India
    Monika Gwalani, Amity University, Jaipur, India
    Meenakshi Bajpai, Arya Mahila PG College, Banaras Hindu University,
        India
    Saad Ullah Khan, Centre for Media and Mass Communiction Studies,
        Jamia Hamdard, New Delhi, India
    Mukta Arora, UN Women ICO, India
    Pranshuta Arora, UNICEF, India

This research paper, "Women-Centric AI: A Step Towards Empowerment," reveals how artificial intelligence can be used in a positive way toward gender equity and greater inclusivity. The critical focus of the study addresses intrinsic bias in the contemporary functioning of AI systems, with particular emphasis on women being the victims of the discriminatory behavior of such AI-based systems. It states that, for the sake of building women-centric AI systems, women's concerns in designing such systems need to be integrated so they remain proactive creators and decision-makers. Qualitative methodology and thematic analysis of expert interviews identify defining features of women-centric AI that can be adapted to meet the challenges specific to women. Targeted AI solutions can drastically reduce the digital gender gap and offer more tailored support in health, education, and employment. Ethical considerations include diverse datasets and gender impact assessments for equitable AI development.

Chapter 6

    Amit Kumar Singh, Amity University, Jaipur, India
    Swapnesh Taterh, Amity University, Jaipur, India
    Arti Kaushik, G.V.M College of Education, Sonepat, India
    Avinash Goswami, The LNM Institute of Information Technology,
       Jaipur, India

This chapter explores the transformative role of Artificial Intelligence (AI) in driving digital transformation across various industries. As organizations navigate the complexities of the digital age, AI emerges as a critical tool for enhancing operational efficiency, improving customer experiences, and fostering innovation. The integration of AI technologies enables businesses to automate routine tasks, analyze vast datasets for actionable insights, and deliver personalized services at scale. However, the adoption of AI is not without challenges; issues such as data quality, ethical considerations, and workforce displacement pose significant hurdles. This paper discusses best practices for AI integration, emphasizing the importance of establishing clear objectives, investing in high-quality data, and prioritizing transparency and ethical governance. Additionally, future trends in AI, including hyper-automation, edge computing, and AI-driven sustainability, are examined as organizations seek to leverage these advancements for competitive advantage.

Chapter 7

    Gunjan Verma, Amity Institute of Information Technology, Amity
       University, Jaipur, India
    Honey Gocher, Amity University, Jaipur, India
    Sweety Verma, KITECH, South Korea
    Yudhveer Singh, Amity Institute of Information Technology, Amity
       University, Jaipur, India
    Arti Kaushik, G.V.M College of Education, Sonepat, India
    Avinash Goswami, The LNM Institute of Information Technology,
       Jaipur, India

Deep learning-based audio classification has transformed the industry with improved speech recognition, genre identification in music, and ambient sound detection. The article explores various approaches, including model architectures, evaluation metrics, and preprocessing techniques. Traditional methods are compared to deep learning techniques, which have enhanced performance. Spectrograms, Mel-Frequency Cepstral Coefficients, and Short-Time Fourier Transform are discussed as preprocessing techniques. The study also evaluates hybrid model architectures, training methods, data augmentation, and transfer learning for better outcomes. The paper emphasises the importance of interpretability, stable datasets, and real-time processing for overcoming challenges in audio classification. It is expected to guide future research and advancements in this field.

    Suyesha Singh, Manipal University Jaipur, India
    Paridhi Jain, Manipal University Jaipur, India

The advancement of brain organoids has revolutionized the fields of biotechnology and neuroscience. Brain organoids are the 3-D structures of the human brain developed from the human's pluripotent cells, which mimic the functionality of the human brain. Organoid intelligence cumulates brain organoids and artificial intelligence, making it possible to develop a biohybrid system with heightened cognitive capacities. This paper aims to critically analyze concerns surrounding brain organoids and organoid intelligence and highlight the scientific progress and ethical considerations, ensuring that new technologies contribute to the community while maintaining fundamental moral principles. The findings of the current work will be useful for decision makers, neuroscientists, ethicists, and the public to promote environmentally conscious innovation and practice organoid intelligence in an ethical way. This work serves as a call to action for addressing ongoing challenges, advancing the field, and utilizing it effectively in the current context.

    Revti Rani Roy, Chanakya National Law University, India
    Ajay Kumar, Chanakya National Law University, India
    Manju Kaushik, Amity University, Jaipur, India
    Aashish Goswami, Amity University Rajasthan, Jaipur, India

There is no shame in acknowledging that profiteering through wealth generation is the most vicious motivator to satiate entrepreneurial aspirations. However, in this globalizing world where resource availability is faced with problem of resource exploitation, whilst economic development bears the cost of climate change, there is a need for a structural change in business behavior to encourage sustainability. The study examines how the state-of-the-art entrepreneurial ventures are insufficient to address the concerns of green economy; and how it can be transformed through simulation of eco-efficient innovation techniques by synergizing determinants of sustainability. The study is an exploration into spectrum of issues surrounding green entrepreneurship and how R&D and regulatory regimes in eco-innovation can contribute to a holistic solution. Using entrepreneurial Responsibility as a variable for sustainable entrepreneurship besides policy interventions- instilling entrepreneurial motivations for firms to go green, will surely stimulate a green economy driven by ethical ecopreneurs.

Chapter 10

Monica Kunte, Symbiosis Centre for Management and Human Resource Development, Symbiosis International University, India
Joseph Mathew, Symbiosis Centre for Management and Human Resource Development, Symbiosis International University, India
Aman Choudhary, Symbiosis Centre for Management and Human Resource Development, Symbiosis International University, India

The research study aims to explore significance of microlearning in modern training initiatives. It seeks to investigate the benefits of microlearning in terms of employee engagement and its effectiveness in achieving concise training targets. This study employs a quantitative as well as qualitative research design to investigate the effectiveness, adaptation and challenges of microlearning in IT sector. Thematic exploration has been utilized to analyse the findings from in depth interviews with Learning and Development Managers and HR managers. The findings suggest that delivering targeted and personalized learning experiences, microlearning can keep employees motivated and continuously updated with relevant skills. The study's originality lies in its in-depth analysis of microlearning's adaptation, challenges, and the potential path ahead in the context of modern organizational training. The research provides managers with valuable insights and practical implications for incorporating microlearning into their organization's learning and development initiatives.

Chapter 11

Rachit Agarwal, Chandigarh University, Mohali, India
Tanya Kumar, Chandigarh Business School of Administration, Mohali, India
Ramneek Ahluwalia, Chandigarh University, Mohali. India

In the evolving landscape of project management, integrating Artificial Intelligence (AI) offers entrepreneurs unprecedented opportunities to enhance operational efficiency, mitigate risks, and achieve strategic goals. This book chapter explores key strategies to optimise return on investment (ROI) through the cost-effective implementation of AI in project management. The case study approach represents how entrepreneurs can achieve flexibility and cost savings by adopting cloud-based AI services for project planning, resource management, and performance analysis.

Furthermore, the chapter explores the role of predictive analytics and machine learning algorithms in optimizing project outcomes and minimizing operational expenses. Moreover, the chapter explores emerging trends and future directions in AI-driven project management, including advancements in natural language processing and autonomous project management systems. In conclusion, this book chapter provides entrepreneurs with practical strategies for achieving cost-effective implementation of AI in project management.

Chapter 12

Antima Sharma, Manipal University Jaipur, India
Vertika Goswami, Manipal University Jaipur, India
Durgesh Batra, Manipal University Jaipur, India
Preeti Nagar, Manipal University Jaipur, India
Arpita Agarwal, Manipal University Jaipur, India
Anadi Trikha, Manipal University Jaipur, India

Corporate Social Responsibility (CSR) has become crucial in modern business due to its potential to drive economic development. This systematic review analyzes the relationship between CSR initiatives and economic development by reviewing empirical studies, theoretical frameworks, and case analyses. It explores how CSR practices, such as addressing climate change and social inequality, contribute to economic growth by enhancing reputation, improving stakeholder relations, and attracting investment. The review identifies emerging trends in CSR, where companies adopt proactive approaches to social and environmental issues, and underscores the role of CSR in fostering sustainable economic development. Through this, the study provides insights into how businesses can leverage CSR to achieve both social responsibility and economic success.

Chapter 13

Mily Lal, Dr. D.Y. Patil School of Science and Technology, Dr. D.Y. Patil
Vidyapeeth, India
S. Neduncheliyan, Department of CSE, School of Computing, Bharath
Institute of Higher Education and Research, India
Arti Kaushik, G.V.M College of Education, Sonepat, India
Avinash Goswami, The LNM Institute of Information Technology,
Jaipur, India

The integration of Artificial Intelligence in healthcare, especially through chatbots, has the potential to transform patient engagement and support. The authors explore how AI-powered chatbots utilize natural language processing, machine learning, and sentiment analysis to offer personalized, accessible, and continuous healthcare. By tailoring interactions to individual needs, these chatbots enhance care quality, promote proactive health management, and improve patient satisfaction and adherence to treatment plans. Despite these benefits, the chapter highlights the importance of a cautious approach, including rigorous testing and ongoing monitoring, to mitigate risks. Successful AI integration will hinge on maintaining trust among patients and healthcare professionals, ensuring AI remains a valuable tool in healthcare's evolving landscape.

Chapter 14

This paper explores the critical role of social capital in enhancing the effectiveness of sales promotional strategies within entrepreneurial ventures. Social capital, defined as the value derived from social networks, provides entrepreneurs with access to essential resources, market information, and trust, all of which are pivotal in crafting successful promotional campaigns. The study reveals a strong positive relation between an entrepreneur's social capital and the success of their sales promotions. Key findings highlight that entrepreneurs with robust social networks can more effectively mobilize resources, gain market insights, and leverage word-of-mouth and viral marketing, ultimately leading to greater customer acquisition and market penetration. This study provides practical insights for entrepreneurs aiming to leverage their social capital for more impactful marketing efforts, offering a pathway to sustained business growth and competitive advantage.

Chapter 15

This study examines the application of human-cantered artificial intelligence (AI) in Islamic finance, with the aim of bridging modern technology and tradition to improve entrepreneurial interaction. In the era of digital transformation, AI is a major force that is transforming various industries including the financial sector. Islamic finance, which prioritizes ethical practices and social justice, has a unique approach in integrating AI according to sharia principles. This research aims to fill the literature gap regarding the application of AI in Islamic finance which is still limited. Through in-depth analysis and a comprehensive approach, the study offers practical and theoretical guidance for Islamic financial institutions to improve operational efficiency, service personalization, and productive and ethical entrepreneurial interactions.

Chapter 16

Artificial intelligence (AI) is progressively transforming service delivery by
executing diverse activities, serving as a significant driver of innovation, but also
posing a danger to human employment. The significance of artificial intelligence
(AI) and how it affects Nigerian telecommunications companies' service quality are
both emphasized in this chapter. The descriptive survey design was used, and the
population of the study comprised staff of listed telecommunication companies in
Nigeria. The findings of the simple linear regression analysis revealed that artificial
intelligence has a positive and significant effect on the service quality of listed
telecommunication companies in Nigeria. This implies that artificial intelligence
is a significant predictor of service quality. The chapter suggests that firms should
provide customers with more personalised services, as this has a significant impact
on their overall experience with the enterprise.

This chapter examines AI's transformative role in humanizing digital customer
experiences, exploring how technologies like natural language processing, machine
learning, and predictive analytics are reshaping customer interactions. It focuses on
hyper-personalization, empathetic conversational AI, and proactive customer service,
emphasizing the balance between AI capabilities and human touch. The chapter
addresses ethical considerations including data privacy, transparency, and fairness
in AI systems. Case studies highlight successes and challenges in implementing
AI-driven customer experience strategies, while also forecasting future trends.
Throughout, it advocates for a human-centric approach where AI augments rather
than replaces human capabilities. This comprehensive overview serves as a valuable
resource for businesses aiming to leverage AI to enhance customer experiences
while maintaining ethical, human-centered practices. The chapter underscores the
potential of AI to create more meaningful and efficient customer interactions when
implemented thoughtfully.

This research introduces a cutting-edge approach to disease prediction using a stacking ensemble model in machine learning. Through meticulous evaluation and testing on data sourced from the National Health Portal of India, the model demonstrated superior accuracy, particularly with the utilization of the random forest ensemble technique. The success of the model lies in its ensemble learning approach, which harnesses the strengths of multiple classification algorithms and incorporates robust hyper parameter selection and cross-validation methods to ensure stability and reliability.

Blended learning is a type of educational learning that incorporates traditional face-to-face learning with online learning. In the hospitality and tourism industry, the blended learning model is utilized for both training and education to ensure workforce competency. The growing interest in blended learning is drawing increased attention from researchers, leading to a rise in publications on the topic within the hospitality and tourism sector. This chapter aims to add to the existing literature with an overview using the bibliometric analysis, a methodology that quantitatively analyzes academic literature and identifies significant trends, patterns, and research publications in the field of hospitality and tourism industry. This analysis was conducted using research articles published between 2014 and 2024, totaling 2,174 samples. The study emphasizes the distinctive insights provided by the bibliometric approach and identifies areas for future exploration. It concludes with a summary of the findings, acknowledges limitations, and offers recommendations for future research.

Foreword

The rapid evolution of technology has brought about significant transformations in how we live, work, and conduct business. At the forefront of this transformation is the field of Artificial Intelligence (AI), which, when coupled with digital entrepreneurship, has reshaped industries and created new opportunities for growth and innovation. The book, Human-Centric AI in Digital Transformation and Entrepreneurship, explores this convergence of AI and entrepreneurship, delving into how human-centric approaches to technology can drive digital transformation across various sectors.

This edited volume presents a diverse range of topics that reflect the multifaceted nature of AI's impact on both business and society. From the exploration of digital entrepreneurship trends in India to the integration of AI in Islamic finance, the chapters cover a rich spectrum of subjects that address both the promises and challenges of AI in the contemporary entrepreneurial landscape. It not only offers cutting-edge research and insights but also highlights the importance of understanding human interactions with AI, ensuring that technology serves to enhance human potential and societal well-being.

The book begins by mapping emerging research trends in digital entrepreneurship, particularly within the Indian context, and moves on to explore pressing contemporary topics such as consumer engagement in the Metaverse, AI's role in customer satisfaction, and the role of AI in driving digital transformation across various industries. Further, it provides deep dives into specialized areas like women-centric AI, the ethical implications of AI in healthcare, and the integration of deep learning in audio classification, demonstrating how AI is creating value in both traditional and emerging fields.

Each chapter offers a unique perspective, shedding light on how AI, when thoughtfully integrated, can provide novel solutions to complex challenges. The volume's emphasis on a human-centric approach serves as a timely reminder that, while technology offers immense potential, it must be designed and implemented with an understanding of human needs, ethics, and values. It is a call to action for entrepreneurs, businesses, and policymakers to collaborate, ensuring that AI's benefits are accessible to all and that its integration is thoughtful, inclusive, and sustainable.

As we witness AI continue to shape the future of entrepreneurship, this book stands as an essential resource for scholars, practitioners, and anyone interested in the intersection of technology, entrepreneurship, and human-centered innovation. It brings together the work of diverse experts, offering a holistic view of how AI is influencing businesses and societies worldwide, and sets the stage for future research and applications in the ever-evolving digital landscape.

It is my hope that this volume will inspire new ideas, spark further inquiry, and encourage a deeper understanding of how AI can be harnessed to create a more inclusive, innovative, and sustainable entrepreneurial ecosystem.

Shandar Ahmad

DBT Bioinformatics Center, School of Computational and Integrative Sciences (SCIS), Jawaharlal Nehru University (JNU), India

Preface

The rapid evolution of technology, coupled with the growing emphasis on innovation and entrepreneurship, has redefined industries and transformed the global economic landscape. This book emerges as a comprehensive exploration of these dynamic shifts, aiming to bridge the gap between theoretical advancements and practical applications in the realms of digital transformation, artificial intelligence, and entrepreneurship.

In an era marked by the pervasive influence of AI, digital platforms, and immersive technologies like the metaverse, organizations and individuals alike are faced with unprecedented opportunities and challenges. Entrepreneurs are harnessing these innovations to create value, foster sustainability, and address societal needs, while businesses are adopting digital tools to enhance efficiency, customer engagement, and competitive advantage. This book delves into these transformative processes through a collection of chapters that examine key trends, challenges, and emerging solutions.

The topics covered range from the impact of generative AI and microlearning to the role of sustainability and green practices in entrepreneurship. Additionally, the book sheds light on the potential of AI in specialized domains such as healthcare, finance, and telecommunication, emphasizing its capacity to solve pressing problems while raising ethical and social considerations. The chapters also explore the evolving landscape of consumer behavior, especially in the metaverse, and provide bibliometric insights into the research trends shaping these domains.

This book is the result of collective efforts by academics, researchers, and practitioners dedicated to advancing knowledge in digital innovation and entrepreneurship. It is designed to serve as a valuable resource for entrepreneurs, business leaders, policymakers, and scholars seeking to understand and navigate the complexities of a rapidly changing technological ecosystem.

As editors, we hope this book inspires readers to think critically, innovate responsibly, and contribute to building sustainable, inclusive, and technologically advanced societies. We are grateful to the contributors whose rigorous research and thought-provoking insights have made this endeavor possible.

We invite you to embark on this intellectual journey and explore the multifaceted impact of technology and innovation on the modern world.

There are 19 chapters in this book project. A brief description of each chapter is provided as follows:

The first chapter titled "Mapping Research Trends in Indian Digital Entrepreneurship: A Bibliometric Analysis" explains the Digital entrepreneurship plays a crucial role in the economic development and innovation landscape of India, propelled by advancements in digital technologies and the widespread availability of the Internet. The chapter addresses a significant research gap by performing a bibliometric analysis of digital entrepreneurship in India, utilizing the Scopus database for the period from 2012 to 2024. By focusing on keywords such as 'digital entrepreneurship' and 'India,' it gathers metadata related to publication titles, authors, dates, journals, and keywords to discern scholarly trends. The analysis delves into academic discussions, revealing patterns in publication, citation networks, authorship, and thematic interests. By employing analytical tools such as R Studio and VOSviewer, it charts the developing research landscape, emphasizing key contributors and emerging themes. This offers a thorough overview of the scholarship surrounding digital entrepreneurship in India and identifies critical areas for future inquiry, providing essential insights into the evolving domain.

The second chapter titled "The Prospects of Consumer Engagement in Metaverse: Identifying the drivers and Barriers" presents the Metaverse has gone extra mile from its origin as a gaming platform to a dynamic digital economy where the users can create, purchase and sell products and services through their avatars. The retailers are increasing brand engagement and awareness by enhancing product visualizations, enabling highly personalized marketing strategies such as virtual try-ons in the virtual world. Metaverse has led to paradigm shift that can leave behind those who lag. Hence, it is important to consider the progress and foresee the future growth. The bibliometric analysis was conducted to get a comprehensive awareness of consumer behavior and consumer engagement in Metaverse. This study examines the research landscape of 274 articles retrieved from Scopus database from 1997 to 2024 to achieve our research objectives. The findings of the study can assist the future researchers in getting in-depth understanding of current trends and themes while identifying the research gap, providing valuable insights for future study directions.

The third chapter titled "Exploring the Influence of Electronic Word of Mouth on Customer Satisfaction: A bibliometric Study" explains the widespread adoption of E-commerce and digital platforms has led to cultural shift towards sharing experiences and feedback as a form of social proof to increase sales. Electronic word of mouth (eWOM) is a significant factor in shaping customer perception, reducing perceived risk and enhancing customer satisfaction while making purchase decisions. The purpose of conducting the bibliometric analysis on eWOM and customer satisfaction is to capture knowledge from published articles to identify trends, themes and gaps. Using Scopus database, a comprehensive analysis was conducted on 265 documents which satisfied the inclusion/exclusion criteria for the period between 2002 to 2024. The analysis focused on the contribution of academic, country, sources and top cited articles and Network analysis for cluster analysis to identify different themes. This research is valuable for assessing the significance, influence, and evolution of research across different fields, providing a quantitative basis for evaluating scientific contributions and guiding future research agendas.

The fourth chapter titled "Leveraging IT for Enhanced Characterization of SiC and Ni Epoxy Composites in High Voltage Applications" presents the Tools have a major impact on the precision, effectiveness, and cost of manufacturing composite structures. In order to provide lightweight, affordable, long-lasting, and useful tools for producing high-performance composite structures, industry and academia must work closely together to leverage and investigate new developments in materials, structures, and tool manufacturing technologies. This is because the performance of composites components is becoming more and more desired. It is frequently problematic to provide insulation at higher voltages (kV range). In this work, we created a variety of epoxy resin samples utilising nickel and silicon carbide, two conducting and non-conducting fillers. Using the centrifugal casting procedure, samples were created. We examined the composite's breakdown voltage, which will eventually offer insulation at greater voltages. The sample with a higher filler concentration had a good breakdown voltage, according to the data.

Chapter 5 titled "Women-Centric AI: A Step Towards Empowerment," reveals how artificial intelligence can be used in a positive way toward gender equity and greater inclusivity. The critical focus of the study addresses intrinsic bias in the contemporary functioning of AI systems, with particular emphasis on women being the victims of the discriminatory behavior of such AI-based systems. It states that, for the sake of building women-centric AI systems, women's concerns in designing such systems need to be integrated so they remain proactive creators and decision-makers. Qualitative methodology and thematic analysis of expert interviews identify defining features of women-centric AI that can be adapted to meet the challenges specific to women. Targeted AI solutions can drastically reduce the digital gender gap and offer more tailored support in health, education, and employment. Ethical considerations include diverse datasets and gender impact assessments for equitable AI development.

Chapter 6 titled "The Role of AI in Digital Transformation" explores the transformative role of Artificial Intelligence (AI) in driving digital transformation across various industries. As organizations navigate the complexities of the digital age, AI emerges as a critical tool for enhancing operational efficiency, improving customer experiences, and fostering innovation. The integration of AI technologies enables businesses to automate routine tasks, analyze vast datasets for actionable insights, and deliver personalized services at scale. However, the adoption of AI is not without challenges; issues such as data quality, ethical considerations, and workforce displacement pose significant hurdles. This paper discusses best practices for AI integration, emphasizing the importance of establishing clear objectives, investing in high-quality data, and prioritizing transparency and ethical governance. Additionally, future trends in AI, including hyper-automation, edge computing, and AI-driven sustainability, are examined as organizations seek to leverage these advancements for competitive advantage.

Chapter 7 titled "A Comprehensive Review on Advancements and Challenges in Audio Classification through Deep Learning" explains that Deep learning-based audio classification has simply reshaped this entire industry by developing reliable solutions for speech recognition, music genre classification, and identification of ambient sounds. Deep learning simplified the extraction of hierarchical features and complicated patterns from audio signals, resulting in accuracy and efficiency that was previously unparalleled. It presents a review of methods applied for audio classification, which includes model topologies, evaluation metrics, and data pretreatment methods in the domain of deep learning. Such comprehensive knowledge is provided in this work by synthesizing findings from many studies on the current state-of-the-art, highlighting remarkable breakthroughs made so far, and the problems that remain. It surveys future research prospects and practical applications of the open challenges faced in constructing stronger data sets, enhanced interpretability of models, and real-time processing capabilities.

Chapter 8 titled "The Neuroethical Nexus of Brain Organoids and AI: Towards the Integration of Organoid Intelligence and Artificial Intelligence" presents the advancement of brain organoids has revolutionized the fields of biotechnology and neuroscience. Brain organoids are the 3-D structures of the human brain developed from the human's pluripotent cells, which mimic the functionality of the human brain. Organoid intelligence cumulates brain organoids and artificial intelligence, making it possible to develop a biohybrid system with heightened cognitive capacities. This paper aims to critically analyze concerns surrounding brain organoids and organoid intelligence and highlight the scientific progress and ethical considerations, ensuring that new technologies contribute to the community while maintaining fundamental moral principles. The findings of the current work will be useful for decision makers, neuroscientists, ethicists, and the public to promote environmentally conscious innovation and practice organoid intelligence in an ethical way. This work serves as a call to action for addressing ongoing challenges, advancing the field, and utilizing it effectively in the current context.

Chapter 9 titled "Fostering Green Economy via Catalyst of Sustainability: Navigating Entrepreneurial Landscape for Eco-Innovation" presents that there is no shame in acknowledging that profiteering through wealth generation is the most vicious motivator to satiate entrepreneurial aspirations. However, in this globalizing world where resource availability is faced with the problem of resource exploitation, whilst economic development bears the cost of climate change, there is a need for a structural change in business behavior to encourage sustainability. The study examines how the state-of-the-art entrepreneurial ventures are insufficient to address the concerns of green economy; and how it can be transformed through simulation of eco-efficient innovation techniques by synergizing determinants of sustainability. The study is an exploration into the spectrum of issues surrounding green entrepreneurship and how R&D and regulatory regimes in eco-innovation can contribute to a holistic solution. Using entrepreneurial Responsibility as a variable for sustainable entrepreneurship, besides policy interventions, instilling entrepreneurial motivations for firms to go green, will surely stimulate a green economy driven by ethical ecopreneurs.

The Chapter 10 titled "Microlearning- its Adaptation, Challenges and Path Ahead" explains the research study aims to explore the significance of microlearning in modern training initiatives. It seeks to investigate the benefits of microlearning in terms of employee engagement and its effectiveness in achieving concise training targets. This study employs a quantitative as well as qualitative research design to investigate the effectiveness, adaptation, and challenges of microlearning in IT sector. Thematic exploration has been utilized to analyse the findings from in-depth interviews with Learning and Development Managers and HR managers. The findings suggest that delivering targeted and personalized learning experiences, microlearning can keep employees motivated and continuously updated with relevant skills. The study's originality lies in its in-depth analysis of microlearning's adaptation, challenges, and the potential path ahead in the context of modern organizational training. The research provides managers with valuable insights and practical implications for incorporating microlearning into their organization's learning and development initiatives.

Chapter 11 titled "Strategies for Cost-Effective Implementation of AI in Project Management, Maximizing ROI for Entrepreneurs" presents that the evolving landscape of project management, integrating Artificial Intelligence (AI) offers entrepreneurs unprecedented opportunities to enhance operational efficiency, mitigate risks, and achieve strategic goals. This book chapter explores key strategies to optimise return on investment (ROI) through the cost-effective implementation of AI in project management. The case study approach represents how entrepreneurs can achieve flexibility and cost savings by adopting cloud-based AI services for project planning, resource management, and performance analysis. Furthermore, the chapter explores the role of predictive analytics and machine learning algorithms in optimizing project outcomes and minimizing operational expenses. Moreover, the chapter explores emerging trends and future directions in AI-driven project management, including advancements in natural language processing and autonomous project management systems. In conclusion, this book chapter provides entrepreneurs with practical strategies for achieving cost-effective implementation of AI in project management.

Chapter 12 titled "The Role of Corporate Social Responsibility in Driving Economic Development A Systematic Review" explains that Corporate Social Responsibility (CSR) has become crucial in modern business due to its potential to drive economic development. This systematic review analyzes the relationship between CSR initiatives and economic development by reviewing empirical studies, theoretical frameworks, and case analyses. It explores how CSR practices, such as addressing climate change and social inequality, contribute to economic growth by enhancing reputation, improving stakeholder relations, and attracting investment. The review identifies emerging trends in CSR, where companies adopt proactive approaches to social and environmental issues, and underscores the role of CSR in fostering sustainable economic development. Through this, the study provides insights into how businesses can leverage CSR to achieve both social responsibility and economic success.

Chapter 13 titled "Healthcare Chatbots Using Artificial Intelligence and Sentiment Analysis" focuses on The integration of Artificial Intelligence in healthcare, especially through chatbots, has the potential to transform patient engagement and support. The authors explore how AI-powered chatbots utilize natural language processing, machine learning, and sentiment analysis to offer personalized, accessible, and continuous healthcare. By tailoring interactions to individual needs, these chatbots enhance care quality, promote proactive health management, and improve patient satisfaction and adherence to treatment plans. Despite these benefits, the chapter highlights the importance of a cautious approach, including rigorous testing and ongoing monitoring, to mitigate risks. Successful AI integration will hinge on maintaining trust among patients and healthcare professionals, ensuring AI remains a valuable tool in healthcare's evolving landscape.

Chapter 14 titled "Social Capital and Sales Promotions: Entrepreneurial Insights into Network-Based Marketing" This paper explores the critical role of social capital in enhancing the effectiveness of sales promotional strategies within entrepreneurial ventures. Social capital, defined as the value derived from social networks, provides entrepreneurs with access to essential resources, market information, and trust, all of which are pivotal in crafting successful promotional campaigns. The study reveals a strong positive relation between an entrepreneur's social capital and the success of their sales promotions. Key findings highlight that entrepreneurs with robust social networks can more effectively mobilize resources, gain market insights, and leverage word-of-mouth and viral marketing, ultimately leading to greater customer acquisition and market penetration. This study provides practical insights for entrepreneurs aiming to leverage their social capital for more impactful marketing efforts, offering a pathway to sustained business growth and competitive advantage.

Chapter 15 titled "Human-Centric AI in Islamic Finance: Bridging Technology and Tradition for Enhanced Entrepreneurial Interactions" This study examines the application of human-cantered artificial intelligence (AI) in Islamic finance, with the aim of bridging modern technology and tradition to improve entrepreneurial interaction. In the era of digital transformation, AI is a major force that is transforming various industries including the financial sector. Islamic finance, which prioritizes ethical practices and social justice, has a unique approach in integrating AI according to sharia principles. This research aims to fill the literature gap regarding the application of AI in Islamic finance which is still limited. Through in-depth analysis and a comprehensive approach, the study offers practical and theoretical guidance for Islamic financial institutions to improve operational efficiency, service personalization, and productive and ethical entrepreneurial interactions.

Chapter 16 titled "Service Quality in The Age of Artificial Intelligence: Evidence from Listed Telecommunication Businesses in Sub-Saharan Africa" focuses on Artificial intelligence (AI) is progressively transforming service delivery by executing diverse activities, serving as a significant driver of innovation, but also posing a danger to human employment. The significance of artificial intelligence (AI) and how it affects Nigerian telecommunications companies' service quality are both emphasized in this chapter. The descriptive survey design was used, and the population of the study comprised staff of listed telecommunication companies in Nigeria. The findings of the simple linear regression analysis revealed that artificial intelligence has a positive and significant effect on the service quality of listed telecommunication companies in Nigeria. This implies that artificial intelligence is a significant predictor of service quality. The chapter suggests that firms should provide customers with more personalised services, as this has a significant impact on their overall experience with the enterprise.

Chapter 17 titled "Humanizing Customer Interaction - AI-Powered Experiences in the Digital Age" This chapter examines AI's transformative role in humanizing digital customer experiences, exploring how technologies like natural language processing, machine learning, and predictive analytics are reshaping customer interactions. It focuses on hyper-personalization, empathetic conversational AI, and proactive customer service, emphasizing the balance between AI capabilities and human touch. The chapter addresses ethical considerations including data privacy, transparency, and fairness in AI systems. Case studies highlight successes and challenges in implementing AI-driven customer experience strategies, while also forecasting future trends. Throughout, it advocates for a human-centric approach where AI augments rather than replaces human capabilities. This comprehensive overview serves as a valuable resource for businesses aiming to leverage AI to enhance customer experiences while maintaining ethical, human-centered practices. The chapter underscores the potential of AI to create more meaningful and efficient customer interactions when implemented thoughtfully.

Chapter 18 titled "A machine learning-based ensemble model for estimating Multiple Disease Prediction" introduces a cutting-edge approach to disease prediction using a stacking ensemble model in machine learning. Through meticulous evaluation and testing on data sourced from the National Health Portal of India, our model demonstrated superior accuracy, particularly with the utilization of the Random Forest ensemble technique. The success of our model lies in its ensemble learning approach, which harnesses the strengths of multiple classification algorithms and incorporates robust hyper parameter selection and cross-validation methods to ensure stability and reliability.

Chapter 19 titled "A Bibliometric Analysis of Blended Learning in Hospitality & Tourism Industry: a Bibliometric Analysis of Blended Learning in Hospitality & Tourism Industry" presents a study on Blended learning, which is a type of educational learning that incorporates traditional face-to-face learning with online learning. In the hospitality and tourism industry, the blended learning model is utilized for both training and education to ensure workforce competency. The growing interest in blended learning is drawing increased attention from researchers, leading to a rise in publications on the topic within the hospitality and tourism sector. This paper aims to add to the existing literature with an overview using the bibliometric analysis, a methodology that quantitatively analyzes academic literature and identifies significant trends, patterns, and research publications in the field of hospitality and tourism industry. This analysis was conducted using research articles published between 2014 and 2024, totalling 2,174 samples. The study emphasizes the distinctive insights provided by the bibliometric approach and identifies areas for future exploration. It concludes with a summary of the findings, acknowledges limitations, and offers recommendations for future research.

CONCLUSION

The book chapters collectively explore the intersection of digital transformation, artificial intelligence, entrepreneurship, and innovation across diverse industries. Through the lens of bibliometric analysis and rigorous research methodologies, this compilation highlights critical themes such as digital entrepreneurship, consumer engagement in the metaverse, generative AI, and human-centric applications of technology. By addressing challenges, identifying emerging trends, and proposing actionable strategies, the chapters contribute to advancing theoretical and practical knowledge in these domains.

The comprehensive review of topics like sustainability, AI in project management, healthcare chatbots, green economy, and microlearning underlines the importance of adapting technological advancements to address pressing societal and environmental challenges. Moreover, the studies emphasize the need for ethical considerations and inclusivity in the design and implementation of AI and digital tools.

Key insights emerge, such as the transformative potential of AI in enhancing operational efficiency and customer experience, the critical role of social capital in entrepreneurial ventures, and the importance of sustainable practices in fostering a green economy. Additionally, the integration of AI in niche areas such as Islamic finance, telecommunication, and hospitality training provides valuable perspectives on how technology can bridge tradition with innovation.

The findings of these studies are not merely academic; they offer practical implications for entrepreneurs, businesses, policymakers, and researchers. By synthesizing bibliometric trends and leveraging interdisciplinary approaches, this book serves as a comprehensive resource for understanding and navigating the complexities of a rapidly evolving digital landscape.

As technology continues to advance, future research must remain focused on addressing ethical challenges, ensuring equitable access, and fostering a balance between technological efficiency and human-centric values. This will be pivotal in creating sustainable, inclusive, and innovation-driven ecosystems that benefit both society and the economy.

Sanjay Misra

Institute of Energy Technology, Halden, Norway

Manju Kaushik

Amity University, Jaipur, India

Amit Jain

Amity University, Jaipur, India

Chitreshh Banerjee

Amity University, Jaipur, India

Chapter 1
Mapping Research Trends in Indian Digital Entrepreneurship:
A Bibliometric Analysis

Pallavi Mishra
https://orcid.org/0000-0002-8315-8473
Amity University, Jaipur, India

Tanushri Mukherjee
https://orcid.org/0000-0001-5120-7982
Amity University, Jaipur, India

Neharshi Srivastava
Amity University, Jaipur, India

ABSTRACT

Digital entrepreneurship plays a crucial role in the economic development and innovation landscape of India, propelled by advancements in digital technologies and the widespread availability of the Internet. The chapter addresses a significant research gap by performing a bibliometric analysis of digital entrepreneurship in India, utilizing the Scopus database for the period from 2012 to 2024. By focusing on keywords such as 'digital entrepreneurship' and 'India,' it gathers metadata related to publication titles, authors, dates, journals, and keywords to discern scholarly trends. The analysis delves into academic discussions, revealing patterns in publication, citation networks, authorship, and thematic interests. By employing analytical tools such as R Studio and VOSviewer, it charts the developing research landscape, emphasizing key contributors and emerging themes. This offers a thorough overview of the scholarship surrounding digital entrepreneurship in India

DOI: 10.4018/979-8-3693-8009-3.ch001

and identifies critical areas for future inquiry, providing essential insights into the evolving domain.

INTRODUCTION

In recent years, the domain of digital entrepreneurship in India has experienced significant growth, propelled by rapid technological advancements, increased internet accessibility, and supportive government initiatives. This evolution has sparked substantial academic interest in digital entrepreneurship, especially as it offers new opportunities for economic development, innovation, and job creation. (Yadav, et al. 2023) Digital entrepreneurship encompasses businesses that operate primarily online, leveraging digital platforms, tools, and data-driven strategies to promote innovation across diverse sectors. Indian entrepreneurs are effectively utilizing digital infrastructures to scale their operations, thereby fostering an environment conducive to academic inquiry into the factors, challenges, and impacts of digital business models within the Indian economy. (Chatterjee, Chaudhuri and Chatterjee 2023)

The rise of digital entrepreneurship in India has been boosted by various government policies, including initiatives like "Digital India" and "Startup India," which aim to enhance digital innovation and entrepreneurship nationwide. These programs have significantly contributed to lowering entry barriers, improving access to technology, and nurturing a vibrant startup ecosystem that flourishes in the digital age (P. Tewari 2023). Additionally, the expansion of venture capital and angel investment within the Indian digital startup landscape has been a vital catalyst for this transformation, allowing entrepreneurs to grow their businesses rapidly and sustainably (S. Mondal, Singh and H. Gupta, 2023).

The objective of this study is to delineate the prevailing research trends in Indian digital entrepreneurship through a bibliometric analysis, which facilitates the assessment of publication trends, influential works, and key research themes.

Bibliometric analysis is a robust quantitative method employed to evaluate and analyze research output systematically. By examining citations, publication trends, and co-authorship networks, this method offers valuable insights into scholarly productivity, published works' influence, and collaboration patterns within and across disciplines. Such evaluations help in identifying significant areas of interest and tracing the trajectory of research in a specific field.

One of the key advantages of bibliometric analysis is its ability to highlight emerging trends, which are crucial for staying updated in rapidly evolving academic and professional landscapes. Additionally, this approach uncovers gaps in existing research, directing future studies toward unexplored or underrepresented topics. (Aria and Cuccurullo, C. 2017) For institutions and researchers alike, these insights are a foundation for setting strategic priorities, optimizing resource allocation, and fostering innovation.

Bibliometric analysis allows stakeholders to assess research performance and enhance decision-making processes systematically. It ensures that efforts are focused on impactful areas, thus contributing to advancements in both academic and practical domains. (Donthu, et al. n.d.) The objective of the study is to perform a thorough bibliometric analysis of academic literature concerning digital entrepreneurship in India, with a particular emphasis on the digital entrepreneurship. Through the examination of research articles, citation networks, and patterns of collaboration, the study seeks to delineate the academic research environment in this domain. The analysis is expected to uncover significant themes, prominent authors, and emerging trends, thereby offering a detailed understanding of the digital entrepreneurship process. By employing bibliometric methodologies, the study not only evaluates the existing research landscape but also pinpoints critical areas for future inquiry, thereby supporting the achievement of its goals and promoting digital entrepreneurship landscape. (Agarwal 2009)

The term bibliometrics is often attributed to Alan Pritchard, who first presented it in his 1969 work titled "Statistical Bibliography or Bibliometrics" (A.Pritchard 1969) The use of bibliometric analysis allows researchers to achieve a comprehensive understanding of their field, identify gaps in knowledge, generate new research ideas, and contextualize their findings within the broader discipline. (Gingras and Yves 2016) The study employs bibliometric techniques for three primary objectives. First, it aids in the examination of large datasets. Second, it applies quantitative approaches, providing an objective evaluation of research. Finally, bibliometrics is an established domain with significant implications across various fields of study.

RESEARCH QUESTIONS

The study utilized bibliometric analysis to study India and digital entrepreneurship by examining publications from 2010 to 2024. The analysis aimed to identify key research themes and influential entities in the field.

Upon reviewing the literature on India and digital entrepreneurship, a significant amount of research was found. The analysis will consider variables such as citations, publications, journals, local impact factors, authors, and countries. Additionally, an examination of keywords indexed in the articles will be conducted to identify research groups and understand the research themes associated with India and digital entrepreneurship.

RQ1 What are the main themes explored in the research on the India and digital entrepreneurship?

RQ2 Has there been a significant change in the number of publications on trends in Indian Digital Entrepreneurship from 2012 to 2024?

RQ3 Which researchers and institutions have been the most productive in terms of publishing on this topic?

RQ4 What are the patterns of Top 10 Studies with highest citation on digital entrepreneurship in India?

RQ5 Which papers have received the highest number of citations in the field?

DATA MINING PROCESS OF SCOPUS PUBLICATIONS

The comprehensive mapping of digital entrepreneurship in India was conducted through a review that encompassed different components. The research publications were analysed in three stages: searching, filtering, and screening.

In the first stage, data was searched using a specific formula that included keywords related to digital entrepreneurship in India in Scopus Database that reflected 58 documents.

In the second stage, filtering was performed three times. Firstly, the key words were added digital entrepreneurship in India along with the search to include research papers, conference papers, book chapters, and books which gave 58 results. The search criteria covered all years from 2012 to 2024 and included all documents to ensure no relevant materials were excluded.

Another filter was English language applied to search the keywords 'digital entrepreneurship in India on Scopus data base and searched within article title, abstract and keywords which gave 58 results in English language.

Finally, in the third stage of screening was done which was extracted in CSV format encapsulating categories like "Title," "Author," "Year," "Journal," "Citations," and "Keywords" that gave total 58 publications as the keywords were digital entrepreneurship in India. This collection of data transferred to VOS viewer analytical software for statistical evaluation, trend visualization, and report creation.

CRITERIA OF DATA MINING ON SCOPUS

Table 1. Criteria of data mining

Keywords	Range	Language
digital entrepreneurship in India	2012-2024	English

Main Information

Figure 1. Main information of the publications of Scopus indexed journals from R Studio

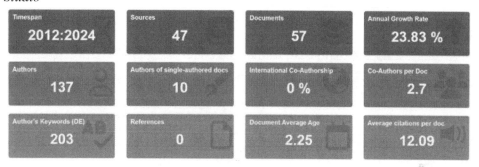

The data visualization provides insights into a bibliometric analysis of a body of research spanning from 2012 to 2024. Here's an interpretation of the key metrics:

Timespan: The analysis covers 12 years, from 2012 to 2024, showcasing the academic trends and output within this period. A total of 57 publications are included in the analysis, reflecting the research productivity during the timespan. The data reflects the annual growth, authors of single-authored documents, international co-authorship co-authors per document, author's keywords, document average age, average citations per document.

Annual Growth Rate (23.83%): A significant annual growth rate of 23.83% suggests a rapidly increasing research output, indicating rising interest or activity in the subject area.

Authors (137): The involvement of 137 unique authors indicates a collaborative and active research community.

Authors of Single-Authored Documents (10): Only 10 publications are single-authored, suggesting that most research in this area is collaborative.

International Co-Authorship (0%): The absence of international collaborations implies that the research is conducted entirely within a domestic or localized context, potentially limiting its global reach and diversity.

Co-Authors per Document (2.7): On average, each document has nearly three co-authors, further emphasizing collaborative efforts in the research process.

Author's Keywords (203): A total of 203 unique keywords were used, reflecting a wide range of topics or concepts covered in the analyzed body of work.

Document Average Age (2.25 years): With an average age of 2.25 years, the documents in the dataset are relatively recent, indicating a focus on contemporary research.

Average Citations per Document (12.09): Each document is cited an average of 12.09 times, which is a strong indicator of the impact and relevance of this research within the academic community.

The research demonstrates robust growth and collaboration but lacks international partnerships. While the high average citations suggest impact, the absence of references and international co-authorship could be areas to improve for broader and more inclusive contributions to the field.

Table 2. Year-wise publications of documents on trends in Indian digital entrepreneurship

Year	Documents	Key Themes
2024	**12**	Digital Entrepreneurial Intention; Capital Theory in Entrepreneurship; Theory of Planned Behavior (TPB); Social Cognitive Career Theory (SCCT); Developing Country Context; Entrepreneurial Education, Business Students: Human, Social, and Financial Capital; Self-Efficacy in Entrepreneurship; Entrepreneurial; Attitudes and Norms; Resource Constraints; Digital Entrepreneurship; Educational Support for Entrepreneurs; Resilience and Adaptability
2023	**17**	3D Printing: Entrepreneurial opportunities, Indian economy, global economy. Digital Entrepreneurship: Indian handicrafts, post-pandemic, challenges, opportunities. Green Entrepreneurship: Circular economy, digitalization, sustainable practices, waste management. Handicraft Industry: Digital technology, innovation, calamities, entrepreneurship. boAt: Marketing strategies, consumer electronics, brand analysis. Family Firms: Technology adoption, entrepreneurial behavior, Indian community. Healthcare Entrepreneurship: Innovative financing, India, primary care sector. Women Entrepreneurs: Challenges, resilience, COVID-19 pandemic. Madhubani Painting: Digital platforms, pandemic sustainability, case study. AI in Marketing: Decision-making, IPL, strategic branding. Digital Entrepreneurship Growth: COVID-19, India, transformation. Digital-Social Entrepreneurship: Sustainable value, technological capabilities, interlinkages. Workplace Flexibility: Paradoxes, organizational analysis, case study. MSMEs: COVID-19 impact, digital transformation, India. Russia-India Cooperation: Trends, economic partnerships, future directions. ICTP: Cooperative trade, economic model, India. G20 Initiatives: Disaster risk reduction, global strategies, implications.
2022	**9**	Digital Entrepreneurship: Business model innovation, online food delivery, India. Relay-as-a-Service: Strategy, innovation, Rivigo, business model analysis. Digital Healthcare: Entrepreneurial opportunities, post-COVID-19, developing countries. Agtech Platforms: Digital connectivity, KisanMitr, participatory action research. COVID-19 Impact: Accelerated digital entrepreneurship, Indian perspective, opportunities. Inclusion: Remote altitudes, 17000 ft Foundation, case study. Retirement to Entrepreneurship: Skills, attitude, technological innovation, transitions. Indian Handicrafts: Marketing challenges, handloom industry, Guthali case study. Smart Agriculture: Entrepreneurship education, needs, challenges, case study in Karnataka.
2021	**7**	E-Government Services: COVID-19 pandemic, demand prediction, SARIMA model. Institutional Challenges: Digital entrepreneurship, developing countries, institutional voids. Startup Ecosystem: Environment, funding activities, India, investment trends. Paper Industry: Resilience, adaptation, "new normal," Indian pulp and paper sector. Fintech and Entrepreneurship: Comparative study, India, Egypt, developing economies. Entrepreneurship Growth: Digital era, key factors, sustainable development, innovation.

continued on following page

Table 2. Continued

Year	Documents	Key Themes
2020	2	ICT4D Conference: Social implications, computers, developing countries, IFIP WG 9.4, ICT4D 2020. Entrepreneurial Ecosystems: Digitization, impact analysis, Indian perspective, globalisation.
2019	4	Science Classrooms: Technology integration, Indian schools, changing nature, science education, perspectives. Digital Divide: India vs. China, digitalization disparities, development issues, comparative analysis. Strategic Innovation: Factors analysis, Indian startups, cross-case study, business advancement. Digital Social Entrepreneurship: Hybridity, digital platforms, social innovation, entrepreneurship models.
2018	2	Zomato Case Study: Foodtech sector, business innovation, growth strategies, market positioning. Social Media and Innovation: Family firms, market segmentation, social media utilization, innovation strategies.
2017	2	Essential Personality Characteristics: Indian millennials, Tobit analysis, personality traits, generational study, behavioral insights. Codezin Case Study: Startup challenges, entrepreneurial resilience, business strategy, emerging markets, case analysis.
2016	0	N/A
2015	1	Service Divide: Digital innovation, healthcare services, service gap, India, technology-enabled solutions. Healthcare Transformation: Digital health, service innovation, patient care improvement, Indian healthcare providers, MIS applications.
2014	0	N/A
2013	0	N/A
2012	1	Entrepreneurship Education: Online learning, computer-based training, distance education, entrepreneurship skills development. Digital Learning: Technology in education, e-learning platforms, entrepreneurship in India, remote education models. Impact Assessment: Education outcomes, learner engagement, effectiveness of distance learning, India-focused study.

The data illustrates the annual generation or recording of documents from 2012 to 2024. The following is an analysis of the observed trends:

ANALYSIS OF THE DATA

Trends and Insights on Digital Entrepreneurship and Related Themes (2012–2024)

The data illustrates a comprehensive examination of themes associated with digital entrepreneurship, entrepreneurial education, and sectoral innovation, particularly within the Indian context, over a decade from 2012 to 2024. The subjects underscore significant shifts in emphasis, shaped by both global and national challenges, including the COVID-19 pandemic, technological advancements, and the changing economic environment.

Key Observations by Year:

2024: The focus is on theoretical frameworks such as Capital Theory and Social Cognitive Career Theory (SCCT), particularly in developing nations. -Key areas of interest include human, social, and financial capital in entrepreneurship, self-efficacy, resource limitations, and educational assistance. -There is a notable emphasis on resilience and adaptability in response to challenges, reflecting post-pandemic trends.

2023: A rich variety of themes emerges, with significant attention given to 3D printing, green entrepreneurship, and healthcare innovation. -The focus is directed towards sectors like handicrafts and family businesses, examining digital transformation and resilience in the aftermath of COVID-19. -Strategic studies encompass AI in marketing, workplace flexibility, and MSMEs, highlighting the increasing influence of technology on entrepreneurial ecosystems. -Enhanced international collaboration is evident through themes such as Russia-India cooperation and G20 initiatives.

2022: The primary emphasis is on business model innovation and sector-specific entrepreneurship in fields such as food delivery, healthcare, and agriculture. -The impact of COVID-19 in accelerating digital entrepreneurship is a recurring theme, alongside the challenges of inclusion in remote regions. -Distinct case studies, including the Guthali case study on handicrafts and KisanMitr in agtech, illustrate grassroots-level digital interventions.

2021: The focus is on institutional challenges and the resilience demonstrated by sectors such as paper and fintech amid the pandemic. The rise of emerging ecosystems, particularly within startups and digital entrepreneurship, underscores the changing landscape of funding and innovation in India. Research also explores sustainable development and the digital transformation of governance in times of crisis.

2019–2020: This period is characterized by investigations into the disparities in digitization between India and China and its implications for entrepreneurship. Initial explorations of digital social entrepreneurship, along with case studies like Zomato, set the stage for more focused research in the following years.

2012–2018: The early themes prioritize entrepreneurship education and digital learning, showcasing the integration of technology in skill enhancement. Innovations across sectors, including the transformation of healthcare in 2015 and case studies such as Codezin in 2017, highlight the emerging significance of digital strategies in entrepreneurship.

Figure 2. Year wise Publications of documents on Trends in Indian Digital Entrepreneurship

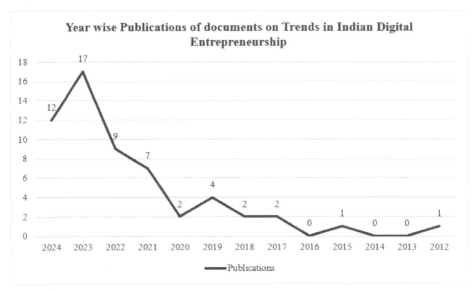

Figure 2 Year wise Publications of documents on Trends in Indian Digital Entrepreneurship

Literature on Digital Entrepreneurship in India (2012-2024)

Table 3. Year-wise documents published in Scopus

Year	Citations
2012	0
2013	0
2014	0
2015	3
2016	12
2017	15
2018	21
2019	22
2020	43
2021	52
2022	89
2023	150
2024	268
Subtotal	**675**

Citation Data

Initial Phase (2012-2014): During this period, no citations were recorded, suggesting either a lack of pertinent publications or insufficient acknowledgment of existing research.

Emerging Recognition (2015-2016): Citations began to emerge in 2015, with a total of 3 citations, indicating the first signs of recognition for the research efforts. The following year, 2016, saw an increase to 12 citations, reflecting an enhanced visibility and impact of prior works.

Progressive Increase (2017-2019): A consistent upward trend is observed, with citations growing from 15 in 2017 to 22 in 2019. This trend signifies a gradual acknowledgment of the contributions made within the academic and research communities.

Notable Growth (2020-2023): The citation count experienced a remarkable increase, starting at 43 in 2020 and rising to 52 in 2021, 89 in 2022, and reaching 150 in 2023. This pattern indicates a phase of heightened impact and broad recognition of the research.

Culminating Impact (2024): In 2024, citations peaked at 268, representing a significant culmination of acknowledgment and relevance of the published works. The total of 675 citations underscores a robust upward trend in academic influence over the years. The rising citation numbers reflect an enhancement in the quality, relevance, and impact of the research output. The exponential growth observed in recent years suggests a bolstered reputation within academic and research spheres, potentially attributed to increased collaboration, improved dissemination strategies, or the exploration of more significant topics.

TYPES OF SCOPUS PUBLICATIONS ON INTERNATIONALIZATION OF HIGHER EDUCATION (2012-2024)

The data mining process revealed a diverse range of Scopus-indexed publications on the internationalization of higher education in India from 2014 to 2024. These publications encompass various types of scholarly work, reflecting the multifaceted nature of research in this area.

Table 4. Types of Scopus publications (2012-2024)

Document Type	Count
Article	34
Book Chapter	13
Conference Paper	6
Conference Review	3
Undefined	1

Figure 3. Types of publication

Types of Publication

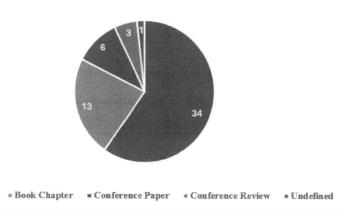

■ Article ■ Book Chapter ■ Conference Paper ■ Conference Review ■ Undefined

Figure 3 Types of Publication

The compilation of Scopus-indexed publications spanning from 2012 to 2024 the data classifies documents into five distinct categories, illustrating their distribution. Articles constitute the largest segment, with 34 entries, representing the most substantial portion of the dataset. Following them, book chapters emerge as the second most common type, with 13 entries. Conference papers consist of six entries, indicating a moderate contribution to the overall dataset. Conference reviews make up three entries, signifying a smaller yet noteworthy presence. Lastly, there is one undefined document, which represents a minimal and unique category. This distribution underscores the dominance of articles and book chapters within the dataset while highlighting the variety of document types included.

Figure 4. VoSviewer map of highest contribution of authors in Scopus publication (2012-2024)

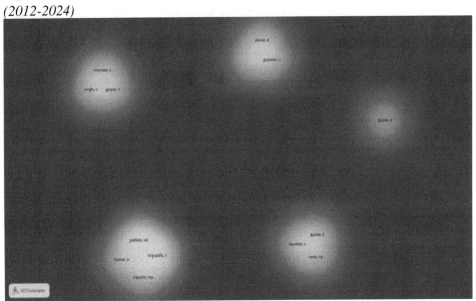

Figure 5. Verify selected authors

 Verify selected authors

Selected	Author	Documents	Citations	Total link strength
☑	tripathi, r	4	30	8
☑	yadav, us	4	30	8
☑	kumar, a	2	12	5
☑	tripathi, ma	2	10	5
☑	david, d	2	12	4
☑	gopalan, s	2	12	4
☑	gupta, h	2	46	4
☑	gupta, s	2	123	4
☑	kamble, s	2	123	4
☑	mondal, s	2	46	4
☑	ramachandran, s	2	12	4
☑	rana, np	2	123	4
☑	singh, s	2	46	4
☑	gupta, a	2	5	0

The analysis conducted using VOS Viewer on Scopus publications from 2012 to 2024 offered a comprehensive evaluation of author contributions, establishing a threshold of at least two documents and a minimum of five citations. Out of the 133 authors examined, 14 authors fulfilled the criteria, resulting in 14 items categorized into five distinct clusters. This analysis underlines the substantial contributions of these authors in terms of both document quantity and citation impact.

Notably, authors Tripathi, R., and Yadav, U.S. each produced four documents, accumulating 30 citations, which reflects their consistent scholarly output. Additionally, Gupta, S., Kamble, S., and Rana, N.P. emerged as the most influential contributors, each achieving 123 citations from their two publications. Gupta, H., Mondal, S., and Singh, S. also exhibited significant influence, each receiving 46 citations for their two documents. Meanwhile, Kumar, A., David, D., Gopalan, S., and Ramachandran, S. consistently contributed two documents, each with 12 citations, while Tripathi, M.A. produced two documents with 10 citations. Gupta, A., although contributing at a lower level, added two documents that garnered five citations.

This analysis yields important insights into the research productivity and citation influence of authors, emphasizing key contributors and their significance within the Scopus publication dataset. The clustering of these authors further reveals collaboration patterns and thematic focuses in their research endeavors.

COUNTRY/TERRITORY WISE SCOPUS PUBLICATIONS 2012-2024

Figure 6. Country/territory-wise Scopus publications

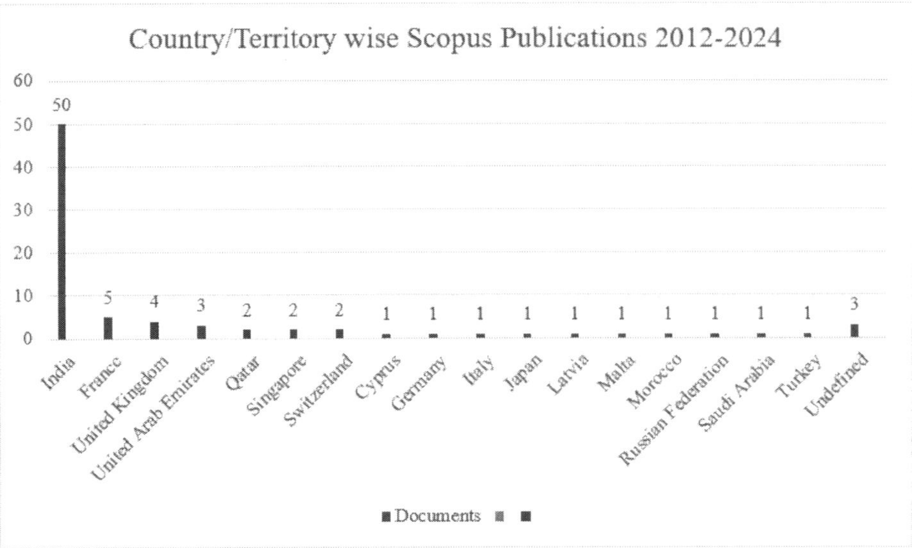

Figure 5 Country/Territory wise Scopus Publications

The analysis of data on publications concerning the internationalization of in India from 2012 to 2024 offers valuable insights into the distribution of scholarly contributions in this field. The data indicates a pronounced concentration of document contributions from India, which stands out with an impressive total of 50 documents, underscoring its preeminent position within the dataset. France follows as the second-largest contributor with 5 documents, while the United Kingdom provides 4, suggesting a moderate level of engagement from these nations. The United Arab Emirates contributes 3 documents, highlighting its growing influence in this domain. Additionally, Qatar, Singapore, and Switzerland each contribute 2 documents, demonstrating their modest yet significant involvement. Several other countries, including Cyprus, Germany, Italy, Japan, Latvia, Malta, Morocco, the Russian Federation, Saudi Arabia, and Turkey, each contribute 1 document, reflecting their smaller but varied participation. Furthermore, there are 3 documents categorized as "Undefined," indicating contributions without specified territorial origins. This distribution illustrates India's dominant contribution and a diverse, albeit uneven, global engagement, characterized by a combination of major contributors and sporadic inputs from other regions.

Top 10 Studies with highest citation on Digital Entrepreneurship in India

Table 6. Top 10 studies on digital entrepreneurship in Indian context

Rank	Title	Authors	Source	Year	Citations
1	Has Covid-19 accelerated opportunities for digital entrepreneurship? An Indian perspective	Modgil, S., Dwivedi, Y.K., Rana, N.P., Gupta, S., Kamble, S.	*Technological Forecasting and Social Change*	2022	122
2	Digital and innovative entrepreneurship in the Indian handicraft sector after the COVID-19 pandemic: challenges and opportunities	Yadav, U.S., Tripathi, R., Tripathi, M.A., Mandal, M., Singh, A.	*Journal of Innovation and Entrepreneurship*	2023	10
3	Achieving Technological Transformation and Social Sustainability: An Industry 4.0 Perspective	Mondal, S., Singh, S., Gupta, H.	*IEEE Transactions on Engineering Management*	2024	10
4	Digital entrepreneurship in developing countries: The role of institutional voids	Soluk, J., Kammerlander, N., Darwin, S.	*Technological Forecasting and Social Change*	2021	82
5	Entrepreneurship in India's Handicraft Industry with the Support of Digital Technology and Innovation During Natural Calamities	Yadav, U.S., Sood, K., Tripathi, R., Grima, S., Yadav, N.	*International Journal of Sustainable Development and Planning*	2023	18
6	Green entrepreneurship and digitalization enabling the circular economy through sustainable waste management - An exploratory study of emerging economy	Mondal, S., Singh, S., Gupta, H.	*Journal of Cleaner Production*	2023	36
7	A study on entrepreneurial opportunities in digital health-care post-Covid-19 from the perspective of developing countries	Khandelwal, R., Kolte, A., Rossi, M.	*Foresight*	2022	16
8	Interlinkages Between Digital-Social Entrepreneurship and Technological Capabilities for Sustainable Value Creation	Sharma, R., Kamble, S., Gupta, S., Rana, N.P., Kumar, K.	*Journal of Global Information Management*	2023	1

continued on following page

Table 6. Continued

Rank	Title	Authors	Source	Year	Citations
9	Evaluation of Factors Affecting Women Artisans as Entrepreneurs in the Handicraft Sector: A Study on Financial, Digital Technology Factors and Developmental Strategies About ODOP in Uttar Pradesh to Boost Economy	Yadav, U.S., Tripathi, R., Kumar, A., Shastri, R.K.	*Journal of the Knowledge Economy*	2024	2
10	Exploring the Interplay of Entrepreneurs' Awareness, Perception, and Intention in Driving Digitalization for Msmes: a Focused Insight Into Sidbi's Role	Shama, Mazhar, S.S., Mittal, P., Khan, F.S., Ur Rehman, A.	*Financial and Credit Activity: Problems of Theory and Practice*	2024	2

Table 6 shows a summary of the ten most referenced academic papers within the realm of higher education, with a specific emphasis on internationalization, higher education, and India. The data presented encompasses a wide array of research articles that examine digital entrepreneurship, including its challenges, opportunities, and effects across various sectors, especially in the aftermath of COVID-19. The leading article, titled "Has Covid-19 accelerated opportunities for digital entrepreneurship? An Indian perspective," authored by Modgil et al. and published in 2022, has achieved the highest citation count of 122, underscoring its substantial impact on the field. This study investigates the ways in which the COVID-19 pandemic has expedited digital entrepreneurship opportunities in India, demonstrating its extensive relevance and influence on the comprehension of digital transformations within the entrepreneurial domain. In contrast, the article "Digital and innovative entrepreneurship in the Indian handicraft sector after the COVID-19 pandemic: challenges and opportunities," authored by Yadav et al. and published in 2023, has received only 10 citations, indicating a more specialized focus on the handicraft sector and its adaptation to digital innovation in the post-pandemic context. Likewise, Achieving Technological Transformation and Social Sustainability: An Industry 4.0 Perspective" by Mondal et al. (2024) also has 10 citations, suggesting that it represents a recent addition to the expanding literature on technological transformations influenced by Industry 4.0 and sustainability.

Significant contributions to the field include the work titled Digital Entrepreneurship in Developing Countries: The Role of Institutional Voids by Soluk et al. (2021), which has garnered 82 citations. This study examines the impact of the lack of institutional frameworks on the development of digital entrepreneurship in emerging economies, highlighting the various challenges and systemic obstacles faced by digital entrepreneurs in settings with inadequate regulatory and infrastructural support.

Another important research piece is Entrepreneurship in India's Handicraft Industry with the Support of Digital Technology and Innovation During Natural Calamities by Yadav et al. (2023), which has received 18 citations. This paper illustrates how digital technology has enabled the handicraft sector in India to overcome the difficulties associated with natural disasters, thereby demonstrating the critical role of technology in enhancing resilience and promoting innovation. Additionally, the study titled Green Entrepreneurship and Digitalization Enabling the Circular Economy through Sustainable Waste Management by Mondal et al. (2023), which has 36 citations, explores the contributions of green entrepreneurship and digital technologies to sustainable waste management practices, a vital aspect of the circular economy in developing regions. Furthermore, Khandelwal et al. (2022) conducted a study on entrepreneurial opportunities in digital healthcare post-Covid-19 from the perspective of developing countries, which has 16 citations. This research investigates the digital transformation of healthcare in the aftermath of the pandemic, identifying new entrepreneurial prospects in these nations.

Other notable studies delve into specialized topics, such as the relationship between digital-social entrepreneurship and technological capabilities for sustainable value creation (Interlinkages Between Digital-Social Entrepreneurship and Technological Capabilities for Sustainable Value Creation, Sharma et al., 2023), as well as the factors influencing women artisans as entrepreneurs in the handicraft industry (Evaluation of Factors Affecting Women Artisans as Entrepreneurs in the Handicraft Sector).

Analysis Of Keywords Occurrence in The Scopus Publications (2012- 2024)

The analysis of keyword occurrences provides valuable insights into the academic and research landscape. It is evident that certain keywords appear frequently, indicating significant trends and focal areas. These keywords include higher education, internationalization, India, education, international education, globalization, students, capacity building, curricula, international cooperation, engineering education, and high education.

Figure 7. Keywords occurrence in the Scopus publications (2012-2024)

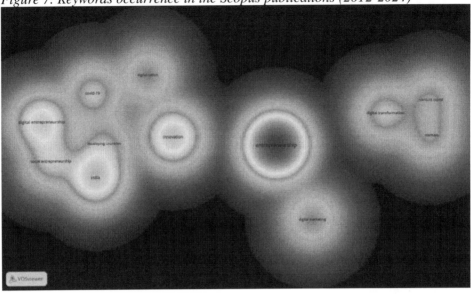

Figure 8. Verify selected key words

 Verify selected keywords

Selected	Keyword	Occurrences	Total link strength
☑	entrepreneurship	24	25
☑	india	8	12
☑	innovation	8	10
☑	startups	3	7
☑	venture capital	3	7
☑	developing countries	3	6
☑	covid-19	4	5
☑	digital transformation	4	5
☑	social entrepreneurship	4	5
☑	digital entrepreneurship	7	3
☑	digitalization	3	3
☑	digital marketing	3	2

The figure illustrates the frequency of occurrences, and the overall link strength associated with various keywords, offering valuable insights into the themes and their interrelations. The keyword "entrepreneurship" is notably the most frequently mentioned, appearing 24 times and exhibiting the highest total link strength of 25.

This indicates that entrepreneurship is a pivotal concept within the subject matter, demonstrating strong connections to other relevant ideas in the dataset.

Following this, "India" ranks as the second most frequent keyword, with 8 occurrences and a total link strength of 12. The relatively high frequency suggests that India is a significant focus of the research, and its link strength implies connections to various important topics, including entrepreneurship, innovation, and startups.

The keyword "innovation" also appears 8 times, with a total link strength of 10, underscoring its relevance in the realms of entrepreneurship and business development. Innovation is intricately linked to the concepts of startups (3 occurrences, 7 link strength) and venture capital (3 occurrences, 7 link strength), both of which are crucial for nurturing new business ideas and scaling operations.

Both "venture capital" and "startups" are mentioned 3 times, each with a link strength of 7, highlighting their vital roles within the entrepreneurial ecosystem, especially during the initial phases of business development. The close relationship between these terms emphasizes the interdependence of capital investment and the formation of new enterprises.

Additionally, the term "developing countries" appears 3 times, with a link strength of 6, indicating the significance of geographical context in entrepreneurial studies. This may reflect discussions surrounding the challenges and opportunities for entrepreneurship in emerging markets, as well as the evolution of innovation and startups in these areas.

The term COVID-19 appears four times, with a cumulative link strength of five. This indicates that the pandemic's influence on entrepreneurship, digital transformation, and business models is a significant subject within the data, although its link strength is comparatively lower than that of other keywords. This suggests that the emphasis may be more on long-term trends rather than the immediate repercussions of the pandemic.

Digital transformation, social entrepreneurship, and digital entrepreneurship are mentioned between four to seven times, each exhibiting moderate link strengths of five or less. These concepts underscore the rising significance of technology and social impact in modern entrepreneurial studies. Notably, digital transformation and digital entrepreneurship highlight the transition towards online business models, while social entrepreneurship emphasizes initiatives aimed at tackling social issues through innovative approaches.

Additionally, digitalization (three occurrences, link strength of three) and digital marketing (three occurrences, link strength of two) are noted with lower frequencies and link strengths. This implies that, although these subjects are relevant, they may not hold as central a position in the data as the more dominant keywords associated with entrepreneurship and innovation.

In summary, the data underscores critical focal points at the convergence of entrepreneurship, innovation, and technology, particularly highlighting the roles of India and developing nations, the effects of COVID-19, and the increasing significance of digital transformation and social entrepreneurship.

CONCLUSION

In conclusion, the field of digital entrepreneurship in India has developed into a vibrant and transformative sector, driven by advancements in technology, greater internet accessibility, and proactive government policies. A bibliometric analysis of research publications spanning from 2012 to 2024 indicates a notable academic interest in exploring the changing trends, challenges, and opportunities within this domain. The systematic methodology utilized in this study not only charts the evolution of research but also emphasizes significant themes, prominent authors, and collaborative networks that influence the conversation surrounding digital entrepreneurship in India. By employing bibliometric methods, the research offers a thorough insight into the current research environment, identifies essential gaps, and lays the groundwork for future investigations. This analysis highlights the vital contribution of digital entrepreneurship to economic growth and innovation, reaffirming its significance in the modern academic and policy context. The bibliometric examination of research conducted from 2012 to 2024 reveals a dynamic and expanding field, characterized by an impressive annual growth rate of 23.83% and a collaborative network of 137 authors contributing to 57 publications. The notable average citations per document (12.09) signify the substantial academic influence of this research. However, the lack of international co-authorship points to a predominantly localized research focus, indicating potential for enhancing global collaborations to achieve greater diversity and wider impact. Furthermore, the dataset's relatively young average document age (2.25 years) and the diverse array of 203 unique keywords reflect a strong emphasis on contemporary and multifaceted subjects. These insights provide a detailed overview of research trends while pinpointing areas for improvement in international partnerships and interdisciplinary collaboration.

REFERENCES

Agarwal, P. (2009). *Indian Higher Education: Envisioning the Future*. SAGE Publications. DOI: 10.4135/9788132104094

Aria, M., & Cuccurullo, C. (2017). Bibliometrix: An R-tool for comprehensive science mapping analysis. *Journal of Informetrics*, *11*(4), 959–975. DOI: 10.1016/j.joi.2017.08.007

Chatterjee, S., Chaudhuri, R., & Chatterjee, S. (2023). Entrepreneurial behavior of family firms in the Indian community: Adoption of a technology platform as a moderator. *Journal of Enterprising Communities*, *17*(2), 433–453. DOI: 10.1108/JEC-08-2021-0122

Donthu, N., S. Kumar, D Mukherjee, N. Pandey, and N., & Lim, W. M. n.d.

Gingras and Yves. 2016. *Bibliometrics and Research Evaluation: Uses and Abuses*. Cambridge: MA: MIT Press.

Mondal, S., Singh, S., & Gupta, H. (2023). Green entrepreneurship and digitalization enabling the circular economy through sustainable waste management: An exploratory study of emerging economy. *Journal of Cleaner Production*, *422*, 422–433. DOI: 10.1016/j.jclepro.2023.138433

Pritchard, A. (1969). Statistical Bibliography or Bibliometrics? *The Journal of Documentation*, *25*, 348–349.

P. Tewari. 2023. Rise of digital entrepreneurship during COVID-19 in India. *In Industry 4.0 and the Digital Transformation of International Business* 135–141.

Yadav, U. S., Sood, K., Tripathi, R., Grima, S., & Yadav, N. (2023). "Entrepreneurship in India's handicraft industry with the support of digital technology and innovation during natural calamities." *International Journal of Sustainable Development and Planning. International Journal of Sustainable Development and Planning*, *18*(6), 1777–1791. DOI: 10.18280/ijsdp.180613

Chapter 2
The Prospects of Consumer Engagement in Metaverse:
Identifying the Drivers and Barriers

Garima Pancholi
https://orcid.org/0000-0002-5039-4998
Amity University, Jaipur, India

Monika Sharma
https://orcid.org/0000-0002-4799-7712
Amity University, Jaipur, India

Abhineet Saxena
https://orcid.org/0000-0003-0467-7900
Amity University, Jaipur, India

Mukesh Kumar Verma
https://orcid.org/0009-0001-7183-6724
University of Rajasthan, India

ABSTRACT

Metaverse has gone the extra mile from its origin as a gaming platform to a dynamic digital economy where the users can create, purchase and sell products and services through their avatars. The retailers are increasing brand engagement and awareness by enhancing product visualizations, enabling highly personalized marketing strategies such as virtual try-ons in the virtual world. Metaverse has led to a paradigm shift that can leave behind those who lag. Hence, it is important to consider

DOI: 10.4018/979-8-3693-8009-3.ch002

progress and foresee future growth. The bibliometric analysis was conducted to get a comprehensive awareness of consumer behavior and consumer engagement in Metaverse. This study examines the research landscape of 274 articles retrieved from Scopus database from 1997 to 2024 to achieve our research objectives. The findings of the study can assist the future researchers in getting in-depth understanding of current trends and themes while identifying the research gap, providing valuable insights for future study directions.

INTRODUCTION

Metaverse is a virtual universe which is created by combining all other virtual worlds. The users can explore Metaverse through Virtual Reality, Augmented Reality and Mixed Reality. Even though it is no longer just a topic of discussion instead it has been attracting a lot of investments from investors around the globe. It is continuously evolving, offering medium for social and business engagement, and innovations in virtual environments. Through Metaverse, the consumers can communicate with the product like never before. It is important for marketers to keep up with these changes and adapt themselves to this new platform. Marketers who are capable of supplying at the intersection of 'what to give' and 'where to give' with solutions that fulfil consumer needs, will have a trailblazer advantage in this ever-changing world. Metaverse is defined as Virtual reality that prevails beyond the realm of real world (Kolesnichenko et al., 2019; Kye et al., 2021), which has capability of matching consumer experiences to the real world (Bale et al., 2022). Metaverse is gaining more popularity as it has potential of giving memorable virtual experiences to their consumers which is hard to achieve in real life. Metaverse Re-tailing is not just a representation of online shopping; instead, it gives opportunity to sense real shopping experience into virtual world. Metaverse can be reached by using a VR headset, or AR goggle, and it helps users to visualize a virtual world and experience bewitching moments. Augmented Reality (AR) and Virtual reality (VR) has potential to transform the communication medium in digital environment (Buhalis et al., 2023). Metaverse not only gives platform for virtual interactions with friends & family but also for providing diverse and innovating services to consumers.

According to the Pew Research Centre, 54% of tech experts believe that by 2040 Metaverse will be in its fully functional state and will be offering immersive experiences. Another research conducted by Metaverse found that there are around 400 million users, among those 80% are people of 16 years of age. The businesses have understood that Metaverse is no longer just a next-generation thing—instead it is platform to create new marketing opportunities in the virtual world. It possesses the potential for engaging consumers through technologies and generating revenue by leveraging new experiences for virtual economy. One of the limitations faced by brick-and-mortar stores is that they are falling short of reaching consumers at all influential moments. This calls out for Metaverse where retailers can interact with customers in much more captivating ways. Metaverse will help retailers by giving a holistic picture of customer behaviour and preferences while making a purchase decision (Chakraborty et al., 2024).

The rising popularity and innovation of Metaverse in retail marketing and its potential effect on Consumer behaviour is beyond the shadow of doubt (Eggenschwiler et al., 2024). Despite this evolution, prior studies lacked the depth and breadth of empirical knowledge of consumer engagement in the Metaverse. Prados-Castillo et al., (2024) have also conducted a bibliometric analysis on Metaverse and customer behaviour but it focused more on tourism sector. The present research study attempts to bridge this gap and attempts to gives comprehensive insights into how consumer engage within the Metaverse.

The bibliometric analysis has been incorporated in the current study, to acknowledge the recent growth of research areas but also serving as a guide for future directions. Bibliometric analysis is also capable of providing a comprehensive knowledge from literature review and evaluating the scope of study related to Metaverse and Consumer engagement. The analysis of current trends and prolific authors, countries and journals will open new frontiers for collaboration, also the identification of most cited articles will help setting a pathway for researcher for analysing the impact of Metaverse on consumer experiences.

The research aims to fulfil and comprehend through this study.

1. The study will identify prominent countries, authors, sources in subject areas through publications and citation analysis.
2. The study will evaluate the current trends and analyse the themes of the subject area.
3. The study will provide frontiers by identifying the gaps and overlooked issues of the subject area.

METHODOLOGY

Data collection

We used the Scopus database to collect data for the present study, which is a renowned abstract and citation resource of important papers from research academics and well-known journals (Van Eck & Waltman, 2014). Bibliometric analysis is a robust and effective method that enables researchers to comprehensively explore and examine extensive amounts of scientific data (Donthu et al., 2021). Scopus was selected above other well-known databases for several reasons (Medias *et al.*, 2023). First, while other databases, like PubMed and Web of Science (WoS) (AlRyalat et al., 2019), are accessible for bibliometric analysis, PubMed is more oriented toward the life sciences and biomedical research (AlRyalat et al., 2019) whereas WoS and Scopus provide multidisciplinary coverage. Secondly, While Scopus has roughly 21,950 journals (22,800 if trade pubs are included), WoS only has 13,100 journals (20,556 if ESCI is included). This suggests that Scopus has more coverage than WoS. Furthermore, WoS only provides 1.6 billion citations, whereas Scopus guarantees access to 1.7 billion. After making this comparison, we choose to move forward with using Scopus for data collection. According to a study conducted by (Tober, 2011), Scopus was found to be highly effective due to its capability of delivering a substantial number of literature explorations.

In the present study, Scopus database was utilized to gather data due to its capability of delivering a comprehensive collection of journals, research articles and patents which allows researchers to identify emerging trends, prominent topics and gaps in literature. From the initial search 669 articles were collected from Scopus database using the search string of TITLE-ABS-KEY, "Articles containing "Metaverse OR Virtual Reality OR Augmented Reality AND Consumer Engagement AND Consumer Behavior AND Customer Experience". These articles were further screened by excluding the subject area other than Business Management and Accounting and duplicates, editorials, conference proceedings and book chapters were also removed along with the articles which were written in other than English language. The screened articles were reduced to include the articles published in journals with suitable impact factor (i.e., impact factor more than 3.00 Based on the 2024 Journal Citation Reports Database (Clarivate, 2024). These articles were then checked for eligibility assessment to remove articles not aligned with our goals (e.g., main focus is not on Consumer Behaviour or satisfaction, use of technology other than Metaverse and VR). Following the systematic review process, our sample comprises 274 articles for further bibliometric analysis.

Table 1. Main information about data

Description	Results
Main Information About Data	
Timespan	1997:2024
Sources (Journals, Books, etc.)	71
Documents	274
Annual Growth Rate %	15.92

continued on following page

Table 1. Continued

Description	Results
Document Average Age	3.78
Average citations per doc	64.44
Document contents	
Keywords Plus (ID)	714
Author's Keywords (DE)	928
AUTHORS	
Authors	751
Authors of single-authored docs	23
Authors Collaboration	
Single-authored docs	24
Co-Authors per Doc	3.31
International co-authorships %	39.05
Document Types	
Article	274

Source: Biblioshiny

Table 1 gives us an overview of the information retrieved from Scopus database. The retrieved articles on the topic of Metaverse and Consumer engagement range from 1997 to 2024. The total number of articles is 274 from 71 journals, 751 authors and 64.44 average citations per document.

TRENDS OF PUBLICATION

The trends in publication can be measured by observing the increase or decrease in publications of articles in the subject area over the period of time. Figure 1 represents the productivity in publications of articles through total publications and Mean Citations per article. The mean citation per article indicates the overall influence or the impact that an article creates within the subject area. Higher mean citation represents relevance or popularity of the article. The first article in the subject area was published in 1997 and till 2016, the number of publications remained static but from 2018 the publications gained momentum and by July' 2024, 54 articles had already been published. The graph represents fluctuations in citations per article, where 2004 has the highest number of citations. In comparison to total publications, the mean citations per article have decreased due to increase in publications since 2018.

Figure 1. Year-wise articles and mean citations per article

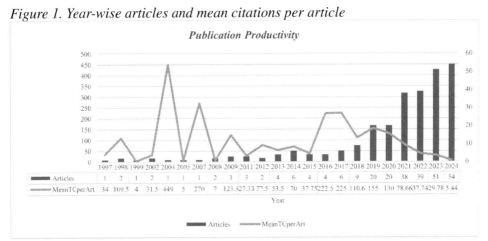

Authors own work

Country Productivity

Figure 2 represents country wise productivity in publications in the subject area of Metaverse and Consumer engagement. For the current study, out of total publishing countries, the top 10 countries are included to represent the publications from 1997 to 2024. The trends reveal that USA has highest number of publications with 124 publications, followed by India and China with 85 and 84 publications respectively. The observation suggests the growing popularity of Metaverse and consumer engagement in the European countries, as 6 out of 10 countries belong to Europe.

Figure 2. Country-wise productivity

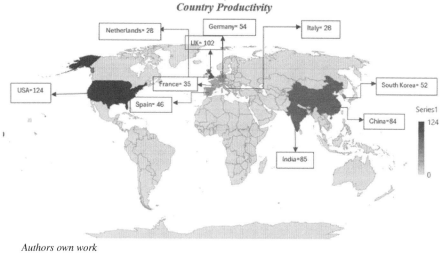

Authors own work

Authors' Productivity

Authors' productivity can be measured by assessing the number of research paper published over a period. Figure 3 represents prolific authors in the field of Metaverse and consumer engagement on the basis of highest article publication and citations. In the present study, Chylinski, M., De Ruyter K., Keeling, Di., and Rauschnabel, P A have highest number of publications, where Rauschnabel, P A has ranked first in terms of citations with 1848 citations and rest all have 1171 citations. Dwivedi, Y K has second highest citations with 1269 citations from 5 publications.

Figure 3. Authors' productivity

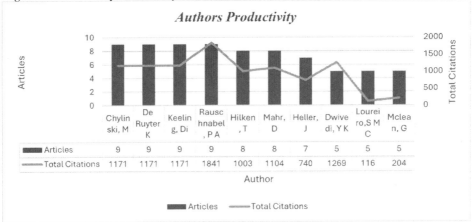

Lotka's Law

Lotka's law is frequently mentioned as Lotka's Inverse Square law of Scientific Productivity, which defines the relationship between the aggregate authors and productivity in terms of publications. It provides a framework for shaping the publication frequency of the author under a specific field. This law states that the aggregate authors making X number of contributions is proportional to the reciprocal of the total sum of authors making a single contribution (Friedman, 2015). It indicates that a lesser number of authors are responsible for a higher number of publications, while a higher number of authors tend to have fewer contributions. Lotka's Law of scientific productivity is shown in the form of an equation:

$Y = C/X^n$

where Y = Interval of authors making X number of contribution
C = constant
X = aggregate publications
N = constant

By applying Lotka's law, researchers gain knowledge about the distribution of authors' productivity and a better grasp of the number of publications in the subject field. Figure 4 represents that in the present study, only 4 authors have published 9 articles which is approximately 0.05% of total authors, while 2 authors have published 8 articles. A total of 670 authors, which is approximately 88.2% of total authors, have only 1 publication. Insights from this analysis will help in encouraging collaborative research, enhance quality and optimal allocation of resources to productive authors.

Figure 4. Authors' productivity through Lotka's law

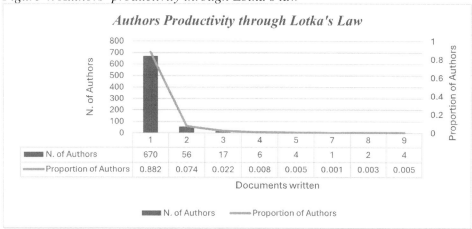

Authors' own work

Source Productivity

The identification of the productivity source in the subject area can be analysed by taking into consideration several factors such as the aggregate sum of articles published, the impact of the sources, and the application of Bradford's law. Bradford's law states that only few core journals have significant research articles while the rest of the journals will have distribution of rest articles. It is also helpful in providing access to relevant and frequently cited articles. According to Figure 5, Journal of Retailing and Consumer Services, Journal of Business Research and Technological Forecasting and Social change are Zone 1 of Core sources as they are having highest number of published articles while the rest are in Zone 2. Core Sources refers to the set of sources which have significant amounts of articles on subject area.

Figure 5. Source productivity through Bradfords law

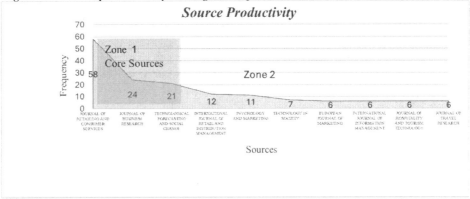

Authors own work

Most Influential Work

Most influential work defined as the published document which is having highest sum of citations. The most cited paper is often recognized as having most contribution to the subject area which will help future researchers to identify and understand the potential research area. Citations are proof of significant contributions and influence of articles in the research community. Table 2 represents the top 10 highest cited documents of the subject area. The article titled, *"Examining the relationship between reviews and sales: The role of reviewer identity disclosure in electronic markets"* which was published in 2008 has received highest citations with 1350 citations and the article titled, *"Setting the future of digital and social media marketing research: Perspectives and research propositions"* published in 2021 has received 812 citations. Both of these articles state the importance of consumer reviews on Virtual platforms for buying behavior of different consumers. The higher number of citations shows the growing importance of dealing with consumer behavior on platforms such as Metaverse and Virtual Reality.

Table 2. Most influential works

Authors	Title	Journal	Year	Total Citations
Forman, C	Examining the relationship between reviews and sales: The role of reviewer identity disclosure in electronic markets	Information System Research	2008	1350
Dwivedi, Y K	Setting the future of digital and social media marketing research: Perspectives and research propositions	International Journal of Information of Information Management	2021	812
Flavián, C	The impact of Virtual, Augmented and mixed reality technologies on the customer experience	Journal of Business Research	2019	700
Kim, M J	Exploring Consumer Behavior in Virtual Reality tourism using an Extended Stimulus-Organism-Response Model	Journal of Travel Research	2020	646
Jiang, Z	Virtual Product experience: Effects of Visual and Functional control of products on perceived diagnosticity and flow in electronic shopping	Journal of Management Information system	2004	449
Buhalis, D	Technological disruptions in services: lessons from tourism and hospitality	Journal of Service management	2019	425
Huang, yc	Exploring the implications of Virtual reality technology in tourism marketing: An integrated research framework	International Journal of Tourism Research	2016	391
Hoyer, W D	Transforming the customer experience through New technologies	Journal of Interactive marketing	2020	390
Javornik, A	Augmented reality: Research agenda for studying the impact of its media characteristics on consumer behavior	Journal of Retailing and Consumer Services	2016	376
Rauschnabel, P A	Augmented reality marketing: How mobile AR apps can improve brand through inspiration	Journal of Retailing and Consumer Services	2019	368

Three Field Plots

The Three field plots are a powerful tool in bibliometric analysis, which showcases the relationship between three dimensions, namely Authors, Countries, Keywords, Journals and affiliation etc. The Three field plot is helpful in visualizing the link or connection between three different fields. In figure 6, left field represents Keywords, middle represents Countries and right field represents Authors. The keyword 'Augmented Reality' has the highest links with UK, and UK in turn, has high links with many Authors. This is followed by 'Virtual Reality' and 'Metaverse' which also has significant links with several countries and Authors.

Figure 6. Three field plot

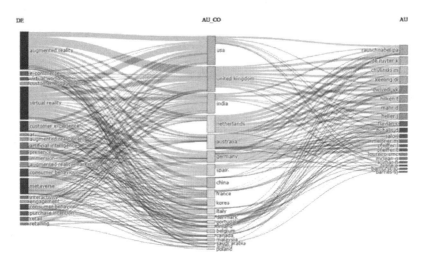

Authors elaboration using Biblioshiny

Thematic Map

A thematic map represents different thematic modules by displaying the relevant author keywords from the bibliographic data. Figure 7 represents a thematic map which is based on author keywords and was mapped into four quadrants of different themes such as Niche theme (Q1), Motor theme(Q2), emerging or declining theme (Q3) and basic themes (Q4). The quadrant 1 representing the niche themes have four clusters: Cluster 1 has Tourism management and tourism market, Cluster 2 has immersion and immersive, cluster 3 has affordances, design and cluster 4 has modeling, mobile augmented reality and research. Quadrant 2 represents motor themes consisting of four clusters, where cluster 1 has mixed reality, behavioral response, consumer perception, cluster 2 has tourism behaviour, research work, tourist destination, cluster 3 has technological development service quality travel behaviour, cluster 4 has purchasing, Metaverses, purchase intention. Quadrant 3, which is based on emerging or declining themes, has 4 clusters the cluster one has managerial implications, cluster 2 has user interfaces cluster 3 has social networking online, and cluster 4 has behavioral research and experimental study. Quadrant 4 has 3 clusters which showcase the basic themes, in this cluster one has sales consumer behaviour, augmented reality, cluster 2 has virtual worlds, interactive computer

36

graphics, human computer interaction, and cluster 3 has virtual reality, consumption behaviour, retailing.

Figure 7. Visualization of thematic mapping

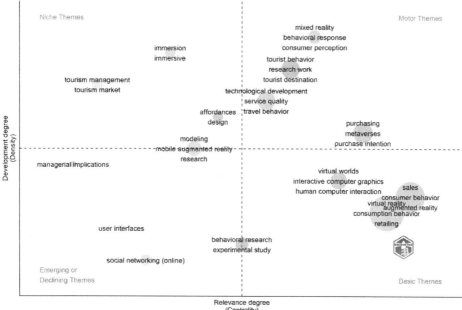

RESEARCH TREND ANALYSIS

The research trend analysis through keywords in bibliometric analysis gives a significant insight into enhancing research quality, developing innovative approaches, anticipating changes by predicting future directions. Figure 8 represents the keywords which have the highest occurrences in the subject area. The size of the keywords is directly related to the number of occurrences. Figure 9 represents the tree map which also shows the occurrences of 52 keywords where Virtual reality has the highest number occurrences (i.e. 78) which holds 13% of total keywords, followed by Consumption behaviour with 65 occurrences. The keywords retailing, marketing and sales are other high occurring keywords. The keywords such as Augmented reality and Metaverses have 18 and 7 occurrences respectively. The analysis indicates that Virtual Reality in the marketing & retail sector is expanding and understanding the consumption behaviour is important to increase the sales.

Figure 8. Word cloud

Authors elaboration using Biblioshiny

Figure 9. Tree map of 50 most frequently used keywords

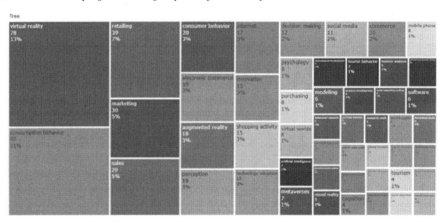

Authors elaboration using Biblioshiny

High Impact Research Topic Analysis

The Graphing keyword analysis is helpful in analysing the fluctuations in the frequency of keywords over a period. The keywords are analysed on the basis of increased usage in literature over longitudinal studies. The visualization of these keywords can help the researchers in identifying the emerging topics. The figure 10 shows the increase in usage of term Augmented Reality after 2012 there is a constant rise in the term. This is followed by Consumer Behaviour and Retailing which also shows increasing trend after 2018.

Figure 10. Word dynamic plot

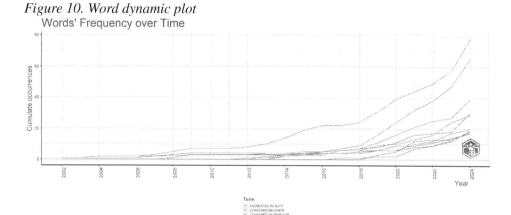

Authors elaboration using Biblioshiny

LITERATURE REVIEW

Metaverse

Metaverse has successfully transformed the real world into the virtual world which has attracted many customers. The term *'Metaverse'* was coined in 1992 by Neal Stephenson in *Snow Crash* which is a science fiction novel. Virtual Reality has paved the way for a new era of consumer engagement which is ahead of traditional online shopping (Dwivedi et al., 2022; Mystakidis, 2022). Metaverse is helping retailers to sell their physical and digital products and consumers are also showing interest in purchasing these products. This collaborative virtual and retail setting is called as "Metaverse Retailing" where consumers can do shopping by interacting and encompassing other customers and salespersons through their digital avatars (Talwar et al., 2022; Yoo et al., 2023). It is important to understand those factors which impact the user adoption of virtual world services. Kumari et al., (2024) have found ten major factors which influence the consumer to use Metaverse services is experience of virtual environment, virtual events, social interaction, social activity, Metaverse ecosystem, buying & selling of assets, possessing virtual assets, other issues related to mobile app, user account, and feature to watch ad-to earn. Caboni et al., (2024) have identified four key motivating approaches for AR usage such as Informational, experiential, social and inspirational which were further classified into intrinsic and extrinsic frameworks. The sensory dimensions are again important in creating immersive phygital (physical + digital) customer experiences in Metaverse which can bridge the gap between physical and digital/virtual world (Batat, 2024). The customised product recommendations, virtual try-on, virtual immersion and interactive productive visualisation can overcome the drawbacks of traditional on-line shopping and influence customer buying behaviours and increase engagement (Nair et al., 2024).

Drivers of Consumer Engagement in Metaverse

Immersive Virtual Reality

Immersive Virtual Reality is a technology which gives the user complete immersion in the virtual world. AR is making shopping more enjoyable and engaging through highly immersive and interactive experiences (Pathak & Prakash, 2023). The IVR can play a crucial role in enhancing customer experience and engagement. The high immersive experience represents immersion of users into 360-degree virtual space through Head mounted display while low immersive experience can be assessed through computer display screens. The study conducted by (Hsiao et al., 2024) has found that low immersive VR has more potential in strengthening customer decision support.

Virtual Influencers

The virtual influencers are also increasing the consumer engagement by showing them product usage behaviour (Yang et al., 2024). The companies introducing sustainable products can also consider them as virtual influencers are good alternative for human influencers as they have low chances of scandals and unethical behaviour which could support green marketing of company (Jiang et al., 2024). Virtual Influencers can be defined as virtual person or Avatar which is created by software for interaction in virtual world who can project human like characteristics and behaviours in order to influence other peoples for marketing or social campaigns (Angmo et al., 2024; Byun & Ahn, 2023; Laszkiewicz & Kalinska-Kula, 2023; Moustakas et al., 2020).

Virtual Product Experience

Virtual Product is a digital representation of a product. The representation can be of a real-world product or from a virtual world. The AR & VR can facilitate consumer engagement through immersive and interactive experience for consumers (Rosário & Dias, 2024). The Virtual Product Experience can facilitate the innovation teams in gathering knowledge for product development by getting insights from consumers. The vividness of customer experience and shopping interaction in virtual environment during purchase triggers the consumer impulsive buying (Chen et al., 2023).

Metaverse/Virtual Fidelity

Visual fidelity means the degree to which visual features in the Metaverse are similar to real world. Higher visual fidelity improves explicit memory of features in visual scene while lower VF increases implicit memory. Gender can play a crucial role for businesses who are aiming to tailor their marketing strategies. In comparison to males, the females are more likely to do preventive shopping which requires browsing than males' counterparts. The intention to visit the store in future has direct relation with fidelity, as decrease in fidelity will have negative impact on female consumers while increasing the fidelity will enhance the male consumers (Frank et al., 2024).

Barries of Consumer Engagement in Metaverse

Privacy Concerns

The marketers use the data generated by consumers by interacting with social media which helps them in offering personalised products (Darmody & Zwick, 2020). The consumers tend to get personalised products according to their preferences to enhance their virtual experience but at the cost of sharing wealth of personal data. To overcome this barrier the retailers should prioritise transparency and facilitate options for sharing personal information.

Data Protection

Establishing products and services in the virtual world and collecting the consumer data is not enough, the protection of collected data is concern (Shahriar, 2024). The consumer's data is being exposed to scrutiny and erosion of their privacy, this calls for protection of their information from third party. It is important for marketers to implement advance cyber security policies and measures to safeguard the data of consumers and their digital assets from theft and fraud.

Technology Adoption Barrier

Metaverse is available for wider and diverse audiences which comes from different demographics and geographies. In order to capture real world consumers, it is important for retailers to make it tech-friendly for all. The marketers can start guiding campaigns by providing comprehensive technical support to consumers to create seamless consumer engagement and experience.

DISCUSSIONS AND MANAGERIAL IMPLICATIONS

Metaverse will take years to reach its full potential, since the entrepreneurs are creating buzz but not producing significant results. The Metaverse has been redefining marketing paradigms, there is still scarcity of research on how content marketing can affect consumer engagement and interaction (Alsoud et al., 2024). In comparison to other online shopping platforms, Metaverse can facilitate more immersive environments for consumers to communicate with each other. Retailers can use this framework intensifying their relationships and engagement with customers which subsequently leads to revenue generation (Mehta et al., 2023).

This study provides an overview of research trends in the influence of Metaverse, Virtual Reality and Augmented Reality on Consumer behavior by extracting 274 articles from Scopus database for the period between 1997 to 2024. The research examines the current trends, collaborations and prominent authors, countries and sources of the field. The major themes and hot topics analysis helps the future researcher in identifying trending topics. The research has gradually started rising from the year 2018. The first paper on the topic of *'Virtual Reality'* was published in 1997 and the term *'Metaverse'* was first coined by Neal Stephenson in 1992 in order to describe 3D Virtual-reality environment. Among all the Countries, USA stands out as highest publishing country with 124 publications, while in Sources, the Journal of retailing and consumer services has highest publications which is 24 publications. Chylinski, M, De Ruyter K, Keeling, Di and Rauschnabel, P A have the highest number of publications with 9 published articles, where Rauschnabel, P A has the highest number of citations with 1848 citations and rest all have 1171 citations. Among highest cited paper, Forman et al., (2008) has citations where they have found that the higher number of reviews having information about the identity of reviewer and sharing same geographical location will have positive impact on the sales of the product. (Dwivedi et al., 2021) has second highest citations, where the authors have discussed about the implications of Social Media Marketing on consumer behavior and engagement, and they have also highlighted challenges from the negative electronic word of mouth. (Flavián et al., 2019) has third highest citations in which the authors have proposed a taxonomy name, EPI where E stands out Embodiment, P for Presence and I for interactivity for classification of technologies. EPI can be a crucial tool for managers who opt for the most suitable technology for market strategies and business goals.

Beyond the theoretical implications discussed above, our study underscores the potential areas for academic and practical fields which can contribute to enhancement of consumer engagement in Metaverse. The drivers and barriers of consumer engagement in Metaverse can prove to be significant for marketers and businessmen. The bibliometric analysis represents the growing interest in the subject area in past few years. This analysis makes it easier for marketers to efficiently observe and visualize the trends and prominent authors, sources and studies which could be assessed in further research. Although Metaverse has been around for a long time, there is scarcity in research. In this context, we propose some direction for future research which could fill the research gap. The future research can be done on Virtual customer integration and intersection of sustainability for virtual customers in Metaverse.

CONCLUSION

Metaverse is revolutionizing the retail marketing paradigms, it is providing numerous opportunities for customer engagement and interaction which can help businesses to set their products and services to meet the requirements of consumers. Analysing the behaviour of customers in Metaverse is significant for achieving its full capacity, paving the way for growth, amplifying customer engagement and ensuring sustainability in business strategies. This study used bibliometric, network and thematic mapping analysis to understand the consumer engagement in Metaverse. A total of 699 articles were retrieved from Scopus database which were further analysed according to research objectives. The total number of articles from 1997 to July 2024 are 274 from 71 journals, 751 authors and 64.44 average citations per document. Bibliometric, network and thematic analysis were performed using Biblioshiny. These findings may have both theoretical and practical significance which can serve as base for future research.

Similar to other studies, current study also has certain limitations. Firstly, this study is based on the articles retrieved from Scopus database only, therefore the article not listed in the database are not considered due to which certain research papers might not have been taken into account for current study. Future researchers can consider using other databases. Secondly, the data collection period was till July 2024, and Metaverse is upgrading day by day, so the researchers should continue literature reviews to provide more concrete guidance based on implications and challenges. Thirdly, the current study has considered the papers written in English language and paper having subject area of business management, so the future researcher can other subject areas and language also. Lastly, the bibliometric analysis should be extended by considering systematic literature review, meta-analysis or region-specific analysis to provide in-depth analysis of previously published literature.

REFERENCES

AlRyalat, S. A. S., Malkawi, L. W., & Momani, S. M. (2019). Comparing Biblio-metric Analysis Using PubMed, Scopus, and Web of Science Databases. *Journal of Visualized Experiments*, *152*(152). Advance online publication. DOI: 10.3791/58494 PMID: 31710021

Alsoud, M., Trawnih, A., Yaseen, H., Majali, T., Alsoud, A. R., & Jaber, O. A. (2024). How could entertainment content marketing affect intention to use the Metaverse? Empirical findings. *International Journal of Information Management Data Insights*, *4*(2), 100258. DOI: 10.1016/j.jjimei.2024.100258

Angmo, P., Mahajan, R., & da Silva Oliveira, A. B. (2024). *Do they look human? Review on virtual influencers*. Management Review Quarterly., DOI: 10.1007/s11301-024-00438-9

Bale, A. S., Ghorpade, N., Hashim, M. F., Vaishnav, J., & Almaspoor, Z. (2022). A Comprehensive Study on Metaverse and Its Impacts on Humans. *Advances in Human-Computer Interaction*, *2022*, 1–11. DOI: 10.1155/2022/3247060

Batat, W. (2024). Phygital customer experience in the Metaverse: A study of consumer sensory perception of sight, touch, sound, scent, and taste. *Journal of Retailing and Consumer Services*, *78*, 103786. DOI: 10.1016/j.jretconser.2024.103786

Buhalis, D., Leung, D., & Lin, M. (2023). Metaverse as a disruptive technology revolutionising tourism management and marketing. *Tourism Management*, *97*, 104724. DOI: 10.1016/j.tourman.2023.104724

Byun, K. J., & Ahn, S. J. (2023). A Systematic Review of Virtual Influencers: Similarities and Differences between Human and Virtual Influencers in Inter-active Advertising. *Journal of Interactive Advertising*, *23*(4), 293–306. DOI: 10.1080/15252019.2023.2236102

Caboni, F., Basile, V., Kumar, H., & Agarwal, D. (2024). A holistic framework for consumer usage modes of augmented reality marketing in retailing. *Journal of Retailing and Consumer Services*, *80*, 103924. DOI: 10.1016/j.jretconser.2024.103924

Chakraborty, D., Polisetty, A., & Rana, N. P. (2024). Consumers' continuance intention towards Metaverse-based virtual stores: A multi-study perspective. *Technological Forecasting and Social Change*, *203*, 123405. DOI: 10.1016/j.techfore.2024.123405

Chen, J. V., Ha, Q.-A., & Vu, M. T. (2023). The Influences of Virtual Reality Shopping Characteristics on Consumers' Impulse Buying Behavior. *International Journal of Human-Computer Interaction*, *39*(17), 3473–3491. DOI: 10.1080/10447318.2022.2098566

Clarivate. (2024). *Journal Citation Report.* https://clarivate.com/academia-government/scientific-and-academic-research/research-funding-analytics/journal-citation-reports

Darmody, A., & Zwick, D. (2020). Manipulate to empower: Hyper-relevance and the contradictions of marketing in the age of surveillance capitalism. *Big Data & Society*, *7*(1), 205395172090411. DOI: 10.1177/2053951720904112

Donthu, N., Kumar, S., Mukherjee, D., Pandey, N., & Lim, W. M. (2021). How to conduct a bibliometric analysis: An overview and guidelines. *Journal of Business Research*, *133*, 285–296. DOI: 10.1016/j.jbusres.2021.04.070

Dwivedi, Y. K., Hughes, L., Baabdullah, A. M., Ribeiro-Navarrete, S., Giannakis, M., Al-Debei, M. M., Dennehy, D., Metri, B., Buhalis, D., Cheung, C. M. K., Conboy, K., Doyle, R., Dubey, R., Dutot, V., Felix, R., Goyal, D. P., Gustafsson, A., Hinsch, C., Jebabli, I., & Wamba, S. F. (2022). Metaverse beyond the hype: Multidisciplinary perspectives on emerging challenges, opportunities, and agenda for research, practice and policy. *International Journal of Information Management*, *66*, 102542. DOI: 10.1016/j.ijinfomgt.2022.102542

Dwivedi, Y. K., Ismagilova, E., Hughes, D. L., Carlson, J., Filieri, R., Jacobson, J., Jain, V., Karjaluoto, H., Kefi, H., Krishen, A. S., Kumar, V., Rahman, M. M., Raman, R., Rauschnabel, P. A., Rowley, J., Salo, J., Tran, G. A., & Wang, Y. (2021). Setting the future of digital and social media marketing research: Perspectives and research propositions. *International Journal of Information Management*, *59*, 102168. DOI: 10.1016/j.ijinfomgt.2020.102168

Eggenschwiler, M., Linzmajer, M., Roggeveen, A. L., & Rudolph, T. (2024). Retailing in the Metaverse: A framework of managerial considerations for success. *Journal of Retailing and Consumer Services*, *79*, 103791. DOI: 10.1016/j.jretconser.2024.103791

Flavián, C., Ibáñez-Sánchez, S., & Orús, C. (2019). The impact of virtual, augmented and mixed reality technologies on the customer experience. *Journal of Business Research*, *100*, 547–560. DOI: 10.1016/j.jbusres.2018.10.050

Forman, C., Ghose, A., & Wiesenfeld, B. (2008). Examining the Relationship Between Reviews and Sales: The Role of Reviewer Identity Disclosure in Electronic Markets. *Information Systems Research*, *19*(3), 291–313. DOI: 10.1287/isre.1080.0193

Frank, D.-A., Peschel, A. O., Otterbring, T., DiPalma, J., & Steinmann, S. (2024). Does Metaverse fidelity matter? Testing the impact of fidelity on consumer responses in virtual retail stores. *International Review of Retail, Distribution and Consumer Research*, *34*(2), 251–284. DOI: 10.1080/09593969.2024.2304810

Friedman, A. (n.d.). *The Power of Lotka's Law Through the Eyes of R Domain-analytical information and knowledge organisation as research infrastructure View project Visual Peer Review View project*. https://www.researchgate.net/publication/280156919

Hsiao, S.-H., Wang, Y.-Y., & Lin, T. L. J. (2024). The impact of low-immersion virtual reality on product sales: Insights from the real estate industry. *Decision Support Systems*, *178*, 114131. DOI: 10.1016/j.dss.2023.114131

Jiang, K., Zheng, J., & Luo, S. (2024). Green power of virtual influencer: The role of virtual influencer image, emotional appeal, and product involvement. *Journal of Retailing and Consumer Services*, *77*, 103660. DOI: 10.1016/j.jretconser.2023.103660

Kolesnichenko, A., McVeigh-Schultz, J., & Isbister, K. (2019). Understanding Emerging Design Practices for Avatar Systems in the Commercial Social VR Ecology. *Proceedings of the 2019 on Designing Interactive Systems Conference*, 241–252. DOI: 10.1145/3322276.3322352

Kumari, V., Bala, P. K., & Chakraborty, S. (2024). A text mining approach to explore factors influencing consumer intention to use Metaverse platform services: Insights from online customer reviews. *Journal of Retailing and Consumer Services*, *81*, 103967. DOI: 10.1016/j.jretconser.2024.103967

Kye, B., Han, N., Kim, E., Park, Y., & Jo, S. (2021). Educational applications of Metaverse: Possibilities and limitations. *Journal of Educational Evaluation for Health Professions*, *18*, 32. DOI: 10.3352/jeehp.2021.18.32 PMID: 34897242

Laszkiewicz, A., & Kalinska-Kula, M. (2023). Virtual influencers as an emerging marketing theory: A systematic literature review. *International Journal of Consumer Studies*, *47*(6), 2479–2494. DOI: 10.1111/ijcs.12956

Mehta, M., Pancholi, G., & Saxena, D. A. (2023). Metaverse changing realm of the business world: A bibliometric snapshot. *Journal of Management Development*, *42*(5), 373–387. DOI: 10.1108/JMD-01-2023-0006

Moustakas, E., Lamba, N., Mahmoud, D., & Ranganathan, C. (2020). Blurring lines between fiction and reality: Perspectives of experts on marketing effectiveness of virtual influencers. *2020 International Conference on Cyber Security and Protection of Digital Services (Cyber Security)*, 1–6. DOI: 10.1109/CyberSecurity49315.2020.9138861

Mystakidis, S. (2022). Metaverse. *Encyclopedia*, *2*(1), 486–497. DOI: 10.3390/encyclopedia2010031

Nair, A. S., & R., D. K. (2024). *ARise to the Occasion* (pp. 184–203). DOI: 10.4018/979-8-3693-2367-0.ch009

Pathak, K., & Prakash, G. (2023). Exploring the role of augmented reality in purchase intention: Through flow and immersive experience. *Technological Forecasting and Social Change*, *196*, 122833. DOI: 10.1016/j.techfore.2023.122833

Prados-Castillo, J. F., Torrecilla-García, J. A., Guaita-Fernandez, P., & De Castro-Pardo, M. (2024). The impact of the Metaverse on consumer behaviour and marketing strategies in tourism. *ESIC Market*, *55*(1), e327. DOI: 10.7200/esicm.55.327

Rosário, A. T., & Dias, J. C. (2024). *Innovative Digital Marketing in Business.*, DOI: 10.4018/979-8-3693-1231-5.ch001

Shahriar, H. (2024). Into the Metaverse: Technological Advances Shaping the Future of Consumer and Retail Marketing. In *The Future of Consumption* (pp. 55–75). Springer International Publishing. DOI: 10.1007/978-3-031-33246-3_4

Talwar, S., Kaur, P., Escobar, O., & Lan, S. (2022). Virtual reality tourism to satisfy wanderlust without wandering: An unconventional innovation to promote sustainability. *Journal of Business Research*, *152*, 128–143. DOI: 10.1016/j.jbusres.2022.07.032

Tober, M. (2011). PubMed, ScienceDirect, Scopus or Google Scholar – Which is the best search engine for an effective literature research in laser medicine? *Medical Laser Application*, *26*(3), 139–144. DOI: 10.1016/j.mla.2011.05.006

van Eck, N. J., & Waltman, L. (2014). Visualizing Bibliometric Networks. In *Measuring Scholarly Impact* (pp. 285–320). Springer International Publishing., DOI: 10.1007/978-3-319-10377-8_13

Yang, D., Zhang, J., Sun, Y., & Huang, Z. (2024). Showing usage behavior or not? The effect of virtual influencers' product usage behavior on consumers. *Journal of Retailing and Consumer Services*, *79*, 103859. DOI: 10.1016/j.jretconser.2024.103859

Yoo, K., Welden, R., Hewett, K., & Haenlein, M. (2023). The merchants of meta: A research agenda to understand the future of retailing in the Metaverse. *Journal of Retailing*, *99*(2), 173–192. DOI: 10.1016/j.jretai.2023.02.002

Chapter 3
Exploring the Influence of Electronic Word of Mouth on Customer Satisfaction:
A Bibliometric Study

Sumit Verma
Chandigarh University, India

Amit Jain
https://orcid.org/0000-0001-7978-1556
Amity University, Jaipur, India

Garima Pancholi
https://orcid.org/0000-0002-5039-4998
Amity University, Jaipur, India

Abhineet Saxena
https://orcid.org/0000-0003-0467-7900
Amity University, Jaipur, India

ABSTRACT

The widespread adoption of E-commerce and digital platforms has led to cultural shift towards sharing experiences and feedback as a form of social proof to increase sales. Electronic word of mouth (eWOM) is a significant factor in shaping customer perception, reducing perceived risk and enhancing customer satisfaction while making purchase decisions. The purpose of conducting the bibliometric analysis on eWOM and customer satisfaction is to capture knowledge from published articles to identify trends, themes and gaps. Using Scopus database, a comprehensive analysis was conducted on 265 documents which satisfied the inclusion/exclusion criteria for the

DOI: 10.4018/979-8-3693-8009-3.ch003

period between 2002 to 2024. The analysis focused on the contribution of academic, country, sources and top cited articles and Network analysis for cluster analysis to identify different themes. This research is valuable for assessing the significance, influence, and evolution of research across different fields, providing a quantitative basis for evaluating scientific contributions and guiding future research agendas.

INTRODUCTION

Digital platforms and Technology have changed the way promotional activities are carried out by companies. Electronic commerce has created a revolution in the marketing world and most of the companies rely on it for marketing their products.(Hendricks & Mwapwele, 2024; Shahriari et al., 2015) As per data by PWC-2018 e-commerce is being used by approximately 462 million internet users in India for online purchase of products. The consumers are not reluctant to write their unvarnished views on the product functioning on online platforms which may act as a pointer for future purchase by other consumers(Zhang et al., 2018). The online reviews have a very strong impact on the attitude formation process for a product or service(Watson & Wu, 2022). Therefore, companies have started paying utmost importance on social and digital marketing in general and e-commerce, in particular. Online platforms help the consumers to interact with their peer group by providing online reviews and feedback to the prospective customers(Dwivedi et al., 2021). These online reviews are also called e-WOM, which influences the purchase decisions. In view of the above it can be inferred that the negative eWOM can have a deleterious impact on the branding of the company. Lim et al., (2012) suggested that online reviews & recommendations help the customer in the evaluation of the product. They also found that in their study there is a significant relationship between online reviews and product evaluation. Social media platforms such as WhatsApp, Facebook, Twitter etc. improve the effectiveness of the company branding through influencer marketing or digital marketing (Khanom, 2023). The advancements in technology have helped the users in exchanging their thoughts and feelings about the product on online platforms like brand communities, review sites and blogs instead of using traditional one-to-one communication. In this advanced understanding reviews and recommendations called as 3R's about the product, which is called Consumer-Generated Content (CGC) helps in the swift flow of communication between consumers, thus increasing the product knowledge. The usage of online marketing by companies such as Walmart and e-tailers like Amazon, Flipkart, snap deal etc., has helped these companies to work upon the feedback given by the consumers and delighting the customers by working on the feedback. Fournier & Avery, (2011) suggested that the concept of negative e-WOM monitors online reviews and develops intervention strategies to counter negative reviews. Monitoring which can be proactive or reactive is an effective tool to dilute the negative e-WOM. Effectiveness of e-WOM management highlights the positive reviews and minimizes the negative reviews which help in customer satisfaction and increased brand equity(Bhat & Chakraborty, 2018). It is pertinent that companies should devise a proper strategy to overcome the challenge of negative e-WOM and can increase consumer loyalty and satisfaction. The web-based complain handling mechanism helps to evaluate the reviews and helps to counter negative eWOM to get back the lost consumers. So, the study discusses theoretically that there is a need for a systematic study to simplify the understanding of eWOM effects on customer satisfaction. The present paper proposes a bibliometric analysis of electronic word of mouth impact on customer satisfaction.

LITERATURE REVIEW

Electronic Word of Mouth

With the rise of digital age, consumers are turning more frequently to electronic word of mouth (e-WOM) as valuable source of information for getting outside perspectives on product and service in purchase decision-making process (Verma & Yadav, 2021). The customers planning for trips are found to perceive the customer generated content such as consumer reviews (Buhalis & Law, 2008; Litvin et al., 2008) more trustworthy in comparison to official destination websites, and travel agents while making decisions on accommodation, visit attractions and dine-out (Dickinger, 2011; Fotis et al., 2012). Feedback from customers is significant for assessing website, gathering input on web designs for facilitating site fabrication according to purpose (Corbitt et al., 2003a). The consumers relying on consumer generated content facilitate crucial turn in marketing strategies, where there is positive influence of preservation of natural tourist attractions, scenic viewpoints, nature-based activities on e-WOM-sharing by tourists (Meenakshy et al., 2024). The published scholarly articles on Electronic Word of Mouth have given preference to urge of buyers who submit feedback (Hennig-Thurau et al., 2004), on the impact of customer reviews on buying intention of other consumers (Senecal & Nantel, 2004) and on the role of customer feedback in affecting sales of several products (Dellarocas, 2003; Godes & Mayzlin, 2002; Zhu & Zhang, 2010). The constructs of perceived ease of use (PEU), perceived usefulness (PU) and perceived value (PV) have a positive influence on electronic word of mouth (e-WOM) which influences the consumers decision-making of visiting and revisiting destinations (Madi et al., 2024). Awad & Ragowsky, (2008a) have found the effect of gender on E-commerce where women's rely heavily on the online reviews before making buying decisions in comparison on men's and also women's tendency to downplay on online reviews in relation to e-WOM quality reflects on aversion to dominating conversations, whereas men's preferences for posting online reflects their inclination towards dominant conversations.

eWOM refers to the sharing of product or service-related information through digital platforms such as social media, online reviews, blogs, forums, and other online channels. eWOM has emerged as a powerful tool for consumers to communicate with each other about brands, products, and services, and has a significant impact on the branding of products and services. The social media applications, web tools, and platforms like Facebook, WhatsApp, LinkedIn and Instagram not only facilitates in collaboration and content sharing but also increase brand awareness, brand image which creates competitive advantage (Faisal & Ekawanto, 2022; Suki et al., 2016). In this literature review, we will explore the concept of eWOM and its impact on branding. One of the earliest studies on eWOM was conducted by Hennig-Thurau et al., (2004) defined eWOM as "any positive or negative statement made by potential, actual, or former customers about a product or company, which is made available to a multitude of people and institutions via the Internet." They found that eWOM had a significant impact on consumer behavior, particularly in the context of online shopping. Consumers rely on eWOM to make purchase decisions, with positive reviews having a stronger influence than negative reviews. The positive eWOM increased brand loyalty, whereas negative eWOM decreased brand loyalty. The study also found that eWOM also plays a significant role in brand crisis management, in the event of a brand crisis, consumers turn to eWOM to gather information and to express their opinions. Another study by (Cheung et al., 2009) investigated the impact of eWOM on brand image and found that positive eWOM significantly enhanced brand image, whereas negative eWOM had a negative impact on brand image. The study also found that the influence of eWOM on brand image was stronger when the source of eWOM was perceived as credible and trustworthy. Research has also shown that eWOM has a significant impact on brand loyalty. A study by (Chen & Xie, 2008) found that effective crisis management communication strategies can help to mitigate the negative impact of eWOM on branding.

Customer Satisfaction

The information collected from online customer behavior has influence on sales, firms' valuation, performance, and developing customer satisfaction (Huang & Crotts, 2019). In the E-commerce and Big Data era, reviews from customers in the form of comments, ratings, and eWOM on the online platforms have substantial business value. The ratings from customers indicate their satisfaction which shows consumption experiences, attributes of customer care, gives comprehensive customer perception which facilitates in designing better feedback system and marketing strategies and to enhance product and services (Zhao et al., 2019a). The paradigm shift towards AI calls for unbiased monitoring implementation of surveys and customer-satisfaction scores (Fry, 2024), since the fake reviews can cause damage to innocent customers (Akhtar et al., 2019). The ratio of fake review ranges from 10% (Hu et al., 2012) to 33% (Salehi-Esfahani & Ozturk, 2018). Fry & Brint, (2025), presents a model for non-fake reviews to estimate the validity and correction while assessing the comprehensive rating for goods or service. The online platforms want to improve their service quality on customer satisfaction, trust, and consumer loyalty level, to increase the retention rate and attract new purchasers through WOM and behavioral loyalty. The consumers belonging to different cultural backgrounds have different reactions to certain factors of customer loyalty (Kassim & Asiah Abdullah, 2010a). Kakeesh et al., (2024) has proposed three dimensions model of Information quality, Interactivity, and ease of use to show impact of e-loyalty by strengthening the customer relationship with current and gain new customers in banking industry.

RESEARCH QUESTIONS

The aim of the present study is to make a substantial contribution in the research field of influence of eWOM on customer satisfaction by identifying the gaps and addressing the underlying research problems. At first, the study contributes to existing literature by yielding extensive construction to subject areas through a bibliometric analysis. This review intends to furnish the authors with the streamlined data on the current state of research. There is an absence of bibliometric analysis on this subject field in existing research work. Therefore, this study contributes to research by addressing the following questions:

RQ1: What is the current trend of scholarly literature of the subject area?

The analysis of the trends of existing literature facilitates the understanding about how the research field has evolved over the period and it encourages the researchers to anticipate the challenges and potential. The insights also hold potential to contribute new knowledge while discovering gaps. For the present study, Scopus database is used to extract the papers published on E-wom and customer satisfaction over the years.

RQ2: What are the high-performing countries, sources, and authors of the subject area?

Identifying the top contributors of research field in terms of countries, and sources provide insights into global advancements. It also facilitates the researchers to benchmark their works through collaboration opportunities with notable authors of the subject area. In this study, the top 10 most contributing countries, sources, affiliations and authors are identified in terms of publications and citations.

RQ3: What is the top 10 highest cited scholarly literature of the subject area?

Exploring research articles with the highest impact in the field provides the direction of the development of emerging research. This can be ascertained by bibliographic coupling, citations per article and publication count.

RQ4: What are the currents themes, gaps of the existing research, and what is the scope for future research?

Exploring the current themes of subject areas not only accentuates the prominent research areas but also paves the way for identifying the niche and emerging themes. Along with that, it also helps in unveiling gaps within the previously published literatures.

RESEARCH METHODOLOGY

The data for current research for bibliometric analysis was extracted from Scopus Database. The first article on E-WOM and customer Satisfaction was published in 2002. Therefore, the time span in this study ranges from 2002 to 2024. The search string for study TITLE-ABS-KEY, "Electronic Word of Mouth OR Ewom AND Customer Satisfaction OR Consumer Satisfaction". The search resulted in 278 articles which were further screened on the basis of inclusion and exclusion criteria. This analysis is based on all emerging and continually evolving research, that's why authors have focused on all sources including conference papers but excluding articles in the press. To enhance the understanding of research, authors have considered English language only, which resulted in 265 articles. The authors began the analysis using Bibliometrix statistical package of Rstudio, which allowed both qualitative and quantitative information such as Documents, authors, countru, sources and word cloud analysis. The authors considered VOSviewer network visualization tool for bibliometric coupling and cluster analysis, which facilitates in identifying trends and gaps.

RESULTS

Bibliometric analysis was performed on 265 documents, which started through showing data set characteristics. The analysis of research publications and citation growth is significant for research as it facilitates in identifying the emerging trends and gaps, along with that it also serves as a guide for future research directions. Therefore, this study presents initial research flow of information, country productivity, authors productivity, sources productivity, thematic and intellectual structure through cluster analysis.

Academic Contribution Metrics

Figure 1 illustrates the output in terms of publications and average citations from scholarly research publications of the subject area for the period ranging from 2002 to 2024. It also helps researchers to determine the areas where there has been stagnant research growth and the areas which are gaining traction and require further investigation. The first article on electronic word of mouth and customer satisfaction was published in 2002. Since then, there has been a static growth in document publications and in terms of average citations, fluctuations can be clearly seen. 2023 marked the highest number of publications which is 31 published documents, while 2003 has highest mean citations per article which 31.77 citations.

Figure 1. Trends in publications and citations impact

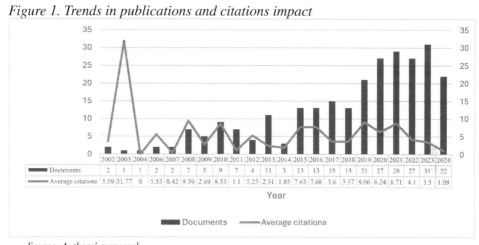

Source: Authors' own work

Academic Contribution by Countries

Figure 2 represents the scholarly research output of various countries in terms of published articles and citations received. The analysis of country-wise publication trends helps researchers to find out leaders and laggards, enhance knowledge transfer, collaboration opportunities to boost research capabilities, and drive socio-economic development for their nation. USA is leading with highest numbers of publications and citations count while 69 and 1514 respectively, followed by China with 112 documents and 594 citations, which is followed by India with 68 documents and 410 citations. Spain is second leading country in citations with 796 citations with only 40 publications, followed by Australia with 761 citations and 19 published documents. This indicates the substantial significance of quality and relevance of research, rather than quantity.

Figure 2. Country-wise trends of publications and citations impact

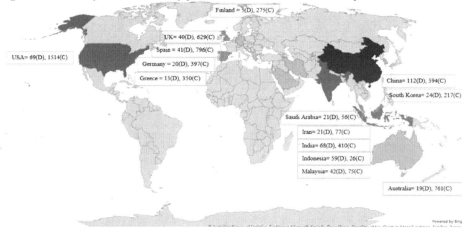

Source: Authors' own work

Academic Contribution by Authors

Figure 3 shows the scholarly contribution by authors in the field of electronic word of mouth and customer satisfaction. Among authors, Xu, X., has highest number of publications and citations with 7 publications and 896 citations over the years, while Srivastava, M., Wang, Y S., and Li, Y., has 4 publications with 100, 135 and 333 citations respectively.

Figure 3. Author-wise trends of publications and citations impact

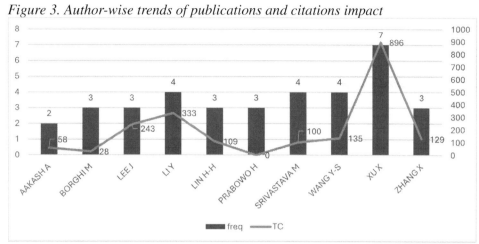

Source: Authors' own work

Academic Contribution by Sources

The analysis of contribution from different sources in the subject area of electronic word of mouth and customer satisfaction facilitates the identification of the most impactful journals to map the research landscape, and foster collaborations. Figure 4 depicts the source productivity using Bradfords law, which describes the scattering and efficiency of scholarly literature withing different journals. Journals of field are divided into 3 parts where each part consists of equal numbers of publications within the research field. Out of top 15 sources, 7 belongs to zone 1 which represents the most influential sources of the subject area which has highest number of published documents. *Sustainability (Switzerland)* has highest number of published documents which is 10 documents, followed by *ACM International Conference Proceeding Series, International Conference on Information Systems (ICIS 2013): Reshaping Society Through Information Systems Design, Journal Of Hospitality And Tourism Technology, Journal Of Retailing And Consumer Services* which is have 5 published documents each.

Figure 4. Source-wise trends of publications

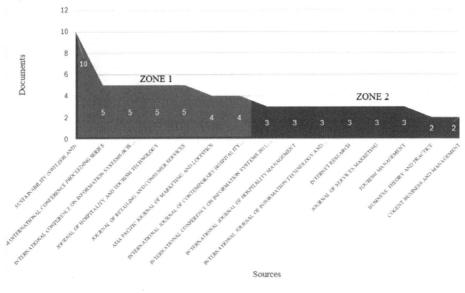

Source: Authors' own work

Key Influential Documents

The most impactful studies on research areas are measured on the basis of citations count as well as the quality, where articles not only answer significant questions but also introduce the new frontiers that influence future research. The citations count is measured by analyzing the articles which are referenced by other articles for research work. Table 1 illustrates documents which have the highest impact in the research field. The document titled, "Trust and e-commerce: a study of consumer perceptions" has highest number of citations, with a total of 699 citations which has 31.77 citations per year, followed by document titled, "Why do travelers trust TripAdvisor? Antecedents of trust towards consumer-generated media and its influence on recommendation adoption and word of mouth" with 579 citations which ha 57.9 citations per year and document titled, "Establishing Trust in Electronic Commerce Through Online Word of Mouth: An Examination Across Genders" with 490 citations and 28.82 citations per year. The analysis shows that there has been substantial research work getting done to understand customer satisfaction through electronic word of mouth.

Table 1. Top-cited documents

Authors	Title	Citations	Total Citations per year
(Corbitt et al., 2003b)	Trust and e-commerce: a study of consumer perceptions	699	31.77
(Filieri et al., 2015)	Why do travelers trust TripAdvisor? Antecedents of trust towards consumer-generated media and its influence on recommendation adoption and word of mouth	579	57.9
(Awad & Ragowsky, 2008b)	Establishing Trust in Electronic Commerce Through Online Word of Mouth: An Examination Across Genders	490	28.82
(Kassim & Asiah Abdullah, 2010b)	The effect of perceived service quality dimensions on customer satisfaction, trust, and loyalty in e-commerce settings	472	31.47
(Zhao et al., 2019b)	Predicting overall customer satisfaction: Big data evidence from hotel online textual reviews	366	61
(Xu & Li, 2016)	The antecedents of customer satisfaction and dissatisfaction toward various types of hotels: A text mining approach	325	36.11
(Casaló et al., 2008)	The role of satisfaction and website usability in developing customer loyalty and positive word-of-mouth in the e-banking services	310	18.24
(Mustak et al., 2021)	The role of satisfaction and website usability in developing customer loyalty and positive word-of-mouth in the e-banking services	244	61
(Pantelidis, 2010)	Electronic Meal Experience: A Content Analysis of Online Restaurant Comments	244	16.27
(Gounaris et al., 2010)	An examination of the effects of service quality and satisfaction on customers' behavioral intentions in e-shopping	213	14.2

Three Field Plot

Sankey diagram is a data visualization tool that shows the flow from one category to another. In Bibliometric analysis, three field plots which is a specific type of Sankey diagram is employed which showcases the connection between 3 categories such as Keywords, countries, authors and sources. Figure 5 shows the relationship between authors, keywords and author countries. The authors have maximum connection with keywords such as Customer satisfaction EWOM and electronic word of mouth and these keywords have maximum connection with China, India and Indonesia. This represents the growing importance of E-WOM & customer satisfaction in Asian countries.

Figure 5. Three field plot

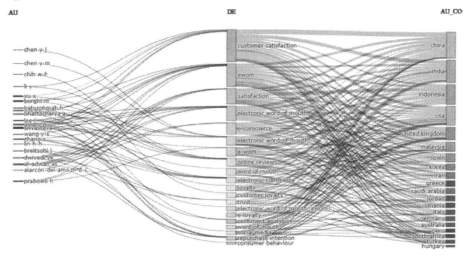

Source: Biblioshiny

Word Cloud

Figure 6 illustrates the word cloud on the basis of the highest occurrences of keywords in the subject area using biblioshiny. The word cloud visually summarizes and interprets the textual data for trend identifications, comparisons and landscape research. The font size of the keywords is directly proportional to its occurrences in the field. The bigger front-size keyword *customer satisfaction* represents its highest number of occurrences, which followed by *electronic commerce, sales* and *electronic word of mouths.*

Figure 6. Word cloud

Source: Biblioshiny

Bibliographic Coupling

Bibliographic Coupling is a widely used metric for identifying conceptual similarity and relation of research in citing a document. Bibliographic couple occurs when two published articles cite the third published article in their research work. It is not only important for identifying the association between two citing documents but also provides insights on the structure of research fonts of different documents, authors and sources. Figure 7 shows the bibliographic coupling of different cited articles using VOS viewer, where (Bhattacharya et al., 2019) has received 36 citations but highest number of link strength as 111. Similarly, (Bhattacharya & Srivastava, 2020), has received 22 citations but has 98 link strength. Furthermore, (Serra-Cantallops et al., 2020), has received 106 citations but it has only 51 link strength.

Figure 7. Bibliometric coupling

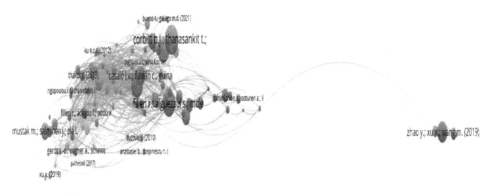

Source: VOS viewer

Table 2. Bibliographic coupling of documents

Document Title	Link Strength	Citations
(Bhattacharya et al., 2019). Customer Experience in Online Shopping: A Structural Modeling Approach, *Journal of Global Marketing (2019)32(1) 3-16*	111	36
(Bhattacharya & Srivastava, 2020). A Framework of Online Customer Experience: An Indian Perspective, *Global Business Review (2020)21(3) 800-817*	98	22
(Ruiz-Alba et al., 2022). Digital platforms: customer satisfaction, eWOM and the moderating role of perceived technological innovativeness, *Information Technology & People (2022)35(7) 2470-2499*	69	30
(Lin & Wang, 2015). Examining E-Commerce Customer Satisfaction and Loyalty: An Integrated Quality-Risk-Value Perspective, *Journal of Organizational Computing and Electronic Commerce (2015)25(4) 379-401*	64	32
(Ismagilova et al., 2020). The effect of characteristics of source credibility on consumer behaviour: A meta-analysis, *Journal of Retailing and Consumer Services (2020)53 101736*	59	101
(Camilleri, 2021). E-Commerce Websites, Consumer Order Fulfillment and After-Sales Service Satisfaction: The Customer Is Always Right, even after the Shopping Cart Check-Out!, *SSRN Electronic Journal (2021)*	58	33
(Izogo & Jayawardhena, 2018). Online shopping experience in an emerging e-retailing market, *Journal of Research in Interactive Marketing (2018)12(2) 193-214*	57	89
(Hsu et al., 2013). Effects of web site characteristics on customer loyalty in B2B e-commerce: evidence from Taiwan, *The Service Industries Journal (2013)33(11) 1026-1050*	52	32

continued on following page

Table 2. Continued

Document Title	Link Strength	Citations
(Serra-Cantallops et al., 2020). Antecedents of positive eWOM in hotels. Exploring the relative role of satisfaction, quality and positive emotional experiences, *International Journal of Contemporary Hospitality Management (2020)32(11) 3457-3477*	51	106
(Srivastava & Sivaramakrishnan, 2021). The impact of eWOM on consumer brand engagement, *Marketing Intelligence & Planning (2021)39(3) 469-484*	49	42

Cluster Analysis

Cluster Analysis holds its significance in elucidating vast academic data for identifying the hidden patterns in data. Similarity and criterion function are important metrics in cluster analysis. In VOSviewer, the formation of clusters or groups are based on co-occurrence of keywords through normalization and community detection algorithms. The group formed through categorized keywords facilitates identification of main themes, niche themes and emerging themes for the research area. In the present study, minimum number of co-occurrences of keywords is set at 7 times and out of 708 keywords, 72 meets threshold. Table 3 illustrates the formation of nine clusters in subject area of eWOM and Customer Satisfaction. Figure 8 shows the keywords are categorized into different clusters and shown in different colors. Cluster one represents the theme of the role of electronic word of mouth in building Loyalty through E-trust, and E-satisfaction and website quality in E-commerce. Cluster two represents the theme of significance of customer loyalty in digital era through perceived risk, quality, value and relationship quality. Cluster three represents the theme of Driving Repurchase intention through social media marketing, brand loyalty, word of mouth and customer satisfaction. Cluster four represents the theme of the role of content analysis by using text mining and user generated content for enhancing customer experience. Cluster five represents the theme of the enhancing the customer engagement in digital tourism through e-commerce, online reviews and relationship marketing. Cluster six represents the theme of Credibility, word of mouth and ease of use as drivers of user engagement. Cluster seven represents the theme of Machine learning and big data for ewom sentiment analysis. Cluster eight represents the theme of Customer involvement for perceived justice and service recovery. Cluster nine represents the theme of Technological influence on brand image and purchase intention.

Table 3. Cluster analysis

Clusters Theme	Keywords	Co-occurrences	Link Strength
Cluster 1 E-WOM building E-trust and E-loyalty through information and website quality in E-commerce	E-Commerce	21	45
	E-Loyalty	8	24
	E-satisfaction	7	24
	E-trust	4	16
	E-wom	43	94
	Electronic Word Of Mouth	28	68
	Information Quality	9	37
	Online Shopping	8	55
	Online Trust	3	12
	Online word of mouth	3	5
	Website quality	3	10
Cluster 2 Perceived risk, value and quality building customer loyalty in digital age	customer loyalty	13	31
	digital marketing	3	10
	loyalty	12	32
	perceived quality	4	9
	perceived risk	5	10
	perceived value	7	18
	product type	3	6
	quality	3	11
	relationship quality	4	9
	reputation	3	11
	satisfaction	32	72
	trust	12	37
Cluster 3 Brand loyalty, Word-of-mouth and Consumer satisfaction influence on Repurchase intention	brand loyalty	3	4
	consumer	5	8
	satisfaction	21	29
	electronic word-of-mouth	3	4
		8	22
	online customer experience	12	26
		3	7
	repurchase intention	3	8
	social media	11	18
	social media marketing		
	social networks		
	word-of -mouth		
Cluster 4 Content analysis through Text mining and user generated content in online reviews for customer reviews	content analysis	3	8
	customer experience	3	4
	customer reviews	3	7
	customer satisfaction	63	122
	electronic word of	9	13
	mouth (ewom)	4	4
	online customer reviews	15	28
	online reviews	5	12
	text mining	3	4
	user generated content		

continued on following page

Table 3. Continued

Clusters Theme	Keywords	Co-occurrences	Link Strength
Cluster 5 Customer Engagement in tourism through electronic commerce, online reviews & relationship marketing	customer engagement consumer behaviour electronic commerce Internet online review relationship marketing tourism	4 7 14 4 3 4 3	7 18 32 13 8 3 5
Cluster 6 Credibility, word of mouth and ease of use drives the user engagement	Credibility Ease of use Engagement sharing economy word of mouth	3 4 3 3 16	7 10 9 6 44
Cluster 7 Machine learning and big data for ewom sentiment analysis	big data Ewom machine learning sentiment analysis	6 43 6 8	13 94 14 18
Cluster 8 Customer involvement for perceived justice and service recovery	Involvement perceived justice service recovery	3 3 7	3 5 11
Cluster 9 Technological influence on brand image and purchase intention	brand image purchase intention technology	3 6 3	11 14 13

Figure 8. Cluster analysis

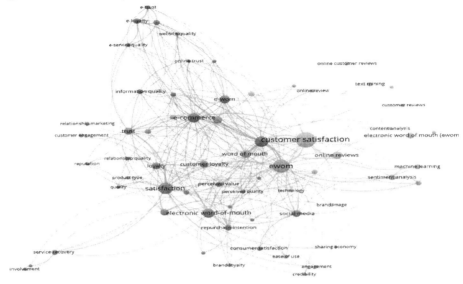

Source: VOS viewer

DISCUSSION

The purpose of this study is to investigate the qualitative and quantitative variables such as Authors, Country, sources, citations and themes. Through that, some developing and niche insights on the subject area of electronic word of mouth and customer satisfaction have emerged. For instance, Xu, X., author have published highest number of articles in the research field of electronic word of mouth and customer satisfaction. Their contribution has substantially focused on customer satisfaction and online reviews among tourists and hotel customers.

Additionally, the present study also provides insights about the sources which has highest number of publish documents which is *"Trust and e-commerce: a study of consumer perceptions"* with 699 citations followed by *"Why do travelers trust TripAdvisor? Antecedents of trust towards consumer-generated media and its influence on recommendation adoption and word of mouth"* and *"Establishing Trust in Electronic Commerce Through Online Word of Mouth: An Examination Across Genders"* with 579 and 490 citations respectively.

From the analysis conducted it was also found that there were substantial publications on electronic word of mouth, customer satisfaction, website quality, relationship marketing, customer engagement and customer behaviour. From this we can say that the central interest of the study is to understand the behaviour of consumer while making a purchase and the significance of electronic word of mouth on purchase decision.

Keyboard analysis on electronic word of mouth and consumer satisfaction provides some insights. Cluster one lays emphasis on significance off electronic word of mouth in developing electronic trust and loyalty by enhancing website quality in electronic commerce. Cluster two is related to finding out the perceived risk, perceived value, and perceived quality for setting customer loyalty through digital marketing. Cluster 3 is related to brand loyalty, word-of-mouth and customer satisfaction influence on repurchase intention. Cluster four shows the significance of content analysis through text mining and user generated content in online reviews for customer reviews. Cluster five is related to customer engagement in tourism through electronic commerce, online reviews and relationship marketing. Cluster six shows of credibility, word of mouth and ease of use driving the user engagement. Cluster seven related to machine learning and big data for electronic word of mouth sentimental analysis. Cluster eight is related to involvement of customers for perceived justice and service recovery. Cluster nine represents the influence of technology on brand image and purchase intention.

However, this study is not free from certain limitations. Firstly, the data collection period is till October 2024, therefore the articles published in subsequent 3 months were not taken into account. Secondly, this considered only articles published in English language from Scopus database, which might increase the chance of missing out some relevant articles. Lastly, Bibliometric analysis entails subjectivity, which influences outcomes of analysis and researchers' judgements. Different researchers might take different decisions to interpret the relationship between nodes which can influence final insights of study.

Implications and Future Propositions

As a result of in-depth research from published documents of the most cited authors, a few challenges and gaps identified to promote future research studies in the field of Electronic Word of mouth and customer satisfaction. This analysis reveals several issues related to electronic word of mouth and customer satisfaction are not fully explored and require extensive examination. Future research can be done on electronic word of mouth and Brand Equity, Brand intention and Brand image. This would help entrepreneurs to understand the influence of eWOM on consumers and strengthen their Brand Image and Equity and enhance satisfaction with current and future consumers.

CONCLUSION

To conclude, our study has utilized science mapping workflow and addressed several research questions to explore the wide academic publications in the subject area of electronic word of mouth and customer satisfaction which was not brought to the forefront of research. This study showcased comprehensive quantitative and qualitative assessment on bibliometric variables and provides theoretical understanding. Electronic word of mouth has turned out to be crucial part of consumer decision making and social proof, where Positive eWOM can enhance brand awareness, improve customer satisfaction, brand reputation, and increase consumer loyalty, while negative eWOM can damage a brand's reputation, reduce customer trust, and result in lost sales. Trust and credibility are crucial elements in the success of eWOM, and social media has become a popular channel for eWOM. Brands must monitor and respond to negative eWOM in a timely and effective manner to mitigate its impact on customer loyalty and its impact on customer satisfaction. Brands need to be aware of the impact of eWOM and develop effective strategies to enhance customer satisfaction and manage their online reputation and to encourage positive eWOM.

REFERENCES

Akhtar, N., Ahmad, W., Siddiqi, U. I., & Akhtar, M. N. (2019). Predictors and outcomes of consumer deception in hotel reviews: The roles of reviewer type and attribution of service failure. *Journal of Hospitality and Tourism Management, 39,* 65–75. DOI: 10.1016/j.jhtm.2019.03.004

Awad, N. F., & Ragowsky, A. (2008a). Establishing Trust in Electronic Commerce Through Online Word of Mouth: An Examination Across Genders. *Journal of Management Information Systems, 24*(4), 101–121. DOI: 10.2753/MIS0742-1222240404

Awad, N. F., & Ragowsky, A. (2008b). Establishing Trust in Electronic Commerce Through Online Word of Mouth: An Examination Across Genders. *Journal of Management Information Systems, 24*(4), 101–121. DOI: 10.2753/MIS0742-1222240404

Bhat, S., & Chakraborty, U. (2018). Online reviews and its impact on brand equity. *International Journal of Internet Marketing and Advertising, 12*(2), 159. DOI: 10.1504/IJIMA.2018.10011683

Bhattacharya, A., & Srivastava, M. (2020). A Framework of Online Customer Experience: An Indian Perspective. *Global Business Review, 21*(3), 800–817. DOI: 10.1177/0972150918778932

Bhattacharya, A., Srivastava, M., & Verma, S. (2019). Customer Experience in Online Shopping: A Structural Modeling Approach. *Journal of Global Marketing, 32*(1), 3–16. DOI: 10.1080/08911762.2018.1441938

Buhalis, D., & Law, R. (2008). Progress in information technology and tourism management: 20 years on and 10 years after the Internet—The state of eTourism research. *Tourism Management, 29*(4), 609–623. DOI: 10.1016/j.tourman.2008.01.005

Camilleri, M. (2021). E-Commerce Websites, Consumer Order Fulfillment and After-Sales Service Satisfaction: The Customer Is Always Right, even after the Shopping Cart Check-Out! SSRN *Electronic Journal.* DOI: 10.2139/ssrn.3853156

Casaló, L. V., Flavián, C., & Guinalíu, M. (2008). The role of satisfaction and website usability in developing customer loyalty and positive word-of-mouth in the e-banking services. *International Journal of Bank Marketing, 26*(6), 399–417. DOI: 10.1108/02652320810902433

Chen, Y., & Xie, J. (2008). Online Consumer Review: Word-of-Mouth as a New Element of Marketing Communication Mix. *Management Science, 54*(3), 477–491. DOI: 10.1287/mnsc.1070.0810

Cheung, C. M. K., Lee, M. K. O., & Thadani, D. R. (2009). *The Impact of Positive Electronic Word-of-Mouth on Consumer Online Purchasing Decision.*, DOI: 10.1007/978-3-642-04754-1_51

Corbitt, B. J., Thanasankit, T., & Yi, H. (2003a). Trust and e-commerce: A study of consumer perceptions. *Electronic Commerce Research and Applications*, *2*(3), 203–215. DOI: 10.1016/S1567-4223(03)00024-3

Corbitt, B. J., Thanasankit, T., & Yi, H. (2003b). Trust and e-commerce: A study of consumer perceptions. *Electronic Commerce Research and Applications*, *2*(3), 203–215. DOI: 10.1016/S1567-4223(03)00024-3

Dellarocas, C. N. (2003). The Digitization of Word-of-Mouth: Promise and Challenges of Online Feedback Mechanisms. SSRN *Electronic Journal*. DOI: 10.2139/ssrn.393042

Dickinger, A. (2011). The Trustworthiness of Online Channels for Experience- and Goal-Directed Search Tasks. *Journal of Travel Research*, *50*(4), 378–391. DOI: 10.1177/0047287510371694

Dwivedi, Y. K., Ismagilova, E., Hughes, D. L., Carlson, J., Filieri, R., Jacobson, J., Jain, V., Karjaluoto, H., Kefi, H., Krishen, A. S., Kumar, V., Rahman, M. M., Raman, R., Rauschnabel, P. A., Rowley, J., Salo, J., Tran, G. A., & Wang, Y. (2021). Setting the future of digital and social media marketing research: Perspectives and research propositions. *International Journal of Information Management*, *59*, 102168. DOI: 10.1016/j.ijinfomgt.2020.102168

Faisal, A., & Ekawanto, I. (2022). The role of Social Media Marketing in increasing Brand Awareness, Brand Image and Purchase Intention. *Indonesian Management and Accounting Research*, *20*(2), 185–208. DOI: 10.25105/imar.v20i2.12554

Filieri, R., Alguezaui, S., & McLeay, F. (2015). Why do travelers trust TripAdvisor? Antecedents of trust towards consumer-generated media and its influence on recommendation adoption and word of mouth. *Tourism Management*, *51*, 174–185. DOI: 10.1016/j.tourman.2015.05.007

Fotis, J., Buhalis, D., & Rossides, N. (2012). Social Media Use and Impact during the Holiday Travel Planning Process. In *Information and Communication Technologies in Tourism 2012* (pp. 13–24). Springer Vienna., DOI: 10.1007/978-3-7091-1142-0_2

Fournier, S., & Avery, J. (2011). The uninvited brand. *Business Horizons*, *54*(3), 193–207. DOI: 10.1016/j.bushor.2011.01.001

Fry, J. (2024). Revisiting student evaluation of teaching during the pandemic. *Applied Economics Letters*, *31*(14), 1259–1263. DOI: 10.1080/13504851.2023.2178623

Fry, J., & Brint, A. (2025). Customer satisfaction scores: New models to estimate the number of fake reviews. *Tourism Management, 106*, 105030. DOI: 10.1016/j.tourman.2024.105030

Godes, D., & Mayzlin, D. (2002). Using Online Conversations to Study Word of Mouth Communication. SSRN *Electronic Journal*. DOI: 10.2139/ssrn.327841

Gounaris, S., Dimitriadis, S., & Stathakopoulos, V. (2010). An examination of the effects of service quality and satisfaction on customers' behavioral intentions in e-shopping. *Journal of Services Marketing, 24*(2), 142–156. DOI: 10.1108/08876041011031118

Hendricks, S., & Mwapwele, S. D. (2024). A systematic literature review on the factors influencing e-commerce adoption in developing countries. *Data and Information Management, 8*(1), 100045. DOI: 10.1016/j.dim.2023.100045

Hennig-Thurau, T., Gwinner, K. P., Walsh, G., & Gremler, D. D. (2004). Electronic word-of-mouth via consumer-opinion platforms: What motivates consumers to articulate themselves on the Internet? *Journal of Interactive Marketing, 18*(1), 38–52. DOI: 10.1002/dir.10073

Hsu, L.-C., Wang, K.-Y., & Chih, W.-H. (2013). Effects of web site characteristics on customer loyalty in B2B e-commerce: Evidence from Taiwan. *Service Industries Journal, 33*(11), 1026–1050. DOI: 10.1080/02642069.2011.624595

Hu, N., Bose, I., Koh, N. S., & Liu, L. (2012). Manipulation of online reviews: An analysis of ratings, readability, and sentiments. *Decision Support Systems, 52*(3), 674–684. DOI: 10.1016/j.dss.2011.11.002

Huang, S., & Crotts, J. (2019). Relationships between Hofstede's cultural dimensions and tourist satisfaction: A cross-country cross-sample examination. *Tourism Management, 72*, 232–241. DOI: 10.1016/j.tourman.2018.12.001

Ismagilova, E., Slade, E., Rana, N. P., & Dwivedi, Y. K. (2020). The effect of characteristics of source credibility on consumer behaviour: A meta-analysis. *Journal of Retailing and Consumer Services, 53*, 101736. DOI: 10.1016/j.jretconser.2019.01.005

Izogo, E. E., & Jayawardhena, C. (2018). Online shopping experience in an emerging e-retailing market. *Journal of Research in Interactive Marketing, 12*(2), 193–214. DOI: 10.1108/JRIM-02-2017-0015

Kakeesh, D., Weshah, G. A., Al, , Ma', N., & Al, . (2024). Building e-loyalty through online banking features: Mediating role of e-trust. *International Journal of Management Practice, 17*(5), 577–599. DOI: 10.1504/IJMP.2024.140867

Kassim, N., & Asiah Abdullah, N. (2010a). The effect of perceived service quality dimensions on customer satisfaction, trust, and loyalty in e-commerce settings. *Asia Pacific Journal of Marketing and Logistics*, *22*(3), 351–371. DOI: 10.1108/13555851011062269

Kassim, N., & Asiah Abdullah, N. (2010b). The effect of perceived service quality dimensions on customer satisfaction, trust, and loyalty in e-commerce settings. *Asia Pacific Journal of Marketing and Logistics*, *22*(3), 351–371. DOI: 10.1108/13555851011062269

Khanom, M. T. (2023). Using social media marketing in the digital era: A necessity or a choice. *International Journal of Research in Business and Social Science (2147- 4478)*, *12*(3), 88–98. DOI: 10.20525/ijrbs.v12i3.2507

Lim, C. H., Chung, J. J., & Pedersen, P. M. (2012). Effects of Electronic Word - of - Mouth Messages. *Chorigia*, *8*(1), 55–76. DOI: 10.4127/ch.2012.0064

Lin, M.-J., & Wang, W.-T. (2015). Examining E-Commerce Customer Satisfaction and Loyalty: An Integrated Quality-Risk-Value Perspective. *Journal of Organizational Computing and Electronic Commerce*, *25*(4), 379–401. DOI: 10.1080/10919392.2015.1089681

Litvin, S. W., Goldsmith, R. E., & Pan, B. (2008). Electronic word-of-mouth in hospitality and tourism management. *Tourism Management*, *29*(3), 458–468. DOI: 10.1016/j.tourman.2007.05.011

Madi, J., Al Khasawneh, M., & Dandis, A. O. (2024). Visiting and revisiting destinations: Impact of augmented reality, content quality, perceived ease of use, perceived value and usefulness on E-WOM. *International Journal of Quality & Reliability Management*, *41*(6), 1550–1571. DOI: 10.1108/IJQRM-10-2023-0314

Meenakshy, M., Prasad, K. D. V., Bolar, K., & Shyamsunder, C. (2024). Electronic word-of-mouth intentions in personal and public networks: A domestic tourist perspective. *Humanities & Social Sciences Communications*, *11*(1), 1226. DOI: 10.1057/s41599-024-03753-4

Mustak, M., Salminen, J., Plé, L., & Wirtz, J. (2021). Artificial intelligence in marketing: Topic modeling, scientometric analysis, and research agenda. *Journal of Business Research*, *124*, 389–404. DOI: 10.1016/j.jbusres.2020.10.044

Pantelidis, I. S. (2010). Electronic Meal Experience: A Content Analysis of Online Restaurant Comments. *Cornell Hospitality Quarterly*, *51*(4), 483–491. DOI: 10.1177/1938965510378574

Ruiz-Alba, J. L., Abou-Foul, M., Nazarian, A., & Foroudi, P. (2022). Digital platforms: Customer satisfaction, eWOM and the moderating role of perceived technological innovativeness. *Information Technology & People, 35*(7), 2470–2499. DOI: 10.1108/ITP-07-2021-0572

Salehi-Esfahani, S., & Ozturk, A. B. (2018). Negative reviews: Formation, spread, and halt of opportunistic behavior. *International Journal of Hospitality Management, 74*, 138–146. DOI: 10.1016/j.ijhm.2018.06.022

Senecal, S., & Nantel, J. (2004). The influence of online product recommendations on consumers' online choices. *Journal of Retailing, 80*(2), 159–169. DOI: 10.1016/j.jretai.2004.04.001

Serra-Cantallops, A., Ramón Cardona, J., & Salvi, F. (2020). Antecedents of positive eWOM in hotels. Exploring the relative role of satisfaction, quality and positive emotional experiences. *International Journal of Contemporary Hospitality Management, 32*(11), 3457–3477. DOI: 10.1108/IJCHM-02-2020-0113

Shahriari, S., Mohammadreza, S., & Gheiji, S. (2015). E-commerce and its impacts on global trend and market. *International Journal of Research -GRANTHAALAYAH, 3*(4), 49–55. DOI: 10.29121/granthaalayah.v3.i4.2015.3022

Srivastava, M., & Sivaramakrishnan, S. (2021). The impact of eWOM on consumer brand engagement. *Marketing Intelligence & Planning, 39*(3), 469–484. DOI: 10.1108/MIP-06-2020-0263

Suki, N. M., Suki, N. M., & Azman, N. S. (2016). Impacts of Corporate Social Responsibility on the Links Between Green Marketing Awareness and Consumer Purchase Intentions. *Procedia Economics and Finance, 37*, 262–268. DOI: 10.1016/S2212-5671(16)30123-X

Verma, S., & Yadav, N. (2021). Past, Present, and Future of Electronic Word of Mouth (EWOM). *Journal of Interactive Marketing, 53*(1), 111–128. DOI: 10.1016/j.intmar.2020.07.001

Watson, F., & Wu, Y. (2022). The Impact of Online Reviews on the Information Flows and Outcomes of Marketing Systems. *Journal of Macromarketing, 42*(1), 146–164. DOI: 10.1177/02761467211042552

Xu, X., & Li, Y. (2016). The antecedents of customer satisfaction and dissatisfaction toward various types of hotels: A text mining approach. *International Journal of Hospitality Management, 55*, 57–69. DOI: 10.1016/j.ijhm.2016.03.003

Zhang, H., Zhao, L., & Gupta, S. (2018). The role of online product recommendations on customer decision making and loyalty in social shopping communities. *International Journal of Information Management*, *38*(1), 150–166. DOI: 10.1016/j.ijinfomgt.2017.07.006

Zhao, Y., Xu, X., & Wang, M. (2019a). Predicting overall customer satisfaction: Big data evidence from hotel online textual reviews. *International Journal of Hospitality Management*, *76*, 111–121. DOI: 10.1016/j.ijhm.2018.03.017

Zhao, Y., Xu, X., & Wang, M. (2019b). Predicting overall customer satisfaction: Big data evidence from hotel online textual reviews. *International Journal of Hospitality Management*, *76*, 111–121. DOI: 10.1016/j.ijhm.2018.03.017

Zhu, F., & Zhang, X. (2010). Impact of Online Consumer Reviews on Sales: The Moderating Role of Product and Consumer Characteristics. *Journal of Marketing*, *74*(2), 133–148. DOI: 10.1509/jm.74.2.133

Chapter 4
Leveraging IT for Enhanced Characterization of SiC and Ni Epoxy Composites in High Voltage Applications

Amit Chaurasia

ⓘ https://orcid.org/0000-0002-4711-7252

Amity University, Jaipur, India

Umesh kumar Dwivedi

Amity University, Jaipur, India

Amita Chaurasiya

ICFAI University, Jaipur, India

ABSTRACT

Tools have a major impact on the precision, effectiveness, and cost of manufacturing composite structures. In order to provide lightweight, affordable, long-lasting, and useful tools for producing high-performance composite structures, industry and academia must work closely together to leverage and investigate new developments in materials, structures, and tool manufacturing technologies. This is because the performance of composites components is becoming more and more desired. It is frequently problematic to provide insulation at higher voltages (kV range). In this work, we created a variety of epoxy resin samples utilising nickel and silicon

DOI: 10.4018/979-8-3693-8009-3.ch004

carbide, two conducting and non-conducting fillers. Using the centrifugal casting procedure, samples were created. We examined the composite's breakdown voltage, which will eventually offer insulation at greater voltages. The sample with a higher filler concentration had a good breakdown voltage, according to the data.

INTRODUCTION

Nowadays requirement of insulation at very high voltages is required for various applications, like electrical circuit breakers in Railway Lines where repairing at a particular distance is being done, that distance require high voltage insulation and at low cost. Researchers have been tried to improve various properties like the thermal conductivity, dielectric strength and mechanical strength (Dwivedi & Hashmi, 2009). But here we used different compound in order to improve the breakdown voltage of the developed composite at a lower cost and to use that composite for high voltage insulation application. Silicon Carbide highly hard, synthetically produced crystalline compound of silicon and carbon. It is a mixture of pure silica sand and carbon in the form of finely ground coke is built up around a carbon conductor within a brick electrical resistance-type furnace. Several investigators have reported very high strengths associated with the Ni epoxy composites for various application and characterization (Zimmerman et al., 2002). This study have taken specifically Silicon Carbide because it contains carbon compounds which is by far the most stable compound known, which makes SiC highly stable. Also the melting point is very high (3100K) which eventually leads to high stability. For the stability purpose other allotropes of carbon could have been used such as Graphite, Diamond etc.. But we preferred SiC over graphite and diamond because diamonds are very costly and graphite is relatively weaker than Silicon Carbide. That's the reason why Silicon Carbide is preferred over Graphite and Diamond (Dwivedi & Hashmi, 2009).

EXPERIMENTAL

Table 1. Materials used during the research

S.No.	Materials Used
1.	Epoxy Resin
2.	Curing Agent/hardner
3.	Commercially available Silicon Carbide(Grit Size: 320) and Nickel

Preparation of Composites

Filler Used: Silicon Carbide and Nickel

In this experiment we use commercially available 320 grit silicon Carbide which finer particle. Blending and Casting method was used to prepare the composites. This method involves some steps and methods for both SIC and nickel are same. They are as follows:

Epoxy Resin and hardener (Curing agent)(10:1) was mixed for around 2-3 minutes manually. After the manual mixing 5wt%, 10wt%, 15wt% & 20wt% of epoxy, silicon carbide was added and they were mixed for 6-7 minutes in order to obtain the homogeneous mixture of the composites.

This was done to increase the viscosity of the solution so that the filler particles can't move freely and the motion of filler particles can be decreased. Although viscosity is a negative parameter, but here viscosity plays an important role in settling down the particles so that after the curing process gradient feature can be extracted from the prepared sample.

The prepared composites were then kept at room temperature for desecration and then the composites were molded in polytetrafloroethylene (PTFE) moulds and moulds were transferred to a clean environment so that no unwanted particles could reach the composite. The composites were then left for curing process by normally cooling temperature. 99% of curing is done in 24 hrs at room temperature, rest 1% takes almost 1 week to cure.

CHARACTERIZATION

Characterization is the basic or fundamental process without which scientific study of material could not be done.

Breakdown Voltage

Breakdown Voltage is that minimum voltage of an insulator which is required to make a portion of insulator conductive. It is also known as striking voltage. It is used w.r.t. insulators. In other word the quantity of electrical force that is required for transforming electrical properties of an object.

- **For measuring breakdown voltage of samples**

In the measuring of the voltage breakdown ASTM standard is used. Breakdown Voltage is measured by filling the oil in the vessel of the testing device and the anode and cathode is connected. A sample is place in between the anode and the cathode and the voltage is continuously applied and at the highest an electric arc is produced and after that the test voltage is switched off automatically. In this way breakdown voltage is tested.

Dielectric Strength

The dielectric can be defined as material's electrical strength as an insulator. It is often measured as the ratio of maximum voltage required to produce a dielectric breakdown. Dielectric Strength is expressed in Volts per unit Thickness. Higher value of voltage indicates higher dielectric strength which individually shows better quality of insulator.

- **For measuring dielectric strength**

It is measured as the ratio of maximum voltage required to produce a dielectric breakdown. Dielectric Strength is expressed in Volts per unit Thickness. Higher value of voltage indicates higher dielectric strength which individually shows better quality of insulator.

MATHEMATICALLY,

$$\text{Dielectric Strength} = \frac{\text{Breakdown Voltage(in Volts)}}{\text{Thickness(in mm)}}$$

RESULT AND DISCUSSION

Table 2. Results of SIC epoxy composites for Breakdown voltage

S.No.	Weight Percentage	Thickness	Voltage Breakdown	Dielectric Strength
1.	0	5 mm	14 kV	2.8 kV/mm
2.	0 (With Centrifuse)	5 mm	15 kV	3 kV/mm
3.	5 (without Centrifuse)	5 mm	14 kV	2.8 kV/mm
4.	10 (without Centrifuse)	5 mm	17.5 kV	3.5 kV/mm

continued on following page

Table 2. Continued

S.No.	Weight Percentage	Thickness	Voltage Breakdown	Dielectric Strength
5.	15 (without Centrifuse)	5 mm	20 kV	4 kV/mm
6.	20 (without Centrifuse)	5 mm	21 kV	4.2 kV/mm
7.	5 (with Centrifuse)	5 mm	14.5 kV	2.9 kV/mm
8.	10 (with Centrifuse)	5 mm	18.5 kV	3.7 kV/mm
9.	15 (with Centrifuse)	5 mm	21 kV	4.2 kV/mm
10.	20 (with Centrifuse)	5 mm	22 kV	4.4 kV/mm

In this performance study various samples using epoxy resin were prepared with different concentrations of silicon carbide used as filler. Various calculations on the samples like size measurement, voltage breakdown measurement, dielectric strength and relative density was done. Half of the samples were centrifused using centrifugal machine and rest were manually mixed. The results have shown that the highest breakdown voltage was found in 20wt% of centrifused silicon carbide sample. This implies that this sample is far most the best samples amongst all the samples and it can be used for high voltage insulation applications. Also, the relative density of the samples has been calculated in order to check the how much the sample is dense.

Table 3. Results of nickel epoxy composites for breakdown voltage

S.No.	Weight Percentage	Thickness	Voltage Breakdown	Dielectric Strength
1.	0	5 mm	14 kV	4.2 kV/mm
2.	0 (With Centrifuse)	5 mm	15 kV	2.9 kV/mm
3.	5 (without Centrifuse)	5 mm	17.4 kV	3.48 kV/mm
4.	10 (without Centrifuse)	5 mm	18.5 kV	3.7 kV/mm
5.	15 (without Centrifuse)	5 mm	18.8 kV	3.76 kV/mm
6.	20 (without Centrifuse)	5 mm	20 kV	4 kV/mm
7.	5 (with Centrifuse)	5 mm	18.4 kV	3.68 kV/mm

continued on following page

Table 3. Continued

S.No.	Weight Percentage	Thickness	Voltage Breakdown	Dielectric Strength
8.	10 (with Centrifuse)	5 mm	17.5 kV	3.5 kV/mm
9.	15 (with Centrifuse)	5 mm	18.5 kV	3.7 kV/mm
10.	20 (with Centrifuse)	5 mm	21 kV	4.2 kV/mm

Figure 1. Graph of dielectric stength of centrifuse and without centrifuse samples

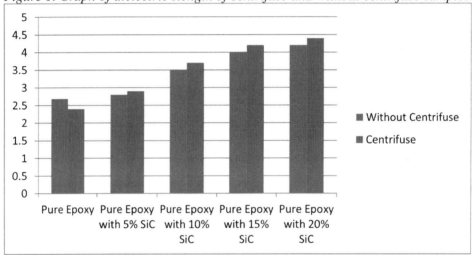

As the concentration increases in the sample dielectric strength is also increasing. Centrifuge sample having more dielectric strength because particle is more uniformly settled down at the bottom due to the centrifugal force applied on it.

Figure 2. Graph of voltage breakdown of centrifuse and without centrifuse sample

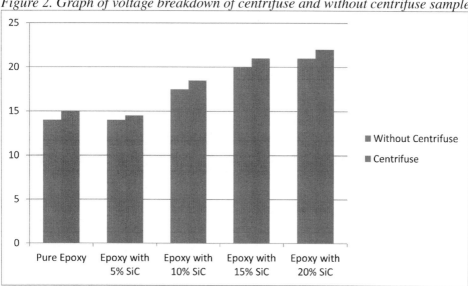

As the concentration increases voltage breakdown of the samples increase because SiC is settled down at the bottom this is the reason of the increase in the breakdown voltage. It follows the low resistance in the without centrifuse sample therefore the voltage breakdown in the without centrifuse sample is less than the centrifuse sample.

Figure 3. Graph of dielectric stength of centrifuse and without centrifuse samples

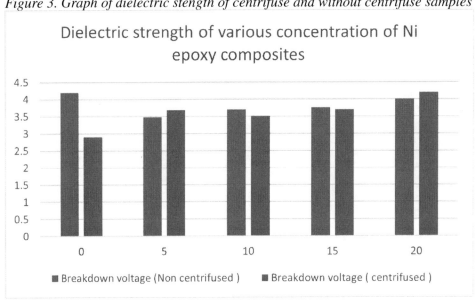

As the concentration increases in the sample dielectric strength is also increasing. Centrifuse sample having more dielectric strength because particle is more uniformly settled down at the bottom due to the centrifugal force applied on it.

CONCLUSION

In this study, 10 various types of samples were prepared for both Sic and Nickel epoxy composites to study the relation between breakdown voltage and thickness of a sample and to find a sample that can suit the high voltage insulation for practical applications. During this study firstly, samples were prepared then the thickness of all those samples was measured. Then, breakdown voltage was found using ASTM method. The result show that the sample having 20wt% of silicon carbide and Nickel has a very good dielectric strength. Meanwhile, we have also found that the break-down voltage of the sample is directly proportional to the thickness of the sample. That means the sample with higher percentage can be suitable for the insulation at higher voltages.To bridge the current divide between business and academia, this study reviews the primary tooling technologies developed for composites in the aerospace industry. It has been investigated how materials, architectures, and tool functions for the production of standard and sophisticated composites have advanced.

A review of the primary and sophisticated tool production processes has been conducted. Lastly, recommendations for tool design and manufacture for advanced composites structures have been provided, along with an overview of current and emerging developments in tooling technology.

REFERENCES

Bhadra, S. Mostafizur Rahaman and P. Noorunnisa Khanam "Electrical and Electronic Application of Polymer–Carbon Composites" , In book: Carbon-Containing Polymer Composites, pp.397-455.DOI: 10.1007/978-981-13-2688-2_12

Chand, N., Dwivedi, U. K., & Sharma, M. K. (2007). Navin Chand*, U.K. Dwivedi, M.K. Sharma "Development and tribological behaviour of UHMWPE filled epoxy gradient composites" [Science Direct.]. *Wear, 262*(1-2), 184–190. DOI: 10.1016/j.wear.2006.04.012

Chaurasia, A., Dwivedi, U. K., Kumari, N., Meena, S., Rathore, D., Hashmi, S. A. R., & Jain, D. (2023). Effect of Graded Dispersion of SiC Particles on Dielectric Behavior of SiC/Epoxy Composite. *Silicon, 15*(2), 913–923. DOI: 10.1007/s12633-022-02057-z

Frechette, M. Trudeau, M. Alamdari, H.D. Boily S."Introductory Remarks on Nano Dielectrics" IEEE Transactions on Dielectrics and Electrical Insulation (Volume: 11, Issue: 5, Oct. 2004)

Imai, T., Sawa, F., Yoshimitsu, T., Ozaki, T., & Shimizu, T. "Preparation and insulation properties of epoxy-layered silicate nanocomposite insulating material applications" The 17th Annual Meeting of the IEEE Lasers and Electro-Optics Society, 2004. LEOS 2004 DOI: DOI: 10.1109/CEIDP.2004.1364272

Ishibe, S. Mori, M. Kozako, M. Hikita M."A New Concept Varistor With Epoxy Microvaristor Composite" IEEE Transactions on Power Delivery (Volume: 29, Issue: 2, April 2014)

Kim, Y. J., Shin, T. S., Do Choi, H., Kwon, J. H., Chung, Y.-C., & Yoon, H. G. (2005). Electrical conductivity of chemically modified epoxy composites. *Carbon, 43*(1), 23–30. DOI: 10.1016/j.carbon.2004.08.015

Kogut, L., & Komvopoulos, K. (2004). Electrical contact resistance theory for conductive rough surfaces separated by a thin insulating film. *Journal of Applied Physics, 95*(2), 576–585. DOI: 10.1063/1.1629392

Sancaktar, E., & Bai, L. (2011). Electrically Conductive Epoxy Composites. *Polymers, 3*(1), 427–466. www.mdpi.com/journal/polymers. DOI: 10.3390/polym3010427

Sancaktar, E., & Bai, L. "Modeling Filler Volume Fraction & Film Thickness Effects on Conductive adhesive Resistivity" *IEEE International Conference on Polymers and Adhesives in Microelectronics* DOI: DOI: 10.1109/POLYTR.2004.1402737

Subodh, G., Deepu, V., Mohanan, P., & Sebestian, M. T. (2009). Dielectric response of high permittivity polymer ceramic composite with low loss tangent. *Applied Physics Letters*, *95*(6), 062903. DOI: 10.1063/1.3200244

Umesh Dwivedi, S. A. R. (2009). Hashmi "SiC dispersed polysulphide epoxy resin based functionally graded material". *Polymer Composites*, *30*(2), 162–168. DOI: 10.1002/pc.20546

Yang, L., Zhaob, Q., Hou, Y., Sun, R., Chengb, M., Shena, M., Zenga, S., Jib, H., & Qiub, J. (2018). High breakdown strength and outstanding piezoelectric performance in flexible PVDF based percolative nanocomposites through the synergistic effect of topological-structure and composition modulations. *Composites. Part A, Applied Science and Manufacturing*, *114*(November), 13–20. DOI: 10.1016/j.compositesa.2018.07.039

Yang, X., Hu, J., & He, J. "Adjusting Nonlinear Characteristics of ZnO-Silicone Rubber Composites by Controlling Filler's Shape and Size" *2016 IEEE International Conference on Dielectrics (ICD)* 23 August 2016. DOI: 10.1109/ICD.2016.7547607

2. Zimmerman, A. F., Palumbo, G., Aust, K. T., & Erb, U. (2002). Mechanical properties of nickel silicon carbide nanocomposites. *Materials Science and Engineering A*, *328*(1–2), 137–146. DOI: 10.1016/S0921-5093(01)01692-6

Chapter 5
Women–Centric AI:
A Step Towards Empowerment

Neharshi Srivastava

https://orcid.org/0000-0002-6828-8907

Amity University, Jaipur, India

Mani Sachdev

Amity University, Jaipur, India

Monika Gwalani

https://orcid.org/0000-0002-6344-1415

Amity University, Jaipur, India

Meenakshi Bajpai

https://orcid.org/0009-0007-1245-0753

Arya Mahila PG College, Banaras Hindu University, India

Saad Ullah Khan

https://orcid.org/0000-0002-3392-7844

Centre for Media and Mass Communiction Studies, Jamia Hamdard, New Delhi, India

Mukta Arora

UN Women ICO, India

Pranshuta Arora

UNICEF, India

ABSTRACT

This research paper, "Women-Centric AI: A Step Towards Empowerment," reveals how artificial intelligence can be used in a positive way toward gender equity and greater inclusivity. The critical focus of the study addresses intrinsic bias in the contemporary functioning of AI systems, with particular emphasis on women being the victims of the discriminatory behavior of such AI-based systems. It states that, for the sake of building women-centric AI systems, women's concerns in designing such systems need to be integrated so they remain proactive creators and decision-makers. Qualitative methodology and thematic analysis of expert interviews identify defining features of women-centric AI that can be adapted to meet the challenges specific to women. Targeted AI solutions can drastically reduce the digital gender gap and offer more tailored support in health, education, and employment. Ethical

DOI: 10.4018/979-8-3693-8009-3.ch005

considerations include diverse datasets and gender impact assessments for equitable AI development.

INTRODUCTION

"When we invest in women and girls, we are investing in the people who invest in everyone else"
Melinda Gates, 2014

Artificial Intelligence has played a very important role in women's empowerment. When we connect women's empowerment with the mainstream of society, we strive for women to stand equally alongside men. Artificial Intelligence has played a very important role in this context. Today women have a strong participation in every field and such a situation, Artificial Intelligence has allowed them to make their mark in every field, not only at home but also in various fields of professional life, with the help of AI they not only do their work but also make a name for themselves in life. Women-centric AI represents a pivotal step towards empowerment, focusing on creating AI systems that specifically address the needs, concerns, and aspirations of women

Artificial Intelligence stands out as one of the determining factors that shape the very core of contemporary society. It influences industries, economies, and daily life in ways unparalleled by other novel technological capabilities. In such rapid transformation, one of the most crucial aspects has become highly one-sided: gender equity. The development and deployment of AI technologies have been guided by very limited perspectives that fail to consider the unique needs and experiences women face. Therefore, with AI progressing further into everyday life, there is an immediate need for shifting towards more inclusive AI systems, truly women-centric, with regard to women's empowerment across social, economic, and cultural spectrums.

Women-centered AI design, development, and deployment aim to provide solutions to cater to the needs of women in specific areas like health, education, economic activities, and justice. This is a paradigm shift that calls for more than just tinkering with existing models; fundamentally, it requires a reimagining of AI from a position of gender equity. By embedding data and perspectives sensitive to gender in the architecture of AI systems, we further an inclusive digital future wherein AI serves and amplifies women's voices and potential, rather than merely serving them.

The impetus toward women-centered AI is gleaned from blatant disparities salient in the landscape of the now-existing digital world. Different studies have demonstrated that AI models-from natural language processing systems to facial recognition technologies-typically harbor biases that put women, especially from marginalized communities, at a disadvantage. These are not technical glitches but strong reflections of those deeply ingrained societal stereotypes and systemic inequalities. For instance, AI hiring algorithms have been identified to accord preference to male applicants over female applicants with equal qualifications, and health monitoring systems often overlook gender-specific symptoms to misdiagnose a disease. These results are indicative of the requirement that AI systems are not only cognizant of the existence of gender but are also designed to be proactive in eradicating it and ensuring empowerment for the gender concerned.

Women-centric AI also goes a long way in mending the wide digital gender divide that still exists across the world. In most parts of the world, women have limited access to digital technologies, digital literacy, and other opportunities related to technology. It would also help design tools and platforms aimed at digital literacy enhancement, adequate education, and training toward opening pathways to the digital economy for women. This is particularly important in strategic areas such as health, where AI might be deployed in offering personalized care, improving maternal health, and maintaining reproductive rights-all areas where many challenges for women are observed.

It also constitutes a somewhat greater opportunity for the nexus of AI and gender equity to serve as a salve to broader societal issues. Indeed, women-centric AI could be a very strong force for change, given the amplification in strength of women's voices both in the technology development process and in policy considerations. Ensuring that women are not just users but creators of AI technologies will ensure the building of more diverse and inclusive AI ecosystems. In this regard, the incorporation of varied perspectives will lead to stronger AI models that will better be placed to serve a heterogeneous population. It is very important to include the voice of a woman in AI development at every level: ideation, design, testing, and implementation. It is only then that such technologies will begin genuinely to reflect, and actually serve, the lived experiences of women.

The Present research discusses how women-centered AI could be a game-changing instrument in strategies of empowerment. Attention is given to the current AI landscape and gender bias, barriers to achieving gender equity in AI, and innovative initiatives that lead from the front for more inclusive AI solutions. While shedding light on these critical issues, the paper advocates for a collaborative, multidisciplinary approach in AI creation-one that understands and champions women's needs and rights around the world. As we forge into an increasingly AI-defined future, making sure it is one serving as a force for equality, rather than an amplifier of existing biases, is not desirable but rather imperative.

REVIEW OF LITERATURE

The concept of women-centric AI has increasingly gathered momentum, with scholars and practitioners tending to show increasing awareness of the need to reduce gender biases within AI systems and tap into the use of AI for empowering women in various spheres. Literature on this aspect addresses various areas that include, among others, discussion on gender bias in AI algorithms, impact of AI on economic and social empowerment of women, and development of ethical frameworks for inclusive AI.

1. Gender Bias in AI Algorithms: A great deal of research has identified that several types of gender biases exist in the AI system, which is largely due to biased data and male-dominated development processes. For example, Buolamwini and Gebru (2018) established that facial recognition technologies have higher error rates for females and people of color, hence indicating a critical gap in how AI systems are trained and validated. Similarly, research conducted by Caliskan, Bryson, and Narayanan (2017) identified that NLP models often reproduce gender stereotypes, associating women with domestic roles and men with professional achievements. These studies show that more diversified data and inclusive AI development are indispensable for achieving fair outcomes.

2. AI and Women's Economic Empowerment: Much literature has focused on the potential that AI may hold for enhancing women's economic empowerment through better access to education, finance, and work. In economies where gendered barriers to women's participation in formal education and/or labor markets either exist or persist, AI-powered platforms can provide them with personalized learning content, skill-building opportunities, and remote employment. A report by the McKinsey Global Institute estimates that AI-driven automation could open up more job opportunities for women, especially in traditionally female-dominated fields like healthcare, education, and social work. However, it warns that lacking targeted interventions, the transition to an AI-driven economy may further exacerbate existing gender disparities, particularly in access to digital skills and high-paying jobs.

3. AI for Social and Health Interventions: There is huge potential for women-centered AI both in healthcare and social services in managing gender-specific needs and improving outcomes. In a study on health care algorithms, Obermeyer et al. (2019) found that most AI models fail to consider symptoms particular to men and women and also treatment responses, leading to misdiagnosis and ineffective treatment. On the other hand, an AI model developed and more conscious of gender issues can put in much effort towards drafting personalized health schemes, especially for maternal health, mental health, and reproductive rights, which solely remain in the domain of women. AI also stands ready to assist social interventions through better access to resources, virtual support systems for victims of domestic violence, enhancement of legal and advocacy efforts.

4. Ethical and Policy Considerations: Therein comes another growing area of priority: the development of ethical guidelines and policy frameworks regarding women-centered AI in the literature. Emphasis is therefore being given to the development of AI systems that are not only technically robust but also ethically sound, correctly orienting toward the principles of gender equity. Noble (2018) argues that AI biases reflect broader societal inequalities and that more critical attention is given to the role of AI in perpetuating discrimination. Similarly, Binns (2018) points to the need for more transparency and accountability in how AI is designed, particularly with respect to data collection, model design, and deployment practices. These studies advocate for inclusive policy-making that calls for gender-sensitive AI practices, including audits to identify biases, diverse teams of developers, and inclusive stakeholder consultations.

5. Women as AI Creators and Leaders: The literature also emphasizes women's involvement at the level of AI research, development, and decision-making. A recent study by West, Kraut, and Chew (2019) finds that women are severely underrepresented in the AI and machine learning fields; hence, there is a lack of diversity in AI solutions. Encouraging more women into the field of AI and its related fields is important to bring inclusiveness to AI systems that genuinely capture diverse perspectives and experiences. For example, there is AI4All and Women in AI that work to do this through mentorship, training, and networking of women. Women-centered artificial intelligence discourse does provide a convincing point for the reevaluation of ways to design and implement AI systems so that the use of AI systems enforces gender equity and empowerment. While there has been considerable progress in the identification of challenges and opportunities related to AI and gender, a great deal remains to be done to achieve appropriate, inclusive, effective, and ethical AI solutions. This review thus demonstrates further research required, policy innovation, and collaboration toward an AI-driven future that truly serves and empowers women worldwide.

SIGNIFICANCE OF THE RESEARCH

The importance of the research on "Women-centric AI: A Step Towards Empowerment" lies in probable transformation of the development and application of artificial intelligence that may bring positive changes regarding gender equity and inclusivity. The attention of special focus will be needed as AI increasingly influences areas as varied as health and education, finance, and social policy; biases and limitations, embedded in most existing AI systems, that often disadvantage women and those coming from the most marginalized communities, will be particularly highlighted. This is an important research feature to underline how AI can purposefully be designed to solve unique needs for women's economic and social empowerment, and reduce the digital gender gap. The study has thus underlined the need for diversity within AI development by advocating for a women-centered AI, so that women do not only play the role of passive users but active creators and decision-makers in AI technologies. Conclusions from this study therefore go a long way to provide valuable inputs for policy makers, developers, and other stakeholders so as to move closer to an AI system that is equitable and ethical. This, therefore, catalyzes the shift towards an inclusive digital future wherein technology becomes a tool of empowerment and not exclusion.

RESEARCH METHODOLOGY

Objectives

- What are some defining characteristics of "women-centric AI," and how might this notion be applied toward the improvement of gender disparities in varied sectors?
- How can AI be engineered or customized to specifically target unique challenges that women face in the realms of health, education, and employment?
- In what ways can AI be designed to ensure women-centric AI systems and what are its ethical considerations and potential biases? How can these systems be developed so that they don't neglect those principles of gender equity?
- Which industries where women are generally underrepresented can AI be used to decrease the gaps between genders, and what should we do next to ensure inclusive AI development?

Research Design

The current study has used a qualitative research design with thematic analysis in order to develop perspectives of AI and IT experts with respect to "women-centric AI" and its consequences for the empowerment of women. This approach is designed to identify and analyze key themes of the structured interviews conducted with industry experts.

Sample

A purposive sample of AI and IT experts with considerable experience in the field was determined for this study. Those being chosen depended on their expertise on AI technologies or their inputs to initiatives that promote gender equity through AI. This means that insights gathered cut across individuals who understand AI and its implications for issues of gender deeply. Thus, these 10 experts belonged to various regions of the country, and it could thus give multifaceted aspects that could be derived from them. They include experiences in developing AI, practicing data science, fair AI practice, and gender-focused technology initiatives. These experiences have enriched the data generated and provided a comprehensive picture for deducing the trend of how AI can be used for the redress of gender imbalances and empower women.

Data Collection

Data collection was done through structured interviews with experts. Questions were set in a way to seek answers on ten specific questions related to the concept of women-centric AI and its implications for women's empowerment.

Interview Questions

1. What does the term "women-centric AI" mean to you, and how do you envision it contributing to women's empowerment?
2. How can AI be specifically designed or adapted to address challenges unique to women in areas such as healthcare, education, and employment?
3. What are some current examples of women-centric AI applications or tools that you believe are making a significant impact?
4. In the development of AI technologies, how can we ensure that women's voices and perspectives are included, particularly in fields where they are underrepresented?
5. How can AI help in bridging the gender gap in industries where women are traditionally underrepresented or marginalized?
6. What ethical considerations should be taken into account when developing AI systems aimed at empowering women, and how can we mitigate potential biases?
7. How do you see AI playing a role in combating issues such as gender-based violence, discrimination, or unequal access to resources?
8. What role do you believe AI should play in shaping the future of work for women, especially considering automation and digital transformation?

9. What can governments, organizations, and educational institutions do to support the development and deployment of AI solutions that promote gender equity and empowerment?
10. Looking ahead, what are the key challenges and opportunities you foresee for women-centric AI, and what steps should be taken to maximize its positive impact?

Data Analysis

Thematic analysis was employed to analyze interview data. The process of analysis is as follows:

> **Transcription:** Interviews were recorded and then transcribed verbatim to accurately represent what the experts said.
> **Coding**: Key phrases and statements were captured through coding, where substantial information relating to every question in the interview could be observed.
> **Development of Themes:** Codes were organized into broader themes reflecting commonalities and patterns across the interviews. Themes are developed based on the responses to each question and are iteratively refined.
> **Analysis:** The identified themes are analyzed for overarching insights and connections among the responses provided by the participants. It looks at ways in which AI can be adapted to help women deal with challenges that are gender-specific.

Thematic analysis was utilized to analyze the interview data. This process began with an initial perusal of the interview transcriptions, thereby having a general overview of the responses. Key phrases, concepts, and ideas were underlined or coded. Preliminary themes were grouped for the codes reviewed and refined so that the themes adequately represented the data and matched the research objectives.

Ethical Considerations

These included informed consent from all of the participants, confidentiality, anonymity, and possible biases that might have occurred during data collection and analysis.

Result: Thematic analysis is a method for identifying, analyzing, and reporting patterns (themes) within data.

Table 1. Themes analysis

Theme	Subtheme	Narratives
Definition and Vision of Women-Centric AI	**Addressing Specific Needs and Challenges**	**Women-centric AI focuses on addressing women's needs in healthcare, education, and employment to reduce inequality.** **- It integrates women's experiences into algorithms, offering equitable access to AI benefits and closing gender gaps. . *(Participant 2, 7)***
	Challenging Societal Biases	It challenges biases in traditional systems by improving access to critical resources like health diagnostics and financial inclusion, fostering women's independence and security. . *(Participant 3, 5, 7)*
Designing AI for Women	**Healthcare Solutions for Women**	**AI can tailor healthcare by using gender-specific data, focusing on overlooked health indicators to detect diseases like breast cancer early. . *(Participant 2, 5)***
	Personalized Education	AI can offer women who missed formal education personalized learning paths, helping them re-enter the workforce and acquire skills.. *(Participant 4, 5, 7)*
	Employment Opportunities and Flexibility	AI platforms can connect women with flexible job opportunities, such as remote or part-time work, aligning with their schedules and responsibilities. . *(Participant 1,3, 7)*
Current Women-Centric AI Applications	**Health and Reproductive AI Tool**	**Clue is an example of an AI-based menstrual tracking app that provides personalized reproductive health insights. . *(Participant 3, 7)***
	Initiatives Promoting Women in AI	AI4ALL encourages young women to pursue careers in AI, bridging the gender gap in technology and empowering women in the industry. She Matters provides mental health support for postpartum depression using AI tools. . *(Participant 5, 6)*
Inclusion of Women in AI Development	**Diversity in AI Development Teams**	**Actively recruiting women in AI development teams and ensuring diversity helps reflect women's perspectives in AI solutions.** **Focus groups with diverse women ensure AI technologies address relevant challenges faced by women. . *(Participant 3, 5, 7)***
	Gender-Impact Assessments	Gender-impact assessments during AI development prevent reinforcing gender biases and ensure women's voices are considered in the final products.. *(Participant 9, 7, 10)*

continued on following page

Table 1. Continued

Theme	Subtheme	Narratives
Bridging Gender Gaps with AI	**AI for Unbiased Recruitment**	**AI can help by creating unbiased recruitment tools that evaluate candidates based on skills and qualifications, opening doors for women in male-dominated industries.** *. (Participant 2, 8)*
	Mentorship and Networking Opportunities	AI-driven mentorship programs can connect women with mentors in underrepresented industries, providing guidance and networking opportunities.. *(Participant 4,3, 7)*
Ethical Considerations in Women-Centric AI	**Bias Mitigation in AI Development**	**AI systems must use diverse datasets to represent women's experiences accurately, ensuring fairness and mitigating biases.** **Regular audits should be conducted to prevent reinforcing harmful gender stereotypes in AI algorithms.** *(Participant 5, 7)*

DISCUSSION

Women-centric AI is a recent development over the last couple of years as a revolutionary approach in reducing disparities caused by gender in health care, education, and employment. As opposed to typical AI systems, which often neglect women's distinct needs and experiences, women-centric AI systems are designed to be more inclusive and equitable, having direct women's perspectives infused within its algorithms. This not only bridges the gap for gender but also serves targeted solutions that empower women to lead healthier, more independent, and economically stable lives (Cummings et al., 2021). Focusing on women's specific challenges, this AI has the potential to create societal change and accessible pathways for women to take advantage of technological advancements.

Indeed, women-centric AI is a vision much broader than equitable access. Herein lies a strategic effort to break the biases inherent in traditional systems. Traditional healthcare structures, finance mechanisms, and employment have historically overlooked or trivialized women's needs, thereby demographically excluding them from critical resources and opportunities. Women-centric AI uses gender-specific data to address such inequalities in striving toward a more balanced and supportive technological landscape, according to Smith and Roberts (2022). For example, AI algorithms trained using gender-specific data enhance early detection and diagnosis of certain diseases like breast cancer with unique characteristics and risk factors for women, such as Kim et al. (2023).

The focus of the discussion will be on some essential aspects of women-centric AI, which includes its role in fulfilling specific needs and challenges, its present applications, and inclusive development practices. With this, it would be easy to comprehend the deep impact that women-centric AI is creating regarding social equity and the empowerment of women across sectors.

The themes identified in this study are as follows:

Definition and Vision of Women-Centric AI

Women-centric AI has come with the purpose of bridging the gaps between genders related to the important aspects of health, education, and employment. The approach will be toward providing equality through the specific requirements of women by directly embedding their experiences into AI algorithms, which will ensure fair access to the benefits of AI and work towards closing the gender gap actively (Cummings et al., 2021; Gupta & Jain, 2023). This inclusion of women's opinions would make the AI technologies more responsive and inclusive, therefore, breaking the long-existing bias in such sectors. According to Sharma and Banerjee, 2022, an inclusive AI development process benefits women's access to technology and is one step forward toward a balanced society since it allows access and equal benefits from AI-driven solutions.

On the other hand, Women-centric AI refers to the approach of AI that especially addresses the different challenges and needs of women in all walks of life, ranging from healthcare, education, work, and social empowerment. This is because traditional AI can unknowingly perpetuate gender biases or may not provide resources to women on a par with men. Women-centric AI addresses these imbalances through algorithms and applications that directly integrate women's views, experiences, and needs into the technology of AI. This vision is aimed at creating an inclusive digital landscape where AI systems would actively contribute to gender equality. By working on gender-responsive frameworks, women-centric AI aims to eliminate obstacles between women and full benefits of technological advancements. For instance, in the health sector, this strategy can tackle the historical dearth of attention provided to women-specific health concerns by exploiting data that reflects unique health risks women face. Such gender-focused applications have already shown promise in diagnosing and managing conditions like endometriosis, breast cancer, and maternal health issues, which have traditionally received limited focus in mainstream healthcare, according to Shah and Gupta (2021), because women-centric AI can come across as a tool for helping women gain access to all sorts of flexible employment or educational resources, which enables women to have enough avenues and skills to be fruitful amidst the ever-changing job scenario. AI-based platforms may empower women who can't work full-time and in a regular setting due to their family obligations or societal pressure, hence participating in the economy on their own terms. According to Henderson et al. (2022), AI-based job platforms for women can enable flexible working conditions that can provide more women with quality employment without sacrificing other parts of their lives.

This is empowering the individual and at the same time striving to bring forth broader social transformation by critically questioning and remodeling older systems and structures. The domain of finance is one such salient example where women are grossly hampered in attaining loans, credits, and other provisions of financial service mainly due to such traditional prejudice. Such AI models, which are being designed and developed as women-centric can work in conjunction with larger financial data models to ensure ease in credit access among women and total financial autonomy. According to Forbes et al. (2021), systems like these hold immense potential for women's empowerment through entrepreneurship by offering a source of investment in businesses, ensuring the assurance of financial security, and ultimately triggering economic growth both individually and at the community level. The last major aspect related to women-centric AI revolves around how it aids bias minimization within technology itself. Traditional AI is often biased by training data that reflects gender stereotypes and ultimately ends up in tools that actually amplify inequalities. Women-centric AI works to challenge such biases through diversified datasets and perspectives in its development, making the resultant tool more accurate to women's experiences. Gender-impact assessments help determine how AI might impact various genders. In the end, developers will make fairer and more reliable technologies if they can make sure the algorithms consider the needs of women and stop hurting stereotypes about them.

In addition, this vision also imagines active engagement by women during AI development processes, that can make AI solutions even more inclusive and relevant. Women working in AI development roles can bring unique insights as well as can be an advocate for features and considerations supporting the needs of female users. According to Taylor and Singh's research in 2022, diverse development teams are able to develop more innovative and inclusive AI solutions because they are in a better position to notice and correct biases that would otherwise be overlooked. This makes gender diversity in tech teams and leadership very crucial for the full realization of women-centric AI. Therefore, women-centric AI is a beckon to the new paradigm that ensures digital spaces are both more representative and equitable. This can then embed needs and experiences of women within AI and ensure that people are afforded all tools, resources, and opportunities that enable full participation in this age of digitization. Dismantling of existing gender biases would yield enormous potential toward building social and economic empowerment in crafting a more balanced and equal future where AI works for all.

Meeting Specific Needs and Challenges

Women-centric AI is distinctive because it solves specific issues in society relating to women, like health diagnostics available easily or financial independence. It offers accessible digital solutions that promote women's autonomy for access to resources meant for security and well-being (Smith & Roberts, 2022). For instance, Forbes et al. (2021) has indicated that AI in application in the area of female financial inclusion positively affects females' ability to handle personal finances and engage in business activities, thus promoting a sense of autonomy. Access and outcomes in preventive health care are changing due to AI-based diagnostics in relation to female-specific diseases in healthcare (Jones et al., 2022). Women-centric AI is a type of AI specifically designed to improve the specific needs and challenges of women in the healthcare, education, and employment sectors where traditional systems have failed. Gender-specific data and insights are used to develop solutions to provide tailored health care with an orientation towards reproductive health and early disease detection, such as breast cancer. It also makes it easier for women to re-enter school, having been out for formal education, and enter the job market with some new skills.

Employment-Based: Women-centric AI will be useful in establishing platforms that will link women with flexible employment opportunities so they can effectively work and at the same time attend to their personal obligations. Besides, it eradicates the hindrances to financial independence by allowing women to access financial services through inclusive algorithms, loans, and credit. Such particular needs do not only benefit individual people but also help society in general in that women will enjoy higher economic and social security.

Designing AI for Women: Healthcare, Education, and Employment

AI-driven Healthcare Solutions for Women: AI-driven healthcare services for women include customized diagnosis services. It aims at gender-specific data particularly during the diagnosis of diseases like breast cancer where early diagnosis remains very crucial. According to Kim et al. (2023), gender-specific algorithms trained on health data boost the accuracy levels during diagnosis of breast cancer and cardiovascular disease among women. Such solutions reduce the likelihood of erroneous diagnoses and timely interventions with respect to diseases usually omitted in other health care systems.

Personalized Education: It will allow AI-based individualized education for women deprived of formal schooling by proposing individual learning pathways, enhancing the capacity of acquiring a new skill set, hence getting back to the working arena. Li and Martin (2021) argued that AI-based educational plans have increased the employability of adult women, enhancing a woman's digital and professional capabilities. Personalized learning has enabled women to bridge their skills gaps in pace and interests, directly answering the questions and issues that may exist or hinder the workforce reentry.

Employment Flexibility and Opportunities: AI Platforms also enable flexible employment. This allows women to have remote work or part-time jobs that better fit their needs, considering that they are also managing family duties. A study notes that the job websites provided by AI can bridge the gap of women representation in industries that are often dominated by males and also let them choose between flexible schedules (Taylor et al., 2023). This will bridge the gap between accessible employment and help balance the demands of work life for women while contributing toward economic stability.

Current Women-Focused AI Applications

The most transformative applications of AI, which are specifically aimed at women's health and wellness, have been their AI-based applications. A perfect example is Clue, a menstrual tracking application. This application uses AI to give users personalized insights related to their reproductive health, hence equipping women with more proactive steps in managing health. Johnson and Lee (2021) studies found that Clue users are more aware and have better management of their reproductive health. AI4ALL, for instance, encourages young women to pursue AI careers and fill the gender gap in technology and develop gender diversity in STEM disciplines (Martinez & Ellis, 2022). She Matters uses AI tools towards the aim of mental health support on the issue of postpartum depression, which is a sensitive area quite often under-resourced (Miller et al., 2022).

Women in AI Development

Increasing more women in development teams about AI will mean higher gender diversity, and the issues related to women and already developed AI solutions would be observed. Utilization of diverse AI teams provides for innovative and unbiased designs in products. The female will uniquely provide different opinions concerning the challenges to be solved through technology. Researchers, Patel and Singh concluded that teams with diversity can do a better job concerning inclusive AI solution design. Research from 2022 points out that such teams would have a better ability to realize the possible biases at algorithmic design. Further developing gender-impact assessments ensures that in the development of AI phases, gender-stereotypical reinforcement is negated, and diversified approaches toward

AI solutions towards women can be ensured. Women in AI development are needed to make the AI technologies more diverse, equitable, and representative. When women are included in the design, development, and deployment of AI, they bring with them critical insights to help shape technology in better meeting the needs of women users. The perspectives reveal gaps in traditional algorithms that could otherwise go unnoticed. The diversity teams develop better, more creative, and more inclusive AI solutions because they are most likely to be in sync with the experiences of all users, not only with those reflected in the data that are male-dominated (Taylor & Singh, 2022). Making AI development more diverse has also been found to encourage the type of working culture that promotes diverse thoughts and leads to creativity. Such an inclusion is of the highest importance for AI development, especially gender-inclusive AI. These women are potentially powerful champions of the cause for conducting gender impact assessments and also for developing AI models in such a manner that is unbiased and accessible to everyone. The search and nurturing of women in AI development add depth to the ethical infrastructure of AI while furthering the cause for industry development of better-fit technologies for diverse human needs.

Closing Gender Gaps With AI

AI for Unbiased Hiring: AI in recruitment ensures that biases in hiring over the years are nullified since the assessment is conducted based solely on skillful candidates, hence promoting chances of women in sectors highly dominated by men. It was evident in Henderson et al.'s 2022 study, which depicted enhanced gender diversity in technology professions through AI-based recruiting technologies, where the tool puts emphasis on skills over some demographic bias.

Mentorship and Networking: The AI-supported mentorship systems enable women to connect with mentors operating in minority fields, such as technology and engineering. Such networking enables women to access additional knowledge and support and professional connections (Kumar & Patel, 2021). Consequently, women tend to make quick career advancement after they gain mentorship targeting them appropriately. On the other hand, With its potential ability to bridge such massive disparities between men and women, AI may be able to eradicate systemic inequalities better with outcomes in employment, educational, and health sectors. Algorithms designed to be unbiased in recruitment will enable the AI to evaluate selections from pure skills and qualification standpoints, hence giving access to women in job occupations viewed to be male-dominated during traditional times. Another major avenue of flexibility open by AI is to provide various remote and part-time platforms women can use that, in turn, allow women to manage both family activities without compromising career development as well. In education, AI-driven learning platforms can provide personalized training and skill-building opportunities for women who missed formal schooling or want to re-skill for new career paths. In healthcare, gender-specific AI applications improve access to diagnostics and early intervention for conditions that disproportionately affect women, such as breast cancer and reproductive health issues. It fills those specific gaps, which allows AI for greater gender balance, making possible personal and professional achievements for women while making society balanced.

Ethics of AI in Women-Centered

To keep AI systems fair and eliminate gender bias, diversified data representing the lives and needs of women must be used. Periodic audits should be carried out on AI models to avoid reinforcing any stereotypes about gender that may lead to harm. As presented by Anderson and Kim in 2023, applying diverse datasets and regular auditing promote the general ethical value of AI in being fair and upholding broader social goals that involve equity and inclusiveness.

Ethics of women-centered AI ensure that technology is promoting equity, inclusivity, and respect for women's diversity of experiences. Ethical development of AI for women thus uses gender-diverse datasets which reflect the variety of women's views to reduce bias and deliver accurate and representative outcomes. Ethical consideration also involves conducting periodic algorithm audits to prevent the formation of harmful stereotypes within the AI system, and also that the technology treats women equitably across all demographic categories.

The other ethical aspect is gender-impact assessments, which allow developers to detect and mitigate unintended implications for women before AI will be widely deployed. It will also require a great deal of transparency along with accountability because when the user understands how AI decides choices for them, it is easier to question the resulting biases. Therefore, in this context, there is a need to consider ethical considerations in its design to promote the ideal of reducing gender gaps and maximizing social equity in the impacts that technology has on women.

In conclusion, women-centric AI is not only a tool for technological progress but also a great enabler of social change. The ability to address specific gender needs, challenge the biased perceptions of society, and provide inclusivity would lead women-centric AI toward making fair systems and promoting sustainable development for all.

CONCLUSION

The women-centric AI turns to be an effective device and tool in addressing this major issue of reducing disparity levels and achieving better balance toward equitable provision of more wholesome resources for health, educational, and job or service opportunities. Through incorporation within AI algorithms, this means unique experiences and the attendant needs that have for years excluded women are gradually becoming well catered to and filled out. The study also discusses how AI of this nature can be used within the transformational change of society, healthcare, and education and for more flexible forms of employment in ways that reflect women's lives. Additionally, gender-impact assessments plus inclusive presence of women in AI development teams are also necessary conditions for fair and inclusive AI technologies.

As women-centric AI arises, it is likely to become a strong catalyst for change in society toward more representation, economic stability, and security for women. Efforts should focus on refining these technologies further to become even more inclusive and ethical ones, ensuring AI becomes a tool for social equity. Targeted interventions, mentorship opportunities, and unbiased recruitment can significantly contribute to the balancing of a technological ecosystem empowering all.

CONTRIBUTION

It will help to add something to the new, slowly emerging discipline of women-centered AI. After all, focusing on how women's experiences can be incorporated into the algorithms for artificial intelligence will directly redress the gender inequality. Thus, through women-centric AI, there may develop better access to personalized health, customized education, and more flexible employment to empower their independence and socio-economic development. Additionally, the research identifies the ethical considerations of AI, including diverse datasets and gender impact assessments, which prevent the occurrence of biases and ensures that AI solutions are closer to responding to women's needs. In this regard, this study provides insights into both existing applications and development practices that may help guide AI designers and policymakers in developing more inclusive and gender-responsive AI systems.

LIMITATIONS

Despite the contribution of this study, there are some limitations. The research work is very much conceptual and has relied just on current instances of AI applications centered on women, not on empirical data that might give a more detailed view of the real-world impact of such technologies. So, the study may miss numerous backgrounds of experience in terms of socio-economic, cultural, and regional aspects which could prove critical in how women experience and benefit from AI. This focus on certain sectors like the health sector, education sector, and employment sector is a limitation since areas of equality, like law enforcement or social services, do not undergo critical scrutiny. This in turn calls for broader empirical studies that cross a much broader spectrum of different contexts and fields.

FUTURE IMPLICATIONS

The future of the study can suggest studying the empirical impacts to assess the real-world efficiency of women-focused AI. Longitudinal studies might be able to establish whether these technologies have an influence over time on women's lives, especially in relation to socio-economic advancement and personal empowerment. Additionally, creating more representative datasets which take into account the diversity of women's experiences across all different demographics and regions will enhance the accuracy and fairness of AI applications. Further research can include its implications on other industries not presented here, which will increase its scope and application. Last but most important, future development requires collaborations between technologists and experts in gender studies coupled with policymakers to push ahead in the advancement of AI on women, ensuring technologies support social equity and aid in the diverse needs of various women around the world.

REFERENCES

Almeida, M., Lima, E., & Lee, D. (2022). Leveraging AI for Women's Entrepreneurship in Developing Economies. *Development Studies Research*, *9*(1), 41–55.

Binns, R. (2018). Fairness in Machine Learning: Lessons from Political Philosophy. *Proceedings of the 2018 Conference on Fairness, Accountability, and Transparency*, 149-159.

Buolamwini, J., & Gebru, T. (2018). Gender Shades: Intersectional Accuracy Disparities in Commercial Gender Classification. *Proceedings of Machine Learning Research*, *81*, 1–15.

Chaudhury, S., & Thakur, A. (2020). Policy Frameworks for Women-Centric AI: Bridging the Gender Digital Divide. *Digital Policy. Regulation & Governance*, *22*(3), 213–231.

Dastin, J. (2018). *Amazon Scraps Secret AI Recruiting Tool That Showed Bias Against Women*. Reuters.

Hunt, V., Prince, S., Dixon-Fyle, S., & Yee, L. (2021). *Diversity Wins: How Inclusion Matters*. McKinsey & Company.

Noble, S. U. (2018). *Algorithms of Oppression: How Search Engines Reinforce Racism*. NYU Press. DOI: 10.18574/nyu/9781479833641.001.0001

Obermeyer, Z., Powers, B., Vogeli, C., & Mullainathan, S. (2019). Dissecting racial bias in an algorithm used to manage the health of populations. *Science*, *366*(6464), 447–453. DOI: 10.1126/science.aax2342 PMID: 31649194

Shin, D. (2020). How VR Can Bridge Gender Gaps in STEM. *Journal of Educational Technology & Society*, *23*(4), 23–35.

Singh, K., Han, A., & Wang, W. (2021). AI-Powered Chatbots for Delivering Telemedicine in Remote Areas. *Journal of Medical Internet Research*, *23*(5), e24732.

West, S. M., Whittaker, M., & Crawford, K. (2019). *Discriminating Systems: Gender, Race, and Power in AI*. AI Now Institute.

Yala, A., Lehman, C., Schuster, T., Portnoi, T., & Barzilay, R. (2020). A Deep Learning Mammography-based Model for Improved Breast Cancer Risk Prediction. *Radiology*, *296*(1), 90–98. PMID: 31063083

Chapter 6
The Role of AI in Digital Transformation

Amit Kumar Singh
https://orcid.org/0000-0002-1325-7329
Amity University, Jaipur, India

Swapnesh Taterh
https://orcid.org/0000-0003-2770-8829
Amity University, Jaipur, India

Arti Kaushik
https://orcid.org/0009-0008-6173-8232
G.V.M College of Education, Sonepat, India

Avinash Goswami
https://orcid.org/0009-0008-7441-9951
The LNM Institute of Information Technology, Jaipur, India

ABSTRACT

This chapter explores the transformative role of Artificial Intelligence (AI) in driving digital transformation across various industries. As organizations navigate the complexities of the digital age, AI emerges as a critical tool for enhancing operational efficiency, improving customer experiences, and fostering innovation. The integration of AI technologies enables businesses to automate routine tasks, analyze vast datasets for actionable insights, and deliver personalized services at scale. However, the adoption of AI is not without challenges; issues such as data quality, ethical considerations, and workforce displacement pose significant hurdles. This paper discusses best practices for AI integration, emphasizing the importance of establishing clear objectives, investing in high-quality data, and prioritizing transparency and ethical governance. Additionally, future trends in AI, including

DOI: 10.4018/979-8-3693-8009-3.ch006

hyper-automation, edge computing, and AI-driven sustainability, are examined as organizations seek to leverage these advancements for competitive advantage.

1. INTRODUCTION

In today's fast-paced digital economy, businesses are under immense pressure to innovate and stay competitive. Traditional business models are being disrupted as organizations across industries undergo digital transformation to adapt to evolving consumer expectations, market conditions, and technological advancements. At the heart of this transformation is Artificial Intelligence (AI)—a technology that enables machines to simulate human intelligence, learn from data, and make decisions that drive efficiency, enhance customer experiences, and open new avenues for growth. AI in Digital Transformation goes beyond automating routine tasks; it allows organizations to leverage vast amounts of data to make data-driven decisions, anticipate customer needs, and optimize complex processes. From personalized customer service in retail to predictive maintenance in manufacturing and enhanced diagnostics in healthcare, AI-driven solutions are revolutionizing industries. AI's capacity to process large datasets, recognize patterns, and generate insights in real-time empowers businesses to operate more dynamically, adapt swiftly, and meet rising demands with precision (Calp, 2020).

As digital transformation reshapes industries, AI has emerged as a foundational technology that enhances every aspect of business. The integration of AI with cloud computing, big data, and IoT further strengthens its potential to revolutionize business functions. Organizations leveraging AI report greater productivity, cost savings, and improved customer experiences—all of which are critical in today's hypercompetitive market (Oyekunle & Boohene, 2024).

1.1 Definition of Digital Transformation

Digital transformation refers to the fundamental rethinking and reorganization of business processes, practices, and customer experiences by leveraging digital technologies. This transformative approach allows organizations to remain competitive, agile, and customer-focused in an ever-evolving digital landscape. While traditional models of transformation might focus solely on process optimization or technology implementation, digital transformation emphasizes a holistic change that affects not just technology but also culture, customer engagement, operational efficiency, and overall business strategy (Chatterjee, Chaudhuri, Vrontis, & Basile, 2022).

Organizations that undergo successful digital transformation often see enhanced customer satisfaction, operational efficiency, and a new culture of innovation. However, without the right technological foundation, even the most ambitious digital transformations are unlikely to reach their full potential. Enter Artificial Intelligence (AI): a technology that has reshaped what's possible within digital transformation initiatives.

1.2 Introduction to Artificial Intelligence (AI)

Artificial Intelligence (AI) is the simulation of human intelligence in machines, enabling them to think, learn, and make decisions autonomously. Broadly, AI encompasses various subfields, including Machine Learning (ML), Natural Language Processing (NLP), Computer Vision, and Robotics Process Automation (RPA). Each of these technologies enables businesses to extract valuable insights from data, automate routine tasks, and develop intelligent systems that can enhance decision-making (Brock & Von Wangenheim, 2019).

AI plays an integral role in digital transformation by providing advanced tools to automate complex tasks, improve customer interaction, and analyze large volumes of data. From predictive analytics in marketing to smart manufacturing processes, AI is increasingly used across industries, transforming them fundamentally (Singh & Taterh, 2023).

1.3 Importance of AI in the Modern Business Landscape

In the contemporary business landscape, AI acts as a catalyst for transformation, helping organizations stay relevant, competitive, and efficient. The capabilities provided by AI go beyond traditional technology solutions, enabling businesses to predict trends, understand customer behavior in real-time, and optimize processes in ways that were previously unattainable.

Consider the impact of AI in industries like healthcare, where it enables rapid diagnostics, or retail, where it personalizes customer experiences. By integrating AI, businesses can transition from merely responding to change to actively shaping their future. Companies that have harnessed the potential of AI often find themselves at the forefront of innovation, with significant advantages in customer satisfaction, revenue generation, and operational agility (Brynjolfsson & McAfee, 2014).

2. UNDERSTANDING DIGITAL TRANSFORMATION AND AI

Digital transformation is the process by which organizations integrate digital technologies across all areas of their business, fundamentally altering how they operate and deliver value to customers. This transformation goes beyond merely adopting new technologies; it requires a strategic shift that embraces agility, data-driven decision-making, and a willingness to challenge traditional business models. At its core, digital transformation involves rethinking processes, fostering a culture of innovation, and realigning organizational structures to respond to evolving customer expectations and a fast-paced digital economy. Organizations undergoing digital transformation strive to enhance operational efficiency, personalize customer experiences, and create new business models that can adapt swiftly to market changes (Davenport & Ronanki, 2018).

Artificial Intelligence (AI) plays a crucial role in digital transformation by providing tools and techniques that allow machines to simulate human intelligence, learn from vast data sets, and perform complex tasks. Through technologies like machine learning, natural language processing, and computer vision, AI allows businesses to automate routine processes, generate insights from large data sets, and make predictions that guide strategic decisions. This ability to harness data-driven insights is critical, as data is the cornerstone of digital transformation (Domingos, 2015). AI's impact extends across industries, from enabling predictive maintenance in manufacturing to enhancing diagnostic accuracy in healthcare and delivering personalized experiences in retail.

When combined, digital transformation and AI create a powerful synergy. AI accelerates digital transformation by making it easier for organizations to analyze data, anticipate customer needs, and innovate in ways that were previously unimaginable. However, successful implementation of AI within a digital transformation strategy requires addressing key challenges such as data quality, skill gaps, and ethical considerations, particularly around issues like data privacy and algorithmic bias. As a result, businesses adopting AI must not only invest in technology but also in training, governance, and ethical frameworks to ensure AI is used responsibly.

Understanding the intersection of digital transformation and AI is essential for any organization aiming to stay competitive. With AI's potential to transform industries and the foundational shift that digital transformation demands, businesses that embrace both can position themselves for sustained growth and relevance in an increasingly digital world (Fountaine, McCarthy, & Saleh, 2019).

2.1 Key Elements of Digital Transformation

Digital transformation extends beyond the adoption of digital tools; it reshapes how organizations operate, serve customers, and approach innovation. This section covers four essential elements crucial to successful digital transformation:

Customer Experience: At the heart of digital transformation lies the objective of creating seamless, intuitive, and personalized customer experiences. AI is central to achieving this goal, offering organizations tools to personalize interactions, analyze customer behaviors, and proactively meet customer needs. By leveraging machine learning and predictive analytics, companies can deliver products and services tailored to individual preferences and behaviors, enhancing engagement and satisfaction (Goodfellow, Bengio, & Courville, 2016).

Operational Agility: Operational agility refers to an organization's ability to respond swiftly and efficiently to market changes, customer demands, and new business opportunities. Digital transformation promotes agility through process automation, real-time data access, and predictive analytics. AI-driven technologies, such as RPA and process optimization algorithms, enable businesses to streamline their workflows, reduce redundancies, and make informed decisions faster (Haenlein & Kaplan, 2019).

Culture and Leadership: Digital transformation isn't limited to technology implementation; it requires a shift in organizational culture. Leadership must foster a mindset that embraces change, innovation, and a willingness to learn. Leaders play a crucial role in integrating AI technologies responsibly and effectively, promoting continuous learning, and aligning the organization's mission with new digital capabilities.

Workforce Enablement: An empowered workforce is essential for the success of digital transformation. By integrating AI into daily operations, companies can automate repetitive tasks, freeing employees to focus on higher-value work that requires creativity, critical thinking, and problem-solving. Furthermore, AI-driven tools can provide real-time insights, augmenting employees' decision-making capabilities and allowing them to contribute more strategically to the organization's goals (Henke, Bughin, Chui, Manyika, Saleh, Wiseman, & Sethupathy, 2016).

2.2 Overview of AI Technologies Supporting Digital Transformation

Artificial intelligence has several key technologies that collectively power digital transformation across industries. Here's an overview of the main AI technologies and their applications:

Machine Learning (ML): Machine Learning is a branch of AI that allows systems to learn from data and improve performance over time. ML is foundational to digital transformation, enabling predictive analytics, recommendation engines, and real-time decision-making. From customer behavior analysis to financial forecasting, ML algorithms are embedded in various business applications to provide personalized, accurate, and adaptive solutions.

Natural Language Processing (NLP): Natural Language Processing enables computers to understand, interpret, and respond to human language. NLP powers chatbots, sentiment analysis, and virtual assistants, helping businesses automate customer service, analyze social media sentiment, and derive insights from text data. By understanding and generating human language, NLP allows companies to interact with customers on a more personal and engaging level (Kaplan & Haenlein, 2020).

Computer Vision: Computer Vision, another AI field, enables machines to interpret and process visual data. Used in sectors like manufacturing, retail, and healthcare, computer vision enhances processes such as quality control, facial recognition, and image analysis. In digital transformation, computer vision provides opportunities for automation in environments where visual inspection or identification is critical.

Robotic Process Automation (RPA): Robotic Process Automation is a technology that uses AI to automate routine and rule-based tasks, such as data entry, invoice processing, and customer onboarding. RPA is particularly valuable in digital transformation because it enhances operational efficiency, minimizes errors, and allows employees to focus on higher-level tasks. RPA also supports scalability, making it easier for organizations to adapt to business growth or changing market conditions (Marr, 2020).

Table 1. Comparison of AI technologies and their use cases

AI Technology	Description	Key Use Cases in Digital Transformation
Machine Learning (ML)	Algorithms that learn from data and improve over time	Predictive analytics, personalized recommendations, anomaly detection
Natural Language Processing (NLP)	AI that understands and processes human language	Chatbots, sentiment analysis, customer support
Computer Vision	AI that interprets and processes visual information	Quality control, image recognition, facial identification
Robotic Process Automation (RPA)	Automates rule-based, repetitive tasks	Data entry, workflow automation, invoice processing

3. IMPACT OF AI ON VARIOUS SECTORS

The integration of Artificial Intelligence (AI) across industries is transforming the way businesses operate, innovate, and deliver value to their customers. In the healthcare sector, AI is revolutionizing diagnostics, patient care, and research. Advanced algorithms help in analyzing medical images with precision, assisting doctors in detecting diseases early, and enabling personalized treatment plans. AI-driven predictive analytics also play a role in anticipating patient needs and optimizing hospital operations, improving patient outcomes, and reducing healthcare costs. In the finance industry, AI is used to enhance security, streamline operations, and improve customer experiences. Algorithms can analyze vast amounts of financial data to detect fraud, assess credit risks, and provide investment recommendations (Mikalef, Boura, Lekakos, & Krogstie, 2019). Furthermore, AI-powered chatbots and virtual assistants are reshaping customer service in finance, allowing institutions to offer 24/7 support and personalized financial advice.

In retail, AI is driving significant changes in inventory management, personalized marketing, and customer engagement. Through data analysis and machine learning, retailers can anticipate demand, optimize stock levels, and reduce wastage. Additionally, AI enables hyper-personalized marketing by analyzing customer behavior and preferences, delivering targeted promotions, and improving customer loyalty. In the manufacturing sector, AI is integral to enhancing production efficiency, predictive maintenance, and quality control. By using AI-powered sensors and IoT data, manufacturers can monitor equipment health in real time, predict potential breakdowns, and reduce downtime, thereby enhancing operational efficiency. Robotics powered by AI are also being used to automate repetitive tasks, allowing human workers to focus on higher-value activities (Rai, Constantinides, & Sarker, 2019).

Education is another field being reshaped by AI. In the education sector, AI is helping to personalize learning experiences, identify students' strengths and weaknesses, and offer targeted educational resources. AI-driven tools can adapt lessons to suit each student's learning pace, providing a more individualized approach to education and improving outcomes. Educators also benefit from AI through automated grading and administrative tasks, allowing them to focus more on teaching. Across these and many other sectors, AI's impact is profound, helping businesses increase efficiency, reduce costs, and offer enhanced, more personalized experiences. As AI technologies continue to evolve, their impact on these sectors will deepen, driving innovation and transforming industry standards.

3.1 AI in Healthcare

In healthcare, AI is rapidly becoming a game-changer by enhancing diagnostic accuracy, optimizing patient care, and accelerating drug discovery. Machine learning algorithms analyze vast datasets to predict patient outcomes, identify disease patterns, and recommend personalized treatment plans (Russell & Norvig, 2020). For example:

- **Diagnostics:** AI-driven diagnostic tools, such as those for detecting early-stage cancers or identifying cardiovascular risks, have improved early diagnosis and treatment.
- **Patient Care:** AI-powered virtual assistants and chatbots support patients by answering questions, setting reminders for medication, and even monitoring vital signs.
- **Drug Discovery:** AI algorithms can analyze biomedical data to identify potential compounds for drug development, significantly reducing the time and cost associated with bringing new drugs to market.

AI in healthcare helps improve patient outcomes, reduce human errors, and optimize resource allocation, enabling a more proactive approach to medicine.

3.2 AI in Finance

The finance sector has embraced AI to improve fraud detection, customer service, and risk management. By analyzing vast amounts of transactional and behavioral data, AI can help financial institutions predict and prevent fraud, as well as enhance customer experiences (Schmidt & Botelho, 2020).

- **Fraud Detection:** Machine learning algorithms identify suspicious patterns in transaction data, detecting and mitigating fraud in real time.
- **Customer Service:** Chatbots and virtual assistants provide 24/7 customer support, answering common questions and helping customers with transactions.
- **Risk Management:** AI models assess credit risks more accurately by analyzing diverse data sources, helping financial institutions make more informed lending decisions.

The integration of AI into finance has not only enhanced security and risk management but has also significantly streamlined operations and improved customer satisfaction.

3.3 AI in Retail

In the retail industry, AI-driven tools are used to personalize shopping experiences, forecast demand, and optimize supply chain management. Retailers rely on AI to enhance customer engagement and streamline operations.

- **Personalization:** AI analyzes customer data to provide personalized product recommendations, improving customer satisfaction and sales.
- **Demand Forecasting:** By predicting demand trends based on historical data, AI helps retailers optimize stock levels, reducing inventory costs and minimizing stockouts.
- **Inventory Management:** AI algorithms optimize inventory, ensuring that popular items are restocked while reducing excess inventory.

Through these applications, AI enables retailers to deliver a tailored shopping experience, reduce operational costs, and improve inventory efficiency.

3.4 AI in Manufacturing

AI is transforming the manufacturing industry by facilitating predictive maintenance, quality control, and supply chain optimization. Manufacturers use AI to minimize downtime, increase productivity, and enhance product quality.

- **Predictive Maintenance:** AI analyzes sensor data from machinery to predict failures before they occur, reducing downtime and maintenance costs.
- **Quality Control:** Computer vision systems inspect products for defects in real time, improving product quality and reducing waste.

- **Supply Chain Optimization:** AI-driven models optimize inventory, forecast demand, and improve logistics, making the supply chain more efficient and resilient.

AI's integration into manufacturing enhances productivity, improves product quality, and reduces costs, helping manufacturers maintain a competitive edge.

3.5 AI in Education

In education, AI supports personalized learning, improves student engagement, and enhances administrative efficiency. Educational institutions use AI to create customized learning experiences, assist students, and streamline operations.

- **Adaptive Learning:** AI-powered platforms assess students' learning styles and adapt content to match individual needs, helping improve learning outcomes.
- **Student Support:** Chatbots provide instant assistance to students, answering questions about enrollment, coursework, and deadlines.
- **Content Creation:** AI tools create or curate course materials, ensuring that content remains relevant and engaging.

By personalizing learning and supporting educators, AI helps institutions improve student satisfaction and operational efficiency.

Table 2. Impact of AI on key sectors

Sector	AI Applications	Benefits and Impact
Healthcare	Diagnostics, patient care, drug discovery	Improved accuracy, early diagnosis, personalized treatment
Finance	Fraud detection, customer support, risk management	Enhanced security, operational efficiency, customer engagement
Retail	Personalization, demand forecasting, inventory management	Increased customer loyalty, cost efficiency
Manufacturing	Predictive maintenance, quality control, supply chain optimization	Reduced downtime, improved productivity, cost savings
Education	Adaptive learning, student support, content creation	Personalized learning, enhanced student experience

4. ENHANCING CUSTOMER EXPERIENCE WITH AI

Artificial Intelligence (AI) is redefining customer experience by enabling businesses to deliver personalized, efficient, and proactive services at scale. AI-driven personalization allows companies to analyze vast amounts of customer data to understand individual preferences, behaviors, and needs. With this information, businesses can create tailored recommendations, targeted promotions, and individualized service offerings, enhancing customer satisfaction and loyalty. For instance, streaming services use AI algorithms to recommend shows and movies based on users' viewing histories, while e-commerce platforms suggest products that align with past purchases or browsing behaviors. AI is also transforming customer support through the use of chatbots and virtual assistants, which provide instant, 24/7 assistance. These AI-powered tools can handle routine inquiries, resolve common issues, and guide customers through transactions, significantly reducing wait times and improving accessibility (Stone, Brooks, Brynjolfsson, Calo, Etzioni, Hager, & Shoham, 2016).

In addition, AI enables companies to anticipate customer needs through predictive analytics. By analyzing patterns in customer data, businesses can proactively address issues, offer timely solutions, and even anticipate future needs. For example, AI can help a telecommunications provider identify potential service disruptions before they impact users, allowing for preemptive action that minimizes inconvenience. This level of service fosters a deeper connection between customers and brands, as it demonstrates attentiveness and a commitment to meeting customer needs in real time. Furthermore, AI can enhance customer experience by optimizing service delivery across channels. From online chat to email support and in-store interactions, AI enables a seamless, Omni channel experience that meets customers where they are, providing consistent and high-quality service across platforms.

Overall, AI empowers businesses to offer customer experiences that are more personalized, responsive, and aligned with customer expectations. As AI technologies continue to advance, the potential for even more immersive and intuitive customer interactions grows, setting new standards for how companies engage with and serve their customer (Verhulst & Young, 2017).

4.1 AI-Driven Personalization

Personalization is one of the most prominent ways AI enhances customer experience. By analyzing customer data—such as browsing behavior, purchase history, and demographic information—AI algorithms deliver highly relevant content and product recommendations to individual users.

- **Product Recommendations:** Online retailers like Amazon and streaming services like Netflix use AI-powered recommendation engines to analyze a user's past behaviors and suggest products or content they are likely to enjoy.
- **Dynamic Content Customization:** AI can dynamically adjust website content based on user preferences, presenting each visitor with a personalized experience.
- **Email and Marketing Personalization:** AI-driven algorithms can create personalized email campaigns, ensuring customers receive offers that match their interests and past behaviors.

Through personalized content and offers, AI strengthens customer loyalty and increases engagement, ultimately driving higher conversion rates and customer satisfaction.

4.2 Chatbots and Virtual Assistants

Chatbots and virtual assistants provide round-the-clock support, helping businesses manage customer inquiries quickly and efficiently. Powered by Natural Language Processing (NLP), these AI-driven tools understand customer requests, provide relevant responses, and even escalate complex issues to human agents when necessary.

- **Automated Customer Support:** Chatbots handle common questions and requests, such as product information, order tracking, and troubleshooting, reducing response times and operational costs.
- **Enhanced Customer Interaction:** Virtual assistants like Siri, Alexa, and Google Assistant not only respond to voice commands but also learn user preferences over time, creating a more engaging and responsive interaction.
- **Multilingual Support:** Advanced AI systems can offer support in multiple languages, making it easier for companies to engage with global audiences and break language barriers.

By reducing wait times and improving service accessibility, chatbots and virtual assistants significantly enhance the customer experience, increasing satisfaction and customer retention.

4.3 Predictive Analytics for Customer Insights

Predictive analytics, powered by machine learning, allows companies to anticipate customer needs and behaviors. By analyzing historical data and identifying patterns, AI-driven predictive models provide valuable insights that help businesses proactively address customer demands.

- **Churn Prediction:** AI models can identify customers at risk of leaving based on behavioral patterns and engagement metrics, allowing companies to take proactive measures to retain them.
- **Customer Lifetime Value (CLV) Prediction:** Predictive models help businesses determine which customers are likely to have a high lifetime value, informing targeted marketing strategies.
- **Real-Time Offers:** AI-driven systems can generate real-time offers and incentives based on a customer's actions, such as browsing a particular product category or adding items to a shopping cart.

With predictive analytics, companies can make data-driven decisions that cater to customers' needs in real time, enhancing engagement and building loyalty.

- **Image Prompt:** Illustration of AI Enhancing Customer Experience: A visually engaging image that depicts AI interacting with customers across multiple devices, including a smartphone, laptop, and smart speaker. In the image, AI is represented by icons such as a chatbot on a website, a virtual assistant on a smart speaker, and personalized content recommendations on the smartphone screen. The background should show a seamless flow of data connecting each device, representing AI-driven personalization and predictive analytics.

5. OPERATIONAL EFFICIENCY AND COST SAVINGS THROUGH AI

Artificial Intelligence (AI) is transforming operational efficiency by automating routine tasks, optimizing processes, and enabling data-driven decision-making across various industries. Through automation, AI takes over repetitive tasks—such as data entry, report generation, and inventory management—freeing up employees to focus on higher-value activities that require creativity and critical thinking. This shift not only enhances productivity but also reduces the risk of human error, leading to more accurate and reliable outcomes. In fields like manufacturing, AI-driven robots handle assembly, packaging, and quality control, performing tasks faster and with greater precision than human workers. Similarly, in the logistics and supply chain sectors, AI algorithms can predict demand trends, optimize delivery routes, and manage stock levels, ultimately minimizing delays, reducing fuel costs, and ensuring products reach customers efficiently.

AI also empowers companies to achieve significant cost savings by improving resource allocation and operational decision-making. Predictive maintenance, for example, uses AI algorithms to analyze equipment performance and forecast potential breakdowns before they happen. By scheduling maintenance only when needed rather than at set intervals, businesses can reduce downtime, prolong equipment lifespan, and cut maintenance costs. In finance and HR departments, AI-based analytics can identify patterns in spending, detect inefficiencies, and suggest strategies for better resource management. These insights allow businesses to make informed choices, reduce waste, and achieve cost-effectiveness without compromising quality or productivity.

Furthermore, AI-driven analytics provide organizations with real-time insights that allow them to adapt quickly to changing conditions, respond to customer demands, and streamline operations. With AI's ability to process and interpret large volumes of data, companies can rapidly identify bottlenecks, forecast inventory needs, and allocate resources more effectively, ensuring smooth operations even during peak demand periods. Overall, AI has become an essential tool for enhancing operational efficiency and realizing cost savings, enabling businesses to stay competitive, agile, and financially sustainable in a dynamic market. AI is increasingly recognized as a powerful tool for enhancing operational efficiency across various industries. By automating routine tasks, optimizing processes, and providing data-driven insights, AI enables organizations to improve productivity and reduce costs. Below, we explore three key areas where AI contributes to operational efficiency: automation, process optimization, and data analysis (Haenlein & Kaplan, 2019).

5.1 Automation of Routine Tasks

One of the most significant impacts of AI is its ability to automate repetitive and time-consuming tasks. This automation can lead to substantial cost savings and allows employees to focus on higher-value activities.

- **Robotic Process Automation (RPA):** RPA technologies automate rule-based tasks across various functions, such as data entry, invoice processing, and report generation. By reducing the manual effort involved in these tasks, organizations can increase throughput and accuracy.

- **AI in Supply Chain Management:** AI systems can automate inventory management, order processing, and logistics planning, streamlining operations and reducing the likelihood of errors.
- **Automated Customer Service:** As mentioned earlier, AI-powered chatbots handle routine customer inquiries, freeing up human agents to tackle more complex issues that require personalized attention.

Through the automation of routine tasks, organizations can significantly enhance efficiency, reduce labor costs, and improve overall productivity.

5.2 Process Optimization

AI algorithms can analyze complex processes to identify inefficiencies and recommend improvements. By optimizing workflows, businesses can achieve higher efficiency and effectiveness in their operations.

- **Predictive Maintenance:** In manufacturing and other industries, AI-driven predictive maintenance identifies equipment issues before they lead to failures, allowing companies to schedule maintenance proactively and avoid costly downtime.
- **Supply Chain Optimization:** AI analyzes historical data and market trends to forecast demand, optimize inventory levels, and streamline logistics. This optimization leads to reduced holding costs and improved delivery times.
- **Quality Control:** AI-powered computer vision systems monitor production lines for defects in real time, ensuring that only high-quality products reach customers while reducing waste and rework costs.

By optimizing processes through AI, organizations can improve service delivery, enhance product quality, and reduce operational costs.

5.3 Data-Driven Insights

AI empowers organizations to harness data for better decision-making. By providing actionable insights based on large datasets, AI helps businesses understand trends, customer behaviors, and market dynamics.

- **Real-Time Analytics:** AI-driven analytics platforms provide real-time insights into operational performance, enabling managers to make informed decisions quickly.

- **Cost Analysis:** AI tools analyze spending patterns, helping organizations identify areas for cost reduction and resource allocation optimization.
- **Strategic Planning:** By predicting future trends based on historical data, AI enables companies to develop strategies that align with market demands, ultimately improving competitiveness and profitability.

With data-driven insights, organizations can make smarter decisions that lead to increased efficiency and lower operational costs.

Table 3. Operational efficiency gains through AI

Area	AI Applications	Efficiency Gains and Cost Savings
Automation of Tasks	Robotic Process Automation, automated customer service	Increased productivity, reduced labor costs
Process Optimization	Predictive maintenance, supply chain optimization	Reduced downtime, improved delivery times, lower waste
Data-Driven Insights	Real-time analytics, cost analysis, strategic planning	Smarter decision-making, better resource allocation

6. CHALLENGES AND ETHICAL CONSIDERATIONS

While AI offers significant benefits for digital transformation, organizations must navigate several challenges and ethical considerations to ensure successful implementation. Addressing these issues is crucial for leveraging AI responsibly and effectively while minimizing potential risks. Below, we explore the primary challenges and ethical considerations associated with AI. While Artificial Intelligence (AI) offers transformative potential across industries, its adoption brings forth significant challenges and ethical considerations. One of the primary challenges is the quality and availability of data. AI systems rely on large, accurate datasets to function effectively, but obtaining clean, representative data can be challenging, especially in sectors with strict privacy regulations like healthcare and finance. Poor data quality can lead to unreliable AI models, and biases in data can skew results, leading to inaccurate predictions or even discriminatory outcomes. Another technical challenge is integration with existing systems. Many organizations operate legacy systems that may not be compatible with advanced AI technologies, making integration complex and costly. Additionally, there is often a shortage of skilled professionals who can manage, develop, and interpret AI solutions effectively, further slowing down AI implementation.

Beyond technical issues, ethical considerations play a crucial role in the responsible deployment of AI. One major ethical concern is bias and fairness. AI algorithms learn from historical data, and if that data contains societal biases, AI can unintentionally perpetuate or even amplify these biases in areas like hiring, lending, and law enforcement. Ensuring that AI models are fair, transparent, and inclusive requires careful oversight, testing, and ongoing adjustments. Privacy is another significant ethical concern. AI systems often process sensitive data, and improper handling of this information can lead to privacy violations. In an era where data breaches and unauthorized data sharing are increasingly common, organizations must prioritize data protection and adhere to regulations like the GDPR to protect user privacy (Mikalef, Boura, Lekakos, & Krogstie, 2019).

Transparency and accountability are also essential considerations, particularly as AI systems make increasingly critical decisions. Many AI models, particularly those using deep learning, operate as "black boxes," where the decision-making process is not easily interpretable. This lack of transparency can erode trust in AI systems, especially in high-stakes applications like healthcare, finance, and criminal justice. To address this, organizations are turning toward Explainable AI (XAI), which aims to make AI decisions more understandable to users and stakeholders. Lastly, responsible use of AI raises questions around workforce displacement. As AI automates more tasks, there are concerns about job losses and the future of work. Organizations must consider how to upskill employees, create new roles, and transition workers into positions that leverage human creativity and judgment.

In summary, while AI has the potential to drive innovation and efficiency, addressing these challenges and ethical considerations is critical for its sustainable and responsible adoption. Organizations must take a proactive approach, implementing ethical frameworks, investing in transparency, and ensuring that AI systems serve all stakeholders fairly and equitably (Russell & Norvig, 2020).

6.1 Challenges of Implementing AI

- **Data Quality and Availability:** AI systems rely heavily on high-quality data for training and decision-making. Organizations often face challenges in collecting, managing, and maintaining accurate data.
- **Data Silos:** Information may be spread across various departments or systems, making it difficult to access and analyze comprehensive datasets.
- **Data Quality:** Inaccurate, incomplete, or biased data can lead to suboptimal AI performance and skewed results, affecting business decisions.

- **Integration with Existing Systems:** Integrating AI technologies with existing IT infrastructures can be complex and resource-intensive. Organizations may encounter challenges in:
- **Legacy Systems:** Many businesses still operate on outdated systems that are not compatible with modern AI technologies.
- **Interoperability:** Ensuring that new AI tools work seamlessly with existing applications and processes is essential for maximizing their value.

Skill Gaps and Change Management- Implementing AI often requires specialized skills that may not be readily available within the organization. Challenges include:

- **Lack of Expertise:** Organizations may struggle to find and retain talent with the necessary AI and data science skills.
- **Change Resistance:** Employees may resist changes brought about by AI implementation, fearing job displacement or disruptions to established workflows.
- By proactively addressing these challenges, organizations can enhance their chances of successfully integrating AI into their digital transformation initiatives.

6.2 Ethical Considerations of AI

As organizations adopt AI technologies, they must also consider the ethical implications of their use. Key ethical considerations include:

Bias and Fairness- AI algorithms can inadvertently perpetuate or amplify biases present in training data, leading to unfair outcomes.

- **Discrimination:** If AI systems are trained on biased datasets, they may make discriminatory decisions in areas such as hiring, lending, and law enforcement.
- **Transparency:** Organizations must strive for transparency in AI decision-making processes to ensure fairness and accountability.

Privacy and Data Security- AI systems often require access to vast amounts of personal data, raising concerns about privacy and security.

- **Data Protection:** Organizations must implement robust data protection measures to safeguard sensitive information and comply with regulations such as GDPR.

- **Informed Consent:** Businesses should ensure that customers are aware of how their data will be used and obtain explicit consent where necessary.

Accountability and Governance- As AI systems make decisions that impact individuals and society, establishing accountability and governance frameworks is crucial.

- **Responsibility:** Organizations should clarify who is responsible for AI-driven decisions, particularly in high-stakes scenarios like healthcare and criminal justice.
- **Regulation:** Businesses must stay informed about evolving regulations surrounding AI use and ensure compliance to avoid legal repercussions.

7. BEST PRACTICES FOR AI INTEGRATION IN DIGITAL TRANSFORMATION

Successfully integrating AI into digital transformation requires careful planning, execution, and ongoing evaluation. Organizations can enhance their chances of successful AI implementation by following best practices that address technical, operational, and cultural aspects of their business. Below are key guidelines to consider when integrating AI technologies. Successfully integrating Artificial Intelligence (AI) into digital transformation efforts requires a well-defined approach that aligns with organizational goals, fosters collaboration, and emphasizes ethical use. One of the first best practices is to establish clear objectives for AI implementation. Organizations should define the specific problems they aim to solve with AI, setting measurable goals that guide project direction and enable tracking of ROI. Starting with a pilot project can be an effective way to assess AI's impact, refine strategies, and build internal support before scaling up. Another best practice is to invest in high-quality data. Data is the foundation of any AI initiative, and ensuring it is clean, relevant, and accessible across departments is crucial. Organizations should implement robust data governance practices and regularly audit data sources to maintain accuracy and relevance.

Building cross-functional teams is essential for a successful AI integration, as it fosters collaboration between IT, data science, and business units. AI initiatives benefit from diverse expertise, ensuring that models are designed with a clear understanding of both technical capabilities and business needs. Additionally, involving end-users early in the process helps align AI solutions with user expectations, improves usability, and accelerates adoption. Establishing strong partnerships with technology providers and AI vendors can also be valuable. These partnerships bring access to advanced tools, technical support, and industry expertise, allowing organizations to stay up-to-date with AI advancements while reducing the burden on internal teams.

Another best practice is to prioritize transparency, ethics, and regulatory compliance throughout AI development and deployment. AI's decision-making processes should be as transparent as possible, and explainable AI (XAI) models can help ensure that users understand AI outcomes, especially in sectors like healthcare and finance. Ethical considerations, such as fairness and accountability, must be integrated from the outset to avoid unintended biases and foster trust. Regular audits of AI models can help identify and correct bias, ensuring that AI systems treat all users equitably. Ongoing monitoring and maintenance are also critical, as AI models can drift over time due to changes in data or market conditions. Regular updates ensure that AI models remain accurate and effective.

Lastly, fostering a culture of continuous learning and adaptation within the organization is key. AI is an evolving field, and organizations need to stay agile to keep pace with technological advancements. Encouraging ongoing training, upskilling employees, and promoting innovation will help businesses maximize AI's potential and remain competitive. By following these best practices, organizations can effectively integrate AI into their digital transformation strategies, driving meaningful results while ensuring ethical and sustainable use of technology.

7.1 Establish Clear Objectives

Before implementing AI, organizations should define clear objectives and align them with their overall digital transformation strategy. This involves:

- **Identifying Use Cases:** Determine specific business problems that AI can solve or areas where AI can enhance efficiency. Prioritize use cases based on potential impact and feasibility.
- **Setting Measurable Goals:** Establish key performance indicators (KPIs) to measure success and track progress. This could include metrics related to customer satisfaction, operational efficiency, or revenue growth.

Having clear objectives ensures that AI initiatives remain focused and aligned with organizational goals.

7.2 Invest in Quality Data

Data is the foundation of effective AI systems. Organizations must prioritize data quality and management to ensure successful AI implementation:

- **Data Collection:** Implement systems for collecting relevant data from various sources, ensuring it is comprehensive and representative of the business landscape.
- **Data Governance:** Establish governance frameworks to ensure data integrity, accuracy, and compliance with regulations. Regularly audit and cleanse data to remove inaccuracies and biases.
- **Data Integration:** Develop strategies for integrating data across departments and systems to create a unified view that enhances AI training and insights.

Investing in quality data not only improves AI performance but also facilitates better decision-making across the organization.

7.3 Foster a Culture of Innovation

Successful AI integration requires a culture that embraces innovation and continuous improvement. Organizations should:

- **Encourage Experimentation:** Promote a mindset of experimentation and learning, where teams feel empowered to explore new ideas and technologies without fear of failure.
- **Invest in Training:** Provide training and development opportunities for employees to build AI-related skills, ensuring that the workforce is equipped to leverage AI tools effectively.
- **Cross-Functional Collaboration:** Foster collaboration among departments—such as IT, operations, and marketing—to leverage diverse perspectives and expertise in AI initiatives.

Cultivating a culture of innovation helps organizations adapt to changing technologies and market conditions.

7.4 Collaborate with Technology Partners

Organizations may benefit from collaborating with technology partners who have expertise in AI development and implementation:

- **Select the Right Vendors:** Evaluate potential technology vendors based on their experience, capabilities, and track record in AI implementation. Look for partners that understand the organization's industry and challenges.

- **Leverage Third-Party Solutions:** Consider using established AI platforms and tools to accelerate development and reduce costs. This can include cloud-based AI services that offer scalability and flexibility.
- **Engage in Knowledge Sharing:** Collaborate with industry peers and research institutions to share insights, best practices, and lessons learned in AI integration.

Partnering with technology experts can help organizations navigate the complexities of AI implementation more effectively.

7.5 Monitor and Evaluate Performance

Once AI systems are implemented, organizations must continuously monitor and evaluate their performance to ensure they are delivering the desired outcomes:

- **Regular Performance Review:** Conduct regular assessments of AI systems against established KPIs to identify areas for improvement. Analyze how AI is impacting business processes and customer experiences.
- **Feedback Mechanisms:** Implement feedback loops that allow users to provide input on AI system performance. This information can help refine algorithms and improve user experience.
- **Adapt and Evolve:** Stay agile and open to change by adjusting AI strategies based on performance data and market dynamics. Be prepared to pivot or scale successful initiatives as needed.

8. FUTURE TRENDS IN AI AND DIGITAL TRANSFORMATION

As technology continues to evolve, the integration of AI into digital transformation strategies will experience significant changes and advancements. Understanding these trends is essential for organizations looking to stay competitive and leverage AI effectively in their digital initiatives. Below are some key trends shaping the future of AI and digital transformation? As Artificial Intelligence (AI) continues to evolve, its role in digital transformation is expanding, shaping new possibilities and setting trends that will redefine industries. One of the most prominent trends is the rise of AI-driven automation. In the coming years, AI will further enhance automation capabilities beyond routine tasks, handling complex processes across various sectors, such as healthcare diagnostics, legal analysis, and financial planning. This level of intelligent automation, often referred to as hyper-automation, will drive even greater operational efficiencies and open up new opportunities for innovation. Another emerging trend is the adoption of Explainable AI (XAI). With AI becoming integral to decision-making in high-stakes industries, there is an increasing demand for transparency and interpretability in AI models. XAI enables organizations to understand and explain how AI systems arrive at specific outcomes, helping to build trust and ensure compliance with ethical and regulatory standards (Schmidt & Botelho, 2020).

The integration of edge AI is also set to transform digital operations. By processing data closer to the source (on devices or local networks), edge AI reduces latency and enhances data privacy, which is especially beneficial in sectors like healthcare, manufacturing, and retail. This trend is driven by advancements in hardware and IoT, allowing organizations to deploy real-time, AI-powered insights directly at the edge, resulting in faster response times and reduced reliance on cloud infrastructure. Another transformative trend is the use of AI for sustainability initiatives. AI is increasingly being leveraged to monitor environmental impact, optimize resource use, and reduce waste. From predictive analytics for energy conservation to intelligent supply chain management, AI is enabling businesses to meet sustainability goals while enhancing operational efficiency.

Personalization at scale is also becoming a key focus in AI-driven digital transformation. As customer expectations for tailored experiences grow, companies are leveraging AI to deliver hyper-personalized services across various channels, predicting customer needs and proactively meeting them. This is particularly evident in e-commerce, entertainment, and finance, where AI analyzes user behavior and preferences to offer products, content, or financial solutions in real time. AI-powered cybersecurity is another critical trend, as organizations face increasingly sophisticated cyber threats. AI systems that detect anomalies, predict vulnerabilities, and respond to attacks autonomously are becoming vital to protecting data and maintaining trust (Stone, Brooks, Brynjolfsson, Calo, Etzioni, Hager, & Shoham, 2016).

Finally, AI ethics and governance are gaining attention as companies and regulators recognize the importance of responsible AI use. Ethical frameworks, transparency standards, and regulatory compliance measures are being developed to guide organizations in using AI responsibly. Future AI systems will likely include built-in ethical safeguards, ensuring fairness, accountability, and minimal bias in decision-making processes. Together, these trends represent a dynamic shift in how AI will drive digital transformation, enabling organizations to innovate responsibly, engage customers more deeply, and operate with greater agility and foresight in an increasingly digital world.

8.1 Increased Adoption of AI-Driven Automation

Automation is expected to become even more prevalent as organizations seek to enhance efficiency and reduce operational costs. Key developments include:

- **Expansion of Robotic Process Automation (RPA):** As RPA technologies improve, businesses will adopt them across a broader range of processes, from finance to HR, leading to increased productivity and accuracy.
- **Intelligent Automation:** The convergence of AI and automation will enable organizations to automate complex tasks that require cognitive capabilities, such as natural language processing and image recognition. This will allow for more sophisticated workflows and decision-making processes.

8.2 Rise of Explainable AI (XAI)

As AI systems become more complex, the need for transparency and interpretability is growing. Explainable AI (XAI) focuses on making AI decision-making processes understandable to users. Key aspects include:

- **Trust and Accountability:** Organizations will prioritize XAI to build trust with customers and stakeholders, ensuring that AI systems provide clear explanations for their decisions.
- **Regulatory Compliance:** As governments and regulatory bodies impose stricter requirements for AI transparency, organizations will adopt XAI solutions to comply with these regulations and mitigate legal risks.

8.3 Integration of AI with Edge Computing

The increasing proliferation of IoT devices and the need for real-time data processing are driving the integration of AI with edge computing. This trend involves:

- **Decentralized AI Processing:** By deploying AI algorithms closer to the data source (i.e., at the edge), organizations can reduce latency, enhance responsiveness, and improve data security.
- **Real-Time Analytics:** Edge AI enables real-time analytics and decision-making, allowing organizations to respond quickly to changing conditions and optimize operations.

8.4 Expansion of AI in Cybersecurity

As digital transformation accelerates, so do the risks associated with cybersecurity threats. AI is becoming an essential tool in the fight against cyber threats through:

- **Proactive Threat Detection:** AI systems can analyze vast amounts of data to identify anomalies and potential security breaches before they escalate.
- **Automated Response Mechanisms:** AI can automate responses to security incidents, enabling organizations to react swiftly and effectively to mitigate risks.

8.5 Enhanced Personalization through AI

The demand for personalized customer experiences will continue to grow, driving innovations in AI technologies. Key developments include:

- **Hyper-Personalization:** Organizations will leverage advanced AI algorithms to create hyper-personalized experiences tailored to individual preferences, behaviors, and real-time context.
- **AI-Driven Content Creation:** AI will play a significant role in generating personalized content, recommendations, and marketing materials, further enhancing customer engagement.

8.6 Greater Focus on Ethical AI Practices

As AI adoption increases, organizations will face heightened scrutiny regarding ethical considerations. Key trends include:

- **Ethical Frameworks:** Businesses will develop and implement ethical frameworks to guide AI development and usage, ensuring responsible practices that prioritize fairness, accountability, and transparency.
- **Diversity and Inclusion:** Companies will focus on building diverse teams to mitigate bias in AI algorithms, promoting fairness and inclusivity in AI-driven decisions.

CONCLUSION

The integration of Artificial Intelligence (AI) into digital transformation strategies is reshaping the way organizations operate, innovate, and compete in the digital era. AI's ability to analyze data, automate processes, enhance customer experiences, and drive operational efficiencies makes it a powerful catalyst for change across industries. By embracing AI, organizations are not only meeting current business demands but also positioning themselves to anticipate future challenges and opportunities. However, the successful adoption of AI in digital transformation requires thoughtful planning, clear objectives, and a commitment to ethical and responsible practices. Issues such as data quality, bias, transparency, and workforce adaptation must be addressed to fully realize AI's potential while minimizing unintended consequences. As AI technology continues to evolve, it will enable organizations to move beyond incremental improvements to achieve transformative changes that redefine customer engagement, optimize business processes, and create new value. Future trends such as edge AI, hyper-personalization, and AI-driven sustainability will further expand AI's impact, offering innovative ways to enhance operations and meet societal needs. By following best practices and staying informed about emerging trends, businesses can harness the full power of AI while upholding ethical standards, creating a sustainable foundation for growth in an increasingly digital world. In this way, AI becomes not just a tool for transformation, but a strategic asset that empowers organizations to thrive, adapt, and lead in the future of business.

REFERENCES

Brock, J. K. U., & Von Wangenheim, F. (2019). Demystifying AI: What digital transformation leaders can teach you about realistic artificial intelligence. *California Management Review*, *61*(4), 110–134. DOI: 10.1177/1536504219865226

Brynjolfsson, E., & McAfee, A. (2014). *The Second Machine Age: Work, Progress, and Prosperity in a Time of Brilliant Technologies*. W.W. Norton & Company.

Calp, M. H. (2020). The role of artificial intelligence within the scope of digital transformation in enterprises. In *Advanced MIS and digital transformation for increased creativity and innovation in business* (pp. 122–146). IGI Global. DOI: 10.4018/978-1-5225-9550-2.ch006

Chatterjee, S., Chaudhuri, R., Vrontis, D., & Basile, G. (2022). Digital transformation and entrepreneurship process in SMEs of India: A moderating role of adoption of AI-CRM capability and strategic planning. *Journal of Strategy and Management*, *15*(3), 416–433. DOI: 10.1108/JSMA-02-2021-0049

Davenport, T. H., & Ronanki, R. (2018). Artificial intelligence for the real world. *Harvard Business Review*, *96*(1), 108–116.

Domingos, P. (2015). *The Master Algorithm: How the Quest for the Ultimate Learning Machine Will Remake Our World*. Basic Books.

Fountaine, T., McCarthy, B., & Saleh, T. (2019). Building the AI-powered organization. *Harvard Business Review*, *97*(4), 62–73.

Goodfellow, I., Bengio, Y., & Courville, A. (2016). *Deep Learning*. MIT Press.

Haenlein, M., & Kaplan, A. (2019). A brief history of artificial intelligence: On the past, present, and future of artificial intelligence. *California Management Review*, *61*(4), 5–14. DOI: 10.1177/0008125619864925

Henke, N., Bughin, J., Chui, M., Manyika, J., Saleh, T., Wiseman, B., & Sethupathy, G. (2016). The age of analytics: Competing in a data-driven world. McKinsey Global Institute. https://www.mckinsey.com/business-functions/mckinsey-analytics/our-insights/the-age-of-analytics-competing-in-a-data-driven-world

Kaplan, J., & Haenlein, M. (2020). Rulers of the world, unite! The challenges and opportunities of artificial intelligence. *Business Horizons*, *63*(1), 37–50. DOI: 10.1016/j.bushor.2019.09.003

Marr, B. (2020). *Artificial Intelligence in Practice: How 50 Successful Companies Used AI and Machine Learning to Solve Problems*. Wiley.

Mikalef, P., Boura, M., Lekakos, G., & Krogstie, J. (2019). Big data analytics capabilities and innovation: The mediating role of dynamic capabilities and moderating effect of the environment. *British Journal of Management*, *30*(2), 272–298. DOI: 10.1111/1467-8551.12343

Oyekunle, D., & Boohene, D. (2024). Digital transformation potential: The role of artificial intelligence in business. *International Journal of Professional Business Review: Int.J. Prof. Bus. Rev.*, *9*(3), 1.

Rai, A., Constantinides, P., & Sarker, S. (2019). Next-generation digital platforms: Toward human–AI hybrids. *Management Information Systems Quarterly*, *43*(1), iii–ix.

Russell, S., & Norvig, P. (2020). *Artificial Intelligence: A Modern Approach* (4th ed.). Pearson.

Schmidt, G., & Botelho, A. (2020). The future of work in the digital era: The rise of the human-machine partnership. *Journal of Business Research*, *115*, 360–367.

Singh, A. K., & Taterh, S. (2023). Exploring the Significance and Obstacles of Adopting Futuristic Technology Perspectives for Entrepreneurship and Sustainable Innovation. In *Futuristic Technology Perspectives on Entrepreneurship and Sustainable Innovation* (pp. 1–10). IGI Global. DOI: 10.4018/978-1-6684-5871-6.ch001

Stone, P., Brooks, R., Brynjolfsson, E., Calo, R., Etzioni, O., Hager, G., & Shoham, Y. (2016). Artificial Intelligence and Life in 2030: One Hundred Year Study on Artificial Intelligence: Report of the 2015–2016 Study Panel. Stanford University. https://ai100.stanford.edu/2016-report

Verhulst, S., & Young, A. (2017). *Open Data in Developing Economies: Toward Building an Evidence Base on What Works and How*. African Minds. DOI: 10.47622/9781928331599

Chapter 7
A Comprehensive Review on Advancements and Challenges in Audio Classification Through Deep Learning

Gunjan Verma
https://orcid.org/0009-0004-4995-3707
Amity Institute of Information Technology, Amity University, Jaipur, India

Honey Gocher
Amity University, Jaipur, India

Sweety Verma
KITECH, South Korea

Yudhveer Singh
Amity Institute of Information Technology, Amity University, Jaipur, India

Arti Kaushik
https://orcid.org/0009-0008-6173-8232
G.V.M College of Education, Sonepat, India

Avinash Goswami
https://orcid.org/0009-0008-7441-9951
The LNM Institute of Information Technology, Jaipur, India

ABSTRACT

Deep learning-based audio classification has transformed the industry with improved speech recognition, genre identification in music, and ambient sound detection. The article explores various approaches, including model architectures, evaluation metrics, and preprocessing techniques. Traditional methods are compared to deep learning techniques, which have enhanced performance. Spectrograms, Mel-Frequency Cepstral Coefficients, and Short-Time Fourier Transform are discussed

DOI: 10.4018/979-8-3693-8009-3.ch007

as preprocessing techniques. The study also evaluates hybrid model architectures, training methods, data augmentation, and transfer learning for better outcomes. The paper emphasises the importance of interpretability, stable datasets, and real-time processing for overcoming challenges in audio classification. It is expected to guide future research and advancements in this field.

1. INTRODUCTION

One of the core tasks with a lot of applications in speech recognition, music genre classification, ambient sound detection, etc., is the classification of an audio signal into meaningful categories. The intricate and hierarchical structure of the audio signal has been limited by the handmade features and shallow learning algorithms used traditionally in most techniques of audio classification (Duan, Yang, & Guo, 2024). Deep learning has completely changed this environment by now allowing models to learn directly from unprocessed audio data, therefore bringing considerable gains in efficiency, robustness, and accuracy.

Two categories of deep learning models that have been useful in many diverse audio classification tasks are convolutional and recurrent neural networks. While RNNs excel in modeling the temporal dependencies existing in sequential data, the CNN models specialize in the extraction of spatial features from spectrograms of audio (Sabha & Selwal, 2024). Furthermore, hybrid models mixing RNNs and CNNs make good use of both architectures and often reach state-of-the-art performance for many diverse tasks in audio classification (Wimalasena & Ranasinghe, 2024). Deep learning applied in audio classification demands deep understanding of both audio signal processing and data preprocessing methods on the one hand, and model architectures, training procedures, and assessment metrics on the other to use it effectively.

This paper aims to review the full context of the state of the art in deep learning audio classification by putting together the research findings from a wide range of studies. A few approaches, applications, and problems are considered in the present research to underline the progress that has been done so far and possible future paths toward better audio categorization (Tarımer & Karadağ, 2024). It will thus be an enlightening read within the reach of many of the variegated scholarship and practitioner mixtures with the use of figures, tables, and code snippets that increase clarity for the complex concepts and approaches covered (Gupta & Kumar, 2024).

2. THEORETICAL BACKGROUND

2.1 Deep Learning in Audio Classification

Accordingly, deep learning models can recognize complex patterns and hierarchies in audio data. Though RNNs are good at modeling temporal dependencies in sequential data, on the other hand, CNNs turn out to be very effective for extracting spatial characteristics from audio spectrograms (Shishkin, Hollosi, Goetze, & Doclo, 2024). Hybrid models ensure the benefits of both architectures by mixing CNN and RNN into one, turning out better results in many applications related to audio classification.

Figure 1. An overview of audio processing pipeline.

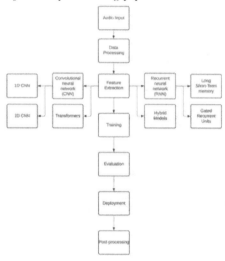

(Esposito et al., 2024)

The figure 1 flowchart includes the developed text architecture and is found in a typical audio processing pipeline where raw audio input is taken first and then, subsequently, using data in the pipeline for better quality. Feature extraction may result from a variety of neural networks, including transformers, hybrid models, and 1D, 2D convolutional neural networks for various features extracted from the audio input (Esposito et al., 2024). The extracted properties are fed for training to machine learning (Gocher, Taterh, & Dadheech, 2023a) models like recurrent neural networks, long short-term memory networks, and gated recurrent units. The trained models are then validated on another dataset to check the efficacy. After that, a model that looks most promising is further post-processed and then implemented on applications. A flowchart has therefore been provided in place of a template for this very purpose (Midavaine, Go, Canez, Simion, & Chatterji, 2024).

3. KEY CONCEPT

3.1 Processing Techniques

Preprocessing is one of the most critical stages in audio classification, as it makes raw audio signals ready for use with deep neural networks (Mei et al., 2024). Common techniques related to preprocessing include feature extraction, enriching data, and normalization.

3.1.1 Normalization

Normalization gives an audio signal its constant range of amplitudes, which is crucial for the stability and convergence of deep learning models (Mohaimenuzzaman, Bergmeir, & Meyer, 2024). Normally in an audio signal, it is done by min-max scaling and mean-variance normalization: techniques that squeeze the amplitude of a signal into a pre-specified range (Vosoughi, Bondi, Wu, & Xu, 2024).

Figure 2. Normalization technique.

(Ye, Ciccarelli, & Kulis, 2024)

140

The impact of normalisation on training images is seen in figure 2. By helping to scale the standard deviation and centre the data around zero, normalisation can enhance the model's convergence (Gocher, Taterh, & Dadheech, 2023b).

4. MODEL ARCHITECTURE

Deep learning is an artificial neural network and is based on the structure and functions of the brain of a human. At the core of a neural network is a neuron, which receives input signals, processes a given signal with the help of weighted connections and produces responses (Ji, Zhang, & Wang, 2024). The deep learning architecture is a topological extension of a multi-layer perceptron using the concept of a sequenced layer of neurons. These networks are deep in the sense of containing many hidden layers, which allows them to extract abstract and sophisticated properties out of unprocessed input (Cappellazzo, Falavigna, & Brutti, 2024).

The major objectives of audio classification are to map any input audio signal against the predefined categories. A few theoretical concepts revolve around this that encompasses feature extraction, optimization, and model training (Ji, Wang, & Sun, 2024). Since the raw audio signals are normally transformed into more interpretive representations such as spectrograms, MFCCs, or log-Mel spectrograms, feature extraction becomes quite essential (Ronchini, Comanducci, & Antonacci, 2024). They capture the rich feature set of deep learning models by showing the frequency content of the audio signal over time.

Convolutional Neural Networks (CNNs): Because they can capture spatial hierarchies, they have been helpful in the handling of picture and audio data. Handling spectrograms in audio categorization, CNNs treat them very much like pictures (Zhang, Chen, Bai, & Wang, 2024). They use filters on convolutional layers to scan over the input spectrogram, picking up local patterns such as phonemes and harmonics. Pooling layers essentially reduce the dimensionality to increase model efficiency but keep the essential characteristics. Since it is hierarchical in nature, deep layers in CNN learn abstract features compared to shallow layers (Lau, Rehman, & Po, 2024).

Figure 3. CNN architecture.

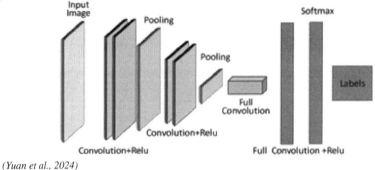

(Yuan et al., 2024)

The standard design of a convolutional neural network (CNN) used for audio categorization is shown in figure 3. It displays the pooling layers, which lower dimensionality, and the convolutional layers, which capture spatial hierarchies, before fully connected layers, which do final classification (Yuan et al., 2024).

Recurrent Neural Networks (RNNs) are the best architectures for tasks that have temporal dependencies in the features, such as speech recognition and music analysis, because they deal with sequential input (Paissan, Della Libera, Ravanelli, & Subakan, 2024). RNNs process the data sequences through recurrent feedback loops, which remember prior inputs. Advanced RNN architectures, including Long Short-Term Memory (LSTM) and Gated Recurrent Units (GRUs), do not suffer from this vanishing gradient problem and hence turn out to be very useful in picking up long-term dependencies (Dinkel, Yan, Wang, Zhang, Wang, & Wang, 2024).

Figure 4. RNN architecture.

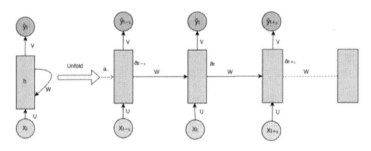

(Lebourdais, Mariotte, Almudévar, Tahon, & Ortega, 2024)

The architecture of a Recurrent Neural Network (RNN) intended for audio classification is shown in figure 4. It has fully linked layers for classification and LSTM layers for processing sequential data (Lebourdais, Mariotte, Almudévar, Tahon, & Ortega, 2024).

Hybrid Models: What is more, the advantages of both architectures are inherited by the RNN-CNN combination. For example, CNNs can make use of spectrophotoms for spatial information extraction, whose output is fed into RNNs to capture temporal dependencies (Yu, Yu, & Wang, 2024). Such a hybrid method provides a more complete understanding of the audio signal and has shown better performance in a variety of audio classification tasks.

Figure 5. Hybrid architecture.

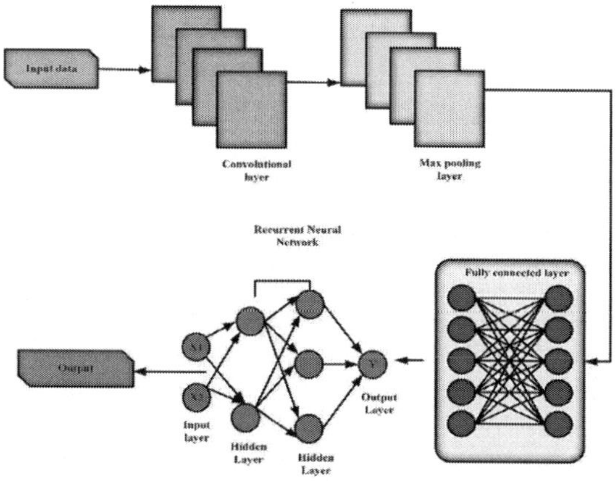

(Chen, Zhou, & Chen, 2024)

The design of a hybrid model for audio categorisation that incorporates CNNs and RNNs is shown in figure 5. Time-distributed convolutional and pooling layers are used to extract spatial data; LSTM layers are then used to capture temporal dependencies; finally, fully connected layers are employed for classification (Chen, Zhou, & Chen, 2024).

5. METHODOLOGIES

Deep learning approaches to audio classification are multi-level complex, with different methods applied at different levels: feature extraction, data preparation, model architecture design, and training plans (Zhao et al., 2024). These approaches will be explained in detail in the next section, including extensive coverage of the steps necessary for the development and deployment of effective deep learning models for audio classification (Bukka, Lalam, Bhatta, & Wright, 2024).

Preprocessing forms the first step in processing raw audio inputs into neural network input formats. This generally comprises techniques such as the extraction of Mel-Frequency Cepstral Coefficients, computation of Short-Time Fourier Transform, and the production of spectrograms (Chen, Jun, Hong, He, & Moon, 2024). STFT will allow the extraction of time-frequency representations by transforming time-domain signals into the frequency domain (Sinnott, 2024). On the other hand, MFCCs summarize the distribution of frequencies over time windows, accounting for essential features required in audio classification.

5.1 Spectrogram Generation

One of the most common ways for the representation of audio signals in deep learning is through spectrograms (Bi, Yu, Jin, & Xu, 2024). A spectrogram is an image that indicates the frequency content of an audio source over time. In creating a spectrogram, STFT is utilized, which means the Short-Time Fourier Transform (Song et al., 2024). It does this by breaking down the audio signal into very short, overlapping segments and applying the Fourier Transform to each segment. The output picture will then be two-dimensional, representing time and frequency.

Figure 6. Spectrogram for audio signals.

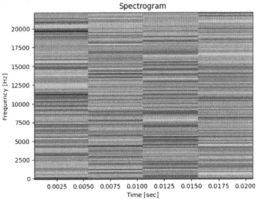

(Chun, Park, & Seo, 2024)

An audio signal's spectrogram is shown in figure 6, which shows how the signal's frequency content changes over time. Time is represented by the x-axis, frequency is represented by the y-axis, and the magnitude of the frequency components is indicated by the color intensity (Chun, Park, & Seo, 2024).

5.2 Mel-Frequency Cepstral Coefficients (MFCCs)

MFCCs are used very often in speech recognition and other speech-related applications (He et al., 2024). First, the log power spectrum of an audio signal is taken, and then a Mel-scale filter bank is applied to get these short-term power spectrums (Hirono et al., 2024). Regarding this, MFCCs are good in capturing relevant perceptual features since the coefficients generated will be very similar in nature to the response of a human ear.

Figure 7. MFCC of an audio signal.

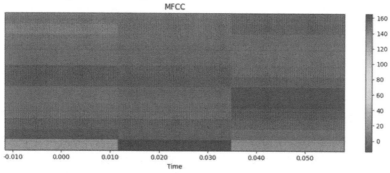

(Singh & Rashmi, 2024)

Figure 7: This figure shows the MFCC representation of an audio stream. The x-axis represents time and the y-axis, the coefficients of the MFCC. The magnitude of the coefficient is represented by different color intensities (Singh & Rashmi, 2024).

5.3 Log-Mel Spectrograms

A log-Mel spectrogram can combine the benefits of spectrograms and MFCCs. It applies first a log transformation and then the Mel-scale filter bank on the spectrogram (Pakkala, Akhila Thejaswi, Rai, & Nagesh, 2024). This forms a better resolution in the low frequency band, which is very critical to most audio classification applications, and captures some perceptually meaningful features (Chen, Meng, Li, & Fang, 2024).

Figure 8. Log-mel spectrogram of an audio signal.

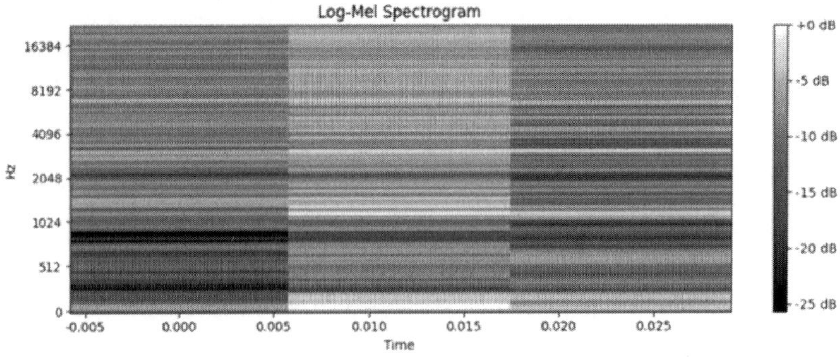

(Damiano, Cramer, Guntoro, & van Waterschoot, 2024)

A log-Mel spectrogram of an audio sample is shown in figure 8. Time is shown by the x-axis, the Mel frequency bands are represented by the y-axis, and the logarithmic magnitude of the frequency components is represented by the colour intensity (Damiano, Cramer, Guntoro, & van Waterschoot, 2024).

6. TRAINING STRATEGIES

6.1 Data Augmentation

These data augmentation techniques artificially inflate a training dataset by applying several modifications to the audio signals. Such methods increase model robustness and generalization since the model is trained using more variations of variables (Kim, Wu, Bondi, & Liu, 2024). Some common examples for the augmentation strategy include time stretching, pitch shifting, random cropping, and addition of background noise.

Figure 9. Data augmentation technique.

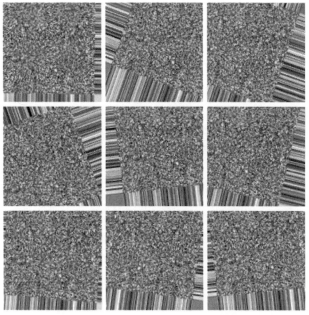

Figure 9 illustrates the various data augmentation methods applied during training, including rotation, width shift, height shift, and horizontal flip. These adjustments further make the model more robust to changes in the input (Diep, Phan, & Truong, 2024).

6.2 Transfer Learning

Transfer learning initializes the model's weights using pre-trained models on large datasets in this case, ImageNet (Sen et al., 2024). This approach will prove useful in scenarios where there is a scarcity of training data, as it allows the model to benefit from experience acquired in the completion of previous tasks (Stankov, 2024). It means that improving the performance of the pre-trained model on the target audio dataset can give very good improvements.

Figure 10. Transfer Learning model for audio classification.

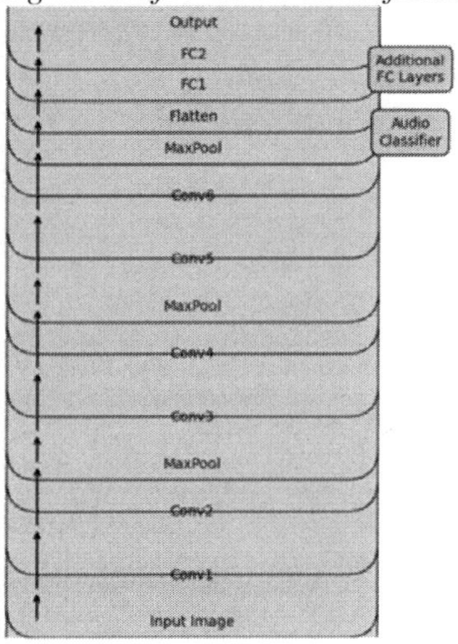

(Zohaib, Asim, & ELAffendi, 2024)

The design of a transfer learning model that makes use of the VGG16 network that has already been trained on ImageNet is shown in figure 10. For audio categorisation, additional fully linked layers are added, and the pre-trained layers are used for feature extraction (Zohaib, Asim, & ELAffendi, 2024).

6.3 Regularization Technique

Regularization techniques like weight decay and dropout constrain model parameters to ensure that the noise is fitted (Rothstein, 2024). Dropout randomly sets a portion of input units while training to zero; thus, the model will not rely too much on any single feature. Weight decay penalizes large weights and helps force the model to pick up more broadly based features (Silva, Whitehead, Lengerich, & Leather, 2024).

Figure 11. Regularization technique for audio classification.

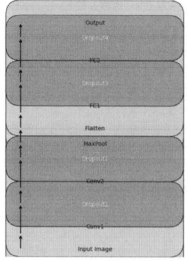

(Tran, Vu, Nguyen, & Nguyen, 2024)

Figure 11 shows how dropout regularisation is used in a deep learning model. To avoid overfitting, dropout layers are included after convolutional and fully linked layers (Tran, Vu, Nguyen, & Nguyen, 2024).

7. APPLICATIONS

Deep learning has influential applications connected with audio classification and affects a variety of industries (Sun, Li, Wang, Xv, & Liu, 2024). This section considers some of the important applications and attempts to highlight different developments taking place in these fields.

7.1 Speech Recognition

Deep learning techniques have been applied with very impressive improvements to ASR. Using deep learning models, the spoken language could be precisely transcript from raw audio data during the extracting process of complicated acoustic properties (Hajihashemi, Alavigharahbagh, Machado, & Tavares, 2024). CNN and RNN especially LSTMs and Transformers are usually used and applied in ASR systems to model the time and space dependencies that are inherently some features of speech signals.

Figure 12. Speech recognition workflow.

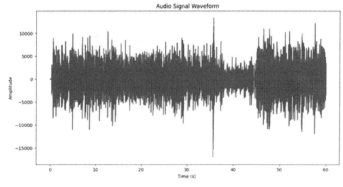

(Belloch et al., 2024)

Figure 12 A deep learning-based system process for speech recognition. The process uses an audio file, understands the spoken words by means of a deep learning model, and transcribes the same (Belloch et al., 2024).

7.2 Music Classification

Music categorization means the sorting of tracks into genres, moods, or instrument categories. Particularly, CNNs demonstrate good results in emulating the harmonic and spectral features of music (Abbas et al., 2024). Temporal dependencies, which exist in musical compositions, can be modeled using RNNs.

Figure 13. Music classification model architecture.

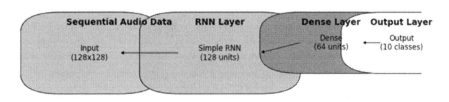

(Jiang, Mutahira, Park, & Muhammad, 2024)

The architecture of a simple Recurrent Neural Network (RNN) for music classification is depicted in figure 12, which also shows how the model processes sequential audio data to classify music into several categories (Jiang, Mutahira, Park, & Muhammad, 2024).

7.3 Environmental Sound Classification

Environmental sound classification is defined as the classification of sounds from all varieties of contexts, which include highway noise, animal sounds, and emergency sirens (Gourisaria, Agrawal, Sahni, & Singh, 2024). Specifically, CNNs have been found to classify ambient noises quite accurately with spectrogram representations for learning discriminative features (Liao et al., 2024).

Figure 14. Environmental sound classification model architecture.

(Mou & Milanova, 2024)

The architecture of a convolutional neural network (CNN) intended for environmental sound classification is depicted in figure 13. Convolutional layers are used for feature extraction, pooling layers are used to reduce dimensionality, and fully linked layers are used for classification (Mou & Milanova, 2024).

8. EVALUATION METRICS

These metrics are designed to judge the performance of audio categorization algorithms. Some commonly used measures as features include the F1-score, accuracy, precision, recall, and area under the receiver operating characteristic curve (Paranayapa, Ranasinghe, Ranmal, Meedeniya, & Perera, 2024). These metrics do convey very important information regarding how algorithms can be used to categorize audio signals (Castro-Ospina, Solarte-Sanchez, Vega-Escobar, Isaza, & Martínez-Vargas, 2024). The common evaluation metrics are shown below in table 1.

Table 1. Common evaluation metrics. (Castro-Ospina, Solarte-Sanchez, Vega-Escobar, Isaza, & Martínez-Vargas, 2024)

Metric	Description
Accuracy	Proportion of correctly classified instances
Precision	Proportion of true positives among predicted positives
Recall	Proportion of true positives among actual positives
F1-score	Harmonic mean of precision and recall
AUC-ROC	Area under ROC curve

Figure 15. A figure to demonstrate common evaluation metrics.

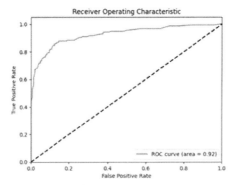

(Ranmal, Ranasinghe, Paranayapa, Meedeniya, & Perera, 2024)

Figure 15: A ROC curve is a graph showing how the true positive rate balances with the false positive rate regarding a model's performance at various levels of classification (Ahammed, Kim, Song, An, & Chen, 2024). These numerical measures, for example, accuracy, precision, recall, F1-score, and AUC-ROC, could all be used to give an overview of the model's performance and the extent to which it can classify instances correctly (Fischer, Orescanin, & Eckstrand, 2024). The greater the discriminations between classes, the higher the AUC-ROC.

9. CONCLUSION AND FUTURE DIRECTION

While deep learning techniques for audio classification have seen impressive progress recently, some obstacles remain. In this respect, we indicate some of the issues and propose possible avenues of investigation.

Large and Diverse Datasets: Deep learning models require, in their training, the presence of large, diverse, and annotated datasets. In the future, research on dataset generation and selection should result in high-quality datasets to span all kinds of audio signals from various surroundings and different conditions.

Model Interpretability: One critical element of developing trust in deep learning models and their decisions is the interpretability of these models. Future research must be focused on visualization techniques of model predictions and their interpretation to get an insight into the characteristics learned and their decision boundaries.

Real-Time Processing: Most audio classification applications, including voice assistants and security systems, require real-time processing. In the future, research must aim at building more efficient models and algorithms that can classify audio in real time at low computing costs and latencies.

Generalization and Robustness: It is a very important task to ensure deep learning models are robust to the variability in audio signals and generalize well to unseen data. Future research should consider investigating techniques that improve model generalization, such as data augmentation, transfer learning, and domain adaptation.

Deep learning for audio classification has greatly improved by neural network topologies, data pretreatment methods, and training strategies. In this paper, there has been an in-depth review of the main concepts, techniques, and applications of the field, highlighting remarkable achievements and remaining challenges. Precisely, this study is going to provide cutting-edge comprehensive overview synthesis of results from many studies, pointing out more reliable datasets, model interpretability, and real-time processing capability. These issues must be brought to the center of future research in audio categorization if it is to open a myriad of new, innovative applications in audio analysis. For those scholars and professionals trying to guide their way through this fast-moving area of deep learning-based audio classification, this review proves to be invaluable.

REFERENCES

Abbas, S., Ojo, S., Al Hejaili, A., Sampedro, G. A., Almadhor, A., Zaidi, M. M., & Kryvinska, N. (2024). Artificial intelligence framework for heart disease classification from audio signals. *Scientific Reports*, *14*(1), 3123. DOI: 10.1038/s41598-024-53778-7 PMID: 38326488

Ahammed, I., Kim, B., Song, S., An, J., & Chen, Z. (2024). Acoustic-based Multitask Construction Equipment and Activity Recognition Using Customized ResNet-18.

Belloch, J. A., Coronado, R., Valls, O., del Amor, R., Leon, G., Naranjo, V., Dolz, M. F., Amor-Martin, A., & Piñero, G. (2024). Urban sound classification using neural networks on embedded FPGAs. *The Journal of Supercomputing*, *80*(9), 1–11. DOI: 10.1007/s11227-024-05947-8

Bi, M., Yu, X., Jin, Z., & Xu, J. (2024). IG-Based Method for Voiceprint Universal Adversarial Perturbation Generation. *Applied Sciences (Basel, Switzerland)*, *14*(3), 1322. DOI: 10.3390/app14031322

Bukka, S. R., Lalam, N., Bhatta, H., & Wright, R. (2024, June). Lab scale demonstration of pipeline third-party damage classification using convolutional neural networks. In *Signal Processing, Sensor/Information Fusion, and Target Recognition XXXIII* (Vol. 13057, pp. 323–330). SPIE. DOI: 10.1117/12.3014005

Cappellazzo, U., Falavigna, D., & Brutti, A. (2024). Efficient Fine-tuning of Audio Spectrogram Transformers via Soft Mixture of Adapters. *arXiv preprint arXiv:2402.00828*. DOI: 10.21437/Interspeech.2024-38

Castro-Ospina, A. E., Solarte-Sanchez, M. A., Vega-Escobar, L. S., Isaza, C., & Martínez-Vargas, J. D. (2024). Graph-Based Audio Classification Using Pre-Trained Models and Graph Neural Networks. *Sensors (Basel)*, *24*(7), 2106. DOI: 10.3390/s24072106 PMID: 38610318

Chen, J., Jun, S. W., Hong, S., He, W., & Moon, J. (2024). Eciton: Very low-power recurrent neural network accelerator for real-time inference at the edge. *ACM Transactions on Reconfigurable Technology and Systems*, *17*(1), 1–25. DOI: 10.1145/3629979

Chen, L., Zhou, X., & Chen, H. (2024, March). Audio Scanning Network: Bridging Time and Frequency Domains for Audio Classification. *Proceedings of the AAAI Conference on Artificial Intelligence*, *38*(10), 11355–11363. DOI: 10.1609/aaai.v38i10.29015

Chen, S., Meng, N., Li, H., & Fang, W.CHEN. (2024). Efficient Deep Neural Network Compression for Environmental Sound Classification on Microcontroller Units. *Turkish Journal of Electrical Engineering and Computer Sciences*, *32*(4), 501–515. DOI: 10.55730/1300-0632.4084

Chun, C., Park, H. J., & Seo, M. B. (2024). Static Sound Event Localization and Detection Using Bipartite Matching Loss for Emergency Monitoring. *Applied Sciences (Basel, Switzerland)*, *14*(4), 1539. DOI: 10.3390/app14041539

Damiano, S., Cramer, B., Guntoro, A., & van Waterschoot, T. (2024). Synthetic data generation techniques for training deep acoustic siren identification networks. *Frontiers in Signal Processing*, *4*, 1358532. DOI: 10.3389/frsip.2024.1358532

Diep, Q. B., Phan, H. Y., & Truong, T. C. (2024). Crossmixed convolutional neural network for digital speech recognition. *PLoS One*, *19*(4), e0302394. DOI: 10.1371/journal.pone.0302394 PMID: 38669233

Dinkel, H., Yan, Z., Wang, Y., Zhang, J., Wang, Y., & Wang, B. (2024). Scaling up masked audio encoder learning for general audio classification. *arXiv preprint arXiv:2406.06992*. DOI: 10.21437/Interspeech.2024-246

Duan, L., Yang, L., & Guo, Y. (2024). SIAlex: Species identification and monitoring based on bird sound features. *Ecological Informatics*, *81*, 102637. DOI: 10.1016/j.ecoinf.2024.102637

Esposito, M., Valente, G., Calaña, Y. P., Dumontier, M., Giordano, B. L., & Formisano, E. (2024). Bridging Auditory Perception and Natural Language Processing with Semantically informed Deep Neural Networks. bioRxiv, 2024-04.

Fischer, J., Orescanin, M., & Eckstrand, E. (2024). VI-PANN: Harnessing Transfer Learning and Uncertainty-Aware Variational Inference for Improved Generalization in Audio Pattern Recognition. *IEEE Access*.

Gocher, H., Taterh, S., & Dadheech, P. (2023). Impact Analysis to Detect and Mitigate Distributed Denial of Service Attacks with Ryu-SDN Controller: A Comparative Analysis of Four Different Machine Learning Classification Algorithms. *SN Computer Science*, *4*(5), 456. DOI: 10.1007/s42979-023-01842-w

Gocher, H., Taterh, S., & Dadheech, P. (2023). *Reinforcing Network Resilience From DDOS: A Review Of Advanced Distributed Denial Of Service (Ddos)*. Attacks And Its Mitigation Techniques.

Gourisaria, M. K., Agrawal, R., Sahni, M., & Singh, P. K. (2024). Comparative analysis of audio classification with MFCC and STFT features using machine learning techniques. *Discover Internet of Things*, *4*(1), 1. DOI: 10.1007/s43926-023-00049-y

Gupta, R., & Kumar, V. Sound Classification in Indian Cities Using Multi-Label Data and Transfer Learning. In *The Second Tiny Papers Track at ICLR2024*.

Hajihashemi, V., Alavigharahbagh, A., Machado, J. J. M., & Tavares, J. M. R. (2024). Novel sound event and sound activity detection framework based on intrinsic mode functions and deep learning. *Multimedia Tools and Applications*, ●●●, 1–29. DOI: 10.1007/s11042-024-19557-2

He, H., Chen, J., Chen, H., Zeng, B., Huang, Y., Zhaopeng, Y., & Chen, X. (2024). Enhancing Insect Sound Classification Using Dual-Tower Network: A Fusion of Temporal and Spectral Feature Perception. *Applied Sciences (Basel, Switzerland)*, *14*(7), 3116. DOI: 10.3390/app14073116

Hirono, Y., Sato, I., Kai, C., Yoshida, A., Kodama, N., Uchida, F., & Kasai, S. (2024). The Approach to Sensing the True Fetal Heart Rate for CTG Monitoring: An Evaluation of Effectiveness of Deep Learning with Doppler Ultrasound Signals. *Bioengineering (Basel, Switzerland)*, *11*(7), 658. DOI: 10.3390/bioengineering11070658 PMID: 39061740

Ji, Q., Wang, Y., & Sun, L. (2024). Mixer is more than just a model. *arXiv preprint arXiv:2402.18007*.

Ji, Q., Zhang, J., & Wang, Y. (2024). ASM: Audio Spectrogram Mixer. *arXiv preprint arXiv:2401.11102*.

Jiang, H., Mutahira, H., Park, U., & Muhammad, M. S. (2024). Scanning dial: The instantaneous audio classification transformer. *Discover Applied Sciences*, *6*(3), 96. DOI: 10.1007/s42452-024-05731-6

Kim, G., Wu, H. H., Bondi, L., & Liu, B. (2024, April). Multi-Modal Continual Pre-Training For Audio Encoders. In *ICASSP 2024-2024 IEEE International Conference on Acoustics, Speech and Signal Processing (ICASSP)* (pp. 691-695). IEEE. DOI: 10.1109/ICASSP48485.2024.10446424

Lau, K. W., Rehman, Y. A. U., & Po, L. M. (2024). AudioRepInceptionNeXt: A lightweight single-stream architecture for efficient audio recognition. *Neurocomputing*, *578*, 127432. DOI: 10.1016/j.neucom.2024.127432

Lebourdais, M., Mariotte, T., Almudévar, A., Tahon, M., & Ortega, A. (2024). Explainable by-design Audio Segmentation through Non-Negative Matrix Factorization and Probing. *arXiv preprint arXiv:2406.13385*. DOI: 10.21437/Interspeech.2024-791

Liao, J., Yi, L., Shi, W., Yang, W., Fang, Y., & Yang, X. (2024). Imperceptible backdoor watermarks for speech recognition model copyright protection. *Visual Intelligence*, *2*(1), 1–10. DOI: 10.1007/s44267-024-00055-w

Mei, X., Meng, C., Liu, H., Kong, Q., Ko, T., Zhao, C., Plumbley, M. D., Zou, Y., & Wang, W. (2024). Wavcaps: A chatgpt-assisted weakly-labelled audio captioning dataset for audio-language multimodal research. *IEEE/ACM Transactions on Audio, Speech, and Language Processing*, *32*, 3339–3354. DOI: 10.1109/TASLP.2024.3419446

Midavaine, N., Go, G. H. T., Canez, D., Simion, I., & Chatterji, S. [Re] On the Reproducibility of Post-Hoc Concept Bottleneck Models. *Transactions on Machine Learning Research*.

Mohaimenuzzaman, M., Bergmeir, C., & Meyer, B. (2024). Deep Active Audio Feature Learning in Resource-Constrained Environments. *IEEE/ACM Transactions on Audio, Speech, and Language Processing*, *32*, 3224–3237. DOI: 10.1109/TASLP.2024.3416697

Mou, A., & Milanova, M. (2024). Performance Analysis of Deep Learning Model-Compression Techniques for Audio Classification on Edge Devices. *Sci*, *6*(2), 21. DOI: 10.3390/sci6020021

Paissan, F., Della Libera, L., Ravanelli, M., & Subakan, C. (2024). Listenable Maps for Zero-Shot Audio Classifiers. *arXiv preprint arXiv:2405.17615*.

Pakkala, P. G. R., Akhila Thejaswi, R., Rai, B. S., & Nagesh, H. R. (2024). Road safety analysis framework based on vehicle vibrations and sounds using deep learning techniques. *International Journal of System Assurance Engineering and Management*, *15*(3), 1086–1097. DOI: 10.1007/s13198-023-02191-w

Paranayapa, T., Ranasinghe, P., Ranmal, D., Meedeniya, D., & Perera, C. (2024). A comparative study of preprocessing and model compression techniques in deep learning for forest sound classification. *Sensors (Basel)*, *24*(4), 1149. DOI: 10.3390/s24041149 PMID: 38400306

Ranmal, D., Ranasinghe, P., Paranayapa, T., Meedeniya, D., & Perera, C. (2024). ESC-NAS: Environment Sound Classification Using Hardware-Aware Neural Architecture Search for the Edge. *Sensors (Basel)*, *24*(12), 3749. DOI: 10.3390/s24123749 PMID: 38931532

Ronchini, F., Comanducci, L., & Antonacci, F. (2024). Synthesizing Soundscapes: Leveraging Text-to-Audio Models for Environmental Sound Classification. *arXiv preprint arXiv:2403.17864*.

Rothstein, M. M. *Data Driven Mel Filter Bank Design for Environmental Sound Analysis* (Doctoral dissertation, Worcester Polytechnic Institute).

Sabha, A., & Selwal, A. (2024). A novel Approach for Audio-based Video Analysis via MFCC Features. *Procedia Computer Science, 235*, 1512–1521. DOI: 10.1016/j.procs.2024.04.142

Sen, A., Rajakumaran, G., Mahdal, M., Usharani, S., Rajasekharan, V., Vincent, R., & Sugavanan, K. (2024). Live event detection for people's safety using NLP and deep learning. *IEEE Access : Practical Innovations, Open Solutions, 12*, 6455–6472. DOI: 10.1109/ACCESS.2023.3349097

Shishkin, S., Hollosi, D., Goetze, S., & Doclo, S. (2024, April). Active Learning for Sound Event Classification Using Bayesian Neural Networks with Gaussian Variational Posterior. In *ICASSP 2024-2024 IEEE International Conference on Acoustics, Speech and Signal Processing (ICASSP)* (pp. 896-900). IEEE. DOI: 10.1109/ICASSP48485.2024.10446970

Silva, D. A., Whitehead, S., Lengerich, C., & Leather, H. (2024). CoLLAT: On adding fine-grained audio understanding to language models using token-level locked-language tuning. *Advances in Neural Information Processing Systems, •••*, 36.

Singh, M. P., & Rashmi, P. (2024). Convolution Neural Networks of Dynamically Sized Filters with Modified Stochastic Gradient Descent Optimizer for Sound Classification.

Sinnott, R. (2024). Predicting and Avoiding Dog Barking Behaviour through Deep Learning. In *Proceedings of the 2024 Australasian Computer Science Week* (pp. 26-35). DOI: 10.1145/3641142.3641176

Song, X., Xiong, J., Wang, M., Mei, Q., & Lin, X. (2024). Combined Data Augmentation on EANN to Identify Indoor Anomalous Sound Event. *Applied Sciences (Basel, Switzerland), 14*(4), 1327. DOI: 10.3390/app14041327

Stankov, I. (2024). *Natural Audio Data Augmentation Techniques* (Master's thesis, Humboldt-Universität zu Berlin).

Sun, Y., Li, J., Wang, L., Xv, J., & Liu, Y. (2024). Deep Learning-based drone acoustic event detection system for microphone arrays. *Multimedia Tools and Applications, 83*(16), 47865–47887. DOI: 10.1007/s11042-023-17477-1

Tarımer, İ., & Karadağ, B. C. Genres Classification of Popular Songs Listening by Using Keras. *Gazi University Journal of Science Part A: Engineering and Innovation, 11*(1), 123-136.

Tran, K. T., Vu, X. S., Nguyen, K., & Nguyen, H. D. (2024). NeuProNet: Neural profiling networks for sound classification. *Neural Computing & Applications, 36*(11), 5873–5887. DOI: 10.1007/s00521-023-09361-8

Vosoughi, A., Bondi, L., Wu, H. H., & Xu, C. (2024, April). Learning Audio Concepts from Counterfactual Natural Language. In *ICASSP 2024-2024 IEEE International Conference on Acoustics, Speech and Signal Processing (ICASSP)* (pp. 366-370). IEEE. DOI: 10.1109/ICASSP48485.2024.10446736

Wimalasena, R. A. L. B., & Ranasinghe, D. D. M. (2024). Audio-Based Vehicle Detection System: Enhancing Safety for Bicycle Riders. *ENGINEER, 57*(01), 85–94. DOI: 10.4038/engineer.v57i1.7613

Ye, Z., Ciccarelli, G., & Kulis, B. (2024, April). Maximum-Entropy Adversarial Audio Augmentation for Keyword Spotting. In *ICASSP 2024-2024 IEEE International Conference on Acoustics, Speech and Signal Processing (ICASSP)* (pp. 10826-10830). IEEE. DOI: 10.1109/ICASSP48485.2024.10446557

Yu, R., Yu, W., & Wang, X. (2024). Kan or mlp: A fairer comparison. *arXiv preprint arXiv:2407.16674.*

Yuan, Y., Chen, Z., Liu, X., Liu, H., Xu, X., Jia, D., . . . Wang, W. (2024). T-CLAP: Temporal-Enhanced Contrastive Language-Audio Pretraining. *arXiv preprint arXiv:2404.17806.* DOI: 10.1109/MLSP58920.2024.10734763

Zhang, D., Chen, J., Bai, J., & Wang, M. (2024). Sound event localization and classification using WASN in Outdoor Environment. *arXiv preprint arXiv:2403.20130.*

Zhao, W., Wang, H., Chen, Y., Pan, X., Zhang, K., & Bai, Z. An Environmental Sound Classification Algorithm Based on Multiscale Channel Feature Fusion.

Zohaib, M., Asim, M., & ELAffendi, M. (2024). Enhancing Emergency Vehicle Detection: A Deep Learning Approach with Multimodal Fusion. *Mathematics, 12*(10), 1514. DOI: 10.3390/math12101514

Chapter 8
The Neuroethical Nexus of Brain Organoids and AI:
Towards the Integration of Organoid Intelligence and Artificial Intelligence

Suyesha Singh
 https://orcid.org/0000-0003-2180-5708
Manipal University Jaipur, India

Paridhi Jain
 https://orcid.org/0009-0002-8233-951X
Manipal University Jaipur, India

ABSTRACT

The advancement of brain organoids has revolutionized the fields of biotechnology and neuroscience. Brain organoids are the 3-D structures of the human brain developed from the human's pluripotent cells, which mimic the functionality of the human brain. Organoid intelligence cumulates brain organoids and artificial intelligence, making it possible to develop a biohybrid system with heightened cognitive capacities. This paper aims to critically analyze concerns surrounding brain organoids and organoid intelligence and highlight the scientific progress and ethical considerations, ensuring that new technologies contribute to the community while maintaining fundamental moral principles. The findings of the current work will be useful for decision makers, neuroscientists, ethicists, and the public to promote environmentally conscious innovation and practice organoid intelligence in an

DOI: 10.4018/979-8-3693-8009-3.ch008

ethical way. This work serves as a call to action for addressing ongoing challenges, advancing the field, and utilizing it effectively in the current context.

INTRODUCTION

The human brain is the most mysterious and complex structure of mankind. Even after rapid advancement in neurological and biological research, various dimensions of the origin and development of the brain and associated illnesses remain unknown. To understand this intricacy of the human brain, scientists developed brain organoids. Brain organoids are an innovative technology that helps understand the human brain's complexity in a laboratory setting. Human pluripotent stem cells can develop the tri-dimensional (3D) structure of the human brain. Brain organoids can mimic the functionality of the human brain. People also refer to them as cerebral organoids or mini brains. The development of brain organoids created a revolution in neuroscience research, and it also provides a newer approach to examining initial neuro-developmental processes, recapitulating neurological diseases, testing the drugs, and disease modeling (Adlakha, 2023). Brain organoids are very helpful for studying important stages in brain development, like neurogenesis and the start of functional neuronal networks with some electrical activity. This is because they can automatically recognize patterns and change them into structures that look like the cortex and other parts of the brain. The major future target of brain organoids research is to regenerate injured brain regions in individuals suffering from any brain injury, strokes, or other neurological disorders by transplanting sections of their brain tissues.

Thus, brain organoids are biological organisms created in-vitro from stem cells, capable of mimicking the normal design and physiology of a human organ. These organisms are prepared in a specialized environment using Matrigel, a protein mixture derived from mouse sarcomas, and an appropriate structure composed of 3D matrix data, as mentioned in Table 1. In some reactors and broths, organoids gather and go through "guided" differentiation and early proliferation from totipotent stem cells to nerve cells (Lavazza & Pizzetti, 2020). Based on the specific culture of brain organoids, their further growth is diversely independent. Despite possessing several anatomical and functional characteristics of a human body organ, they are smaller, simpler counterparts, which is why they are referred to as 'Organoids'. Presently accessible organoids can mimic several physiological functions, such as the retina, gut, kidney, pancreas, thyroid, inner ear, liver, and more (Little, 2017). Several sources, including embryonic cells and induced pluripotent stem cells converted from mature cells, can provide the stem cells or pluripotent cells needed to create brain organoids (Romito & Cobellis, 2015). These pluripotent stem cells can transform into any cells, such as glial cells or neurons. The genetic factors of the individual and the environmental factors interact intricately to direct the process of transformation of pluripotent cells into different cells. Furthermore, researchers employ a distinct transcription element to either activate or deactivate specific genes, or expose cells to multiple messenger molecules to mimic the conditions observed in a specific tissue (Andrews & Nowakowski, 2019). These cells can be meticulously manipulated to develop into a vast variety of cells, ranging from neurons to heart cells. As mentioned earlier, brain organoids can function just like the human brain. Prior to this development, it was impossible to study human neurodevelopment using conventional animal models such as mice, which have very distinct brain architectures and processes. Unlike other animal models, brain organoids can replicate key aspects of the human brain, including cell movement, differentiation, and interaction. Thus, they are an essential resource for researching crucial illnesses and disabilities such as schizophrenia and autism spectrum disorders (Andrews & Nowakowski, 2019).

Figure 1 presents the major features and descriptions of brain organoids. However, human brain organoids appear to raise more pertinent ethical questions.

Figure 1. Major features of brain organoids

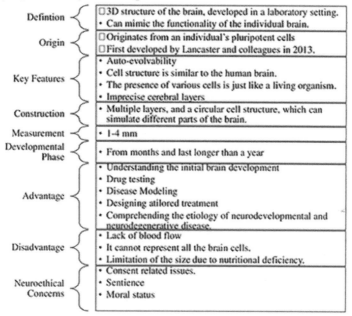

FORMATION OF BRAIN ORGANOIDS

The process of developing brain organoids begins with the process known as cerebral organoid differentiation, which imitates the physiological process that happens during the early brain development in the embryo of a person (as mentioned in Fig. 1). Initially, pluripotent stem cells are obtained from initial stage embryos or induced pluripotent stem cells (Lancaster & Knoblich, 2014). For the optimal growth of these cells, they are cultivated in an ideal and safe environment. Whenever neural-induced stimuli stimulate these cells, they unite to form embryoid bodies that further initiate neural progenitor cells (Xue et al., 2018). Afterward, these progenitors grow and start taking the shape of organized layers, which eventually leads to the development of brain organoids. After their development, these organoids take a few weeks to months to shape into the human brain and its various cells. Before being applied, these brain organoids are first reviewed to make sure that they reflect important arenas of the human brain.

Figure 2. depicts the formation process of brain organoids.

ORGANOID INTELLIGENCE

Organoid Intelligence (OI) has proposed in recent years that brain organoids are evolving, potentially serving as a learning mechanism rather than a static model. These studies have been conducted to increase the computing capacity of brain-like structures. It describes the combination of artificial intelligence (AI) and machine learning (ML) along with brain organoids that can produce biohybrid systems that will have the capacity to carry out complex cognitive functions. These hybrid technologies can revolutionize multiple domains, including neuroscience, neurocomputing, and machine learning (As mentioned in Table 2). Additionally, they can potentially contribute to the development of unique AI-based architectures influenced by the information-processing system of the human brain along with complex neural interfaces and brain-machine interfaces. Organoid intelligence is the biological-electronic computer technology, and it has numerous potential applications. Therefore, this technology can contribute to our understanding of the pathological and physiological outcomes of infectious diseases, intoxication states, degenerative and developmental disabilities, as well as the bodily processes of cognition, memory, and learning. Organoid intelligence could also expand the use of a new "neuro-mimetic AI algorithm" and help the development of new brain-computer interfaces (Ballav et al., 2024).

It is crucial to address the significant ethical concerns associated with the development of organoid intelligence in order to foster environmentally conscious innovation. The present work envisions the creation of biohybrid systems, which fuse brain organoids with electrical equipment, to build new computer frameworks or learn more about cognitive disease and its functionality. These kinds of developments are noteworthy from a scientific standpoint, but they also raise important neurological issues.

Figure 3. Major characteristics of organoid intelligence

Table 2. Characteristics of organoid intelligence (OI)

The presented work aims to explore the neuroethical issues of rapidly changing subjects as scientists keep expanding the limits of the possibilities achievable through brain organoids and organoid intelligence. To successfully navigate the upcoming field of neuroscience and biohybrid technology, this paper emphasizes the need for thorough neuroethical regulation and discourse. This work also seeks to assess the challenging equilibrium between the quest for knowledge, innovation, and the moral obligation that precedes such breakthroughs.

DEVELOPMENT & CURRENT STATUS OF BRAIN ORGANOIDS

Only about a decade ago did the term 'organoids' emerge in scientific research, despite researchers' long-standing efforts to transform 2D structures into 3D cellular formations (Sato et al., 2009). Following the initial instances of partially or substantially developing human organs, researchers successfully cultivated a collection of human nerve cells in 3D in 2013. The development occurred by the same biochemical mechanism that forms the brain during pregnancy (Lancaster et al., 2013). Lancaster & his colleagues conducted ground-breaking research in the field, starting with the artificial pluripotent stem cell conversion of human skin cells to create an organoid brain as an experimental model for microcephaly. Researchers used a patient's microcephaly cells to create brain organoids that closely resembled the patient's brain. Subsequently, the identification of the protein thought to be faulty in microcephaly allowed the researchers to develop organoids that appeared to be slightly immune to this condition. Lancaster's research findings indicate the existence of distinct but interdependent brain arenas, each with internally connected neural connections and a greater degree of cellular similarity (Paşca et al., 2015). The study used organoids of approximately 4 mm in size, which varied in gene expression, to stimulate in-vitro development in vivo, at least until the late mid-fetal stage of gestation.

The research conducted by Qian et al. (2019) produced promising results in the study of organoids, revealing neurons in all six layers of the cerebral cortex, despite the lack of fully established and mature circuits. Not only do organoids face developmental issues, but they also lack blood vessels, which presents another challenge. Lancaster et al. encased each organoid inside a layer of nutritionally enriched material and a nutrient bath, whereby the organoids spun concurrently to absorb maximum amounts of food to feed the interior cells. However, due to a lack of vascularity, the nucleus of cells in the majority of organoids decays quickly. They also lack meninges, immunological cells, glial cells, and surrounding embryonic structures. The high degree of heterogeneity among organoids poses a challenge to organoid models.

The adult mouse brain has been first attempted to implant brain organoids. Of their growth as a method for achieving the vascularization of these kinds of structures. The development of the brain and blood vessel network in human brain tissue was achieved through the fusion of the human organoid and the mouse host tissue within the prostheses (Mansour et al., 2018).

Notwithstanding these challenges, Lancaster & Knoblich have reported a procedure for producing three-dimensional brain tissue that closely resembles the natural developmental program. "In a month or two, using this methodology in a typical tissue culture space can result in the development of the cerebral cortex, choroid plexus, ventral telencephalon, and retinal identities, among other structures" (Lancaster & Knoblich, 2014). However, the hippocampus and spinal cord are also there, along with layers of cortex. Furthermore, scientists have demonstrated the ability to maintain organoids alive for over a year. According to Kelava & Lancaster, scientists can use human pluripotent stem cells to generate organoids capable of replicating the initial stages of human embryonic and foetal brain development, focusing on genetic production and cellular biology.

Despite efforts to address plumbing and scaffolding issues, which provide oxygen and nutrients and enable the development of brain organoids beyond their millimetre size, these organoids still face certain limitations in their ability to replicate an in-vivo brain in an in-vitro setting. It is important to consider that in-vivo organs are constantly created into their final configuration through the expansion, restructuring, and differentiation of cell components. Genes regulate these processes, and they also self-regulate epigenetically based on the activation and deactivation of specific genes' biochemical signals from the environment. This means that a mini brain that develops in-vitro, separated from a whole embryo and without interacting with the outside world, may not be able to fully develop as real brains do.

The major goal of developing brain organoids is to replicate the properties of the human brain, which has advanced significantly since the initial research. However, there are still certain obstacles to overcome, such as the lack of blood circulation that prevents layers of brain organoids from nourishing (Pham et al., 2018). However, current research on brain organoids has indicated certain significant characteristics of the nervous system. Brain organoids can connect the spinal cord and transfer the nerve signals, which can cause muscle contraction (Quadrato et al., 2017). It can also represent specialized and independent electrical activity, or neuronal communication, as reported by Giandomenico et al. in 2019.

According to research by Trujillo et al. (2019), brain organoids developed from induced pluripotent cells have the ability to generate their own periodic and regular oscillatory network electrical activity, a pattern that matches the EEG patterns of neonatal babies. This further suggests that 10-month-old brain organoids can grow through specific genetic programs, reflecting brain activity even in the absence of external neuronal inputs. The unique feature of a healthy human brain is the ring rate, which may reach 2-3 beats per second, and the types of waves such as alpha, gamma, and delta. Furthermore, organoids' age can be predicted through a machine learning model that was developed on the EEG features of neonatal babies, using the organoid's electrical signals as a basis.

Compared to the adult human brain, brain organoids are significantly smaller, lack spatial structure, and differ in terms of neuronal quantity, intricacy, maturity, and biological component deficiency, all of which are characteristics of a living creature. Additionally, they are unable to exchange input and output with the outside world. Brain organoids, on the contrary, can respond to sensory stimuli, as mentioned earlier. Nevertheless, brain organoids can only mimic certain areas of the brain, such as the cerebellum or forebrain. The electrical activity in the brain organoids is similar to that in the newborn infant's brain.

Recent studies have demonstrated synced neuronal activity, a hallmark of all major brain processes, particularly memory, in the lab-grown cerebral cortex model (Sakaguchi et al., 2019). Furthermore, the neurons in the brain organoids expand naturally, suggesting that even this kind of nerve cell cultured in a laboratory setting exhibits the same activity as human neurons, permitting the growth and formation of new pathways. Every aspect of this may give rise to ethical issues based on the potential to produce sentient humans, beginning with moral status. However, in addition to these ethical concerns, it's crucial to take into account the application and specific ethical considerations of brain organoids in research.

USES & APPLICATIONS OF BRAIN ORGANOID

Brain organoid and organoid intelligence is an emerging field and has gained neuroscientists' attention in the past few years. The uses and applications of brain organoid is described below.

Understanding the development of the brain: Understanding the development of the brain is crucial because, as previously mentioned, the brain is the most complex structure in the human body, and the mechanisms underlying certain parts of the brain remain unknown. These brain organoids can help understand the intricacy of brain regions, neuronal networks, and cell differentiation (Modefferi et al., 2021).

Modeling the disease: The use of brain organoids can aid in our understanding of complex psychological and neurological illnesses, including schizophrenia, Alzheimer's, dementia, autism spectrum disorder, and microcephaly. Scientists can develop these brain structures in in-vitro settings, which can aid in understanding the cellular and neuronal development of the brain in this disease and in designing effective therapeutic interventions (Lu et al., 2022; Eichmüller & Knoblich, 2022).

Tailored medical treatment: Scientists can use brain organoids developed from an individual stem and the pluripotent cell to understand the effectiveness of treatment. When there is a significant degree of individual heterogeneity, we can design this treatment modality (Plummer et al., 2019).

Assessing the Side Effects of Toxins: Brain organoids can be efficacious in evaluating the neurotoxic effects of several chemicals, contaminants, and environmental poisons (Zhong et al., 2020).

Evaluation of the medicinal safety: Brain organoids can be utilized to evaluate the safety of new drugs before clinical trials; they are effective in lowering the side effects on human beings.

Machine learning and artificial intelligence: The major aim of organoid intelligence is to develop an artificial intelligence-based system using brain organoids as physiological substrates. The goal is to develop a bio-hybrid system that possesses the cognitive ability just like human beings, which is impossible in traditional silicon-based artificial intelligence (Smirnova et al., 2023).

Promoting ethical use of artificial intelligence: By investigating the limitations of artificial intelligence, brain organoids can guide the development of ethical guidelines for artificial intelligence that emulates or interacts with human cognitive processes.

Usefulness in genomic investigations: Brain organoids are an effective tool for studying genomic and epigenetic investigation because they enable the modification of certain genes to explore their involvement in the development of the brain and its functions. These usages reflect that brain organoids and organoid intelligence research can change the discovery of medicine, advance the boundaries of artificial intelligence, and completely change the knowledge of the brain.

NEUROETHICAL ISSUES RELATED TO BRAIN ORGANOIDS

The use of brain organoids has been on the rise in recent years, but it also presents several ethical issues (Hartung et al., 2024). Some core issues are:

Consciousness/Sentience: One of the major neuroethical issues of brain organoid research is that it raises questions regarding the moral standing of these things. Establishing even a minimal level of consciousness in brain organoids could challenge established ethical frameworks, given that in-vitro models typically lack subjective experience. In order to prevent the illicit production of pain-sensing brain organoids, it is crucial to carefully consider the current legal and ethical implications, informed consent, and research limitations. As this technology advances, society must engage in extensive discourse to effectively address the challenges it presents.

Greely (2021) expressed concerns regarding the ethical and legal status of human brain organoids, highlighting the potential blurring of the line between a real human brain and a human brain organoid. Contradictorily, Hoppe et al. (2023) disagree with Greely's perspective and state that, despite years of consciousness research conducted on adult human brains, the concept remains unclear. Therefore, it is unlikely human brain organoids can attain consciousness without interaction with the external environment, and furthermore, it will be difficult for researchers to identify any subjective sensations they may experience.

Data Privacy: Data privacy is a crucial consideration in brain organoid ethical issues, particularly when these organoids integrate with artificial intelligence and machine learning. Whenever researchers use these organoids with any technology, there is a significant risk of theft or misuse of sensitive material. It is crucial to safeguard the privacy of individuals whose data or stem cells have been used to create these brain organoids, as well as to avoid any future misfortune, abuse, or prejudice. As these technologies are rapidly developing, it is crucial to develop and maintain stricter ethical guidelines so that even the donor has this belief that their data is managed securely, anonymised, and will be used only with consent.

Hartung et al. (2024) highlighted the potential ethical concerns with privacy and issues regarding intellectual rights when working with induced pluripotent stem cells (iPSC). They stated some of the questions raised in this scenario are how much of what is identified when studying the donor's cells, including their potential health issues, is reported back to them and if the donor's rights to the cells extend beyond the donation.

Obtaining informed consent from decisional-competent adults is a relatively straightforward process, requiring them to understand the study's potential risks and benefits. However, appropriate legal guardians must obtain informed consent from young children, adults with cognitive deficits, or adults with brain disorders. The process entails educating the donors about the creation of a genetically matching brain organoid using their donated biomaterials, which could potentially address their health concerns (Hyun et al., 2020).

Animal Research: Even though brain organoids offer a more precise representation of human brain growth and disease compared to traditional animal research, the problem persists because these organoids are unable to accurately replicate the intricate brain activities observed in living animals. Careful consideration is necessary to strike a balance between the advancement of brain organoids research and the ethical treatment of animals, while minimizing the use of animals.

Despite the existence of regulations and the principles of replacement, refinement, and reduction, the question of justifying animal use in research remains relevant, given that animals are sentient beings capable of experiencing pain and discomfort. In their 2018 study, Mansour et al. refer to Gage's work in which 40–50-day-old human brain organoids were grafted into immunodeficient mice and watched for 0.5–8 months to see if any of these missing parts could be established even though the brains did not have the vasculature, microenvironment, or neuronal circuits that exist in vivo. They found neuronal differentiation, maturation, gliogenesis, the integration of microglia, and axon growth in multiple regions of the mouse host brain, indicating the presence of synaptic connections. They assessed the spatial abilities of grafted and ungrafted mice using the Barnes maze, and though they found no significant difference in performance, ungrafted mice performed slightly better. Another important finding was that grafted mice showed neither any observable benefits nor problems. Grafting remains permissible due to these potential benefits, provided that strict guidelines and regulations adhere to them. Furthermore, there is an acknowledgement of risk in the creation of new deficits and defects due to human-to-animal chimerism, with the challenges of the chimerism increasing when more larger and complex brains of rodents are used (Hyun, 2013).

Inequality in Access: The world's distribution of financial assets is uneven. This raises serious questions about the uneven access to technologies related to brain organoids. The development of brain organoids and their sustainability over a limited period requires a significant financial investment. Only well-funded institutions can conduct this type of research. This might lead to unequal access to innovative therapeutics, tailored medicine, and advances in the knowledge of neurological illnesses. To address these disparities, policymakers need to promote inclusive policies and make sure that disadvantaged communities also have similar access to prospective healthcare and technological breakthroughs.

Ravn et al. (2023) indicated that one of the potential concerns of brain organoid research is the widening of existing health inequality. The participants of the study worry the research on organoids will strengthen the hierarchies of illness that influence how much funding it receives for research and treatment resources and hence, with the knowledge, technology, and treatments using brain organoids, be easily accessible to promote health equality.

Commercialization: The commercialization of brain organoid technology raises concerns about the potential exploitation and commoditization of individual biological assets. As these technologies are developing, there is rapid advancement in the testing of drugs, tailored treatments, and other uses, and there is a chance that these commercial interests can take precedence over ethical issues. That can result in issues such as unequal healthcare access, keeping business interests before patient care, and the improper monetization of genetic material or sensitive data. Further, this commercialization could propel technological development and implementation quickly without sufficient regulatory bodies to monitor and punish unethical practices. Stringent laws are required to regulate this commercialization.

As the research on brain organoids progresses, both public and private stakeholders, such as pharmaceutical businesses, are increasingly interested in obtaining, modifying, and producing biomaterial (De Jongh et al., 2022; Hoppe et al., 2023). These concerns about the commercialization of brain organoid research benefits have been associated with unequal access to health care. Participants in the study expressed concern that if the public doesn't have access to the findings of brain organoid studies, it could lead to the commercialization of these findings through overpriced treatments, thereby exacerbating the disparity in health care access.

Lack of Regulatory Bodies: The absence of a specific regulatory body for brain organoid research poses a serious ethical issue. Without having a specific legislative framework, brain organoid technology will poorly address important issues such as informed consent, data privacy, and the susceptibility of these organoids to becoming sentient. Due to the lack of regulation, there is a possibility that unethical practices include unrestricted experimentation, exploitation, and uneven access to this technology. The creation of regulatory authority specific to brain organoids and organoid intelligence can promote ethical practices.

The National Academies of Sciences, Engineering, and Medicine (2021) identified two ethically dubious practices that high-income countries (HIC), with more stringent research regulatory bodies, employ. These two practices include ethical dumping and helicopter research. Ethics dumping is a practice where HIC exports the research to low- to middle-income countries (LMIC) where the ethical standards are less stringent or robust and have difficulty implementing them due to lower resources. On the other hand, helicopter research refers to a practice where HIC conducts research in low- to middle-income countries (LMIC) with minimal or no involvement from local authorities or marginalized communities.

Lee and Sawai (2023) assert that while regulatory bodies exist to prevent ethical and moral transgressions during research, the implementation of these rules faces obstacles due to jurisdictional limitations or private actors issuing them, resulting in no legally binding effects on the parties involved.

Informed consent: Informed consent is a crucial ethical concern in brain organoid research, as it entails getting express authorization from people whose biological components, such as stem cells, are used to make organoids. Donors must be fully aware of the breadth, hazards, and prospective uses of their contribution given the intricacy and potential repercussions of the research, including the possibility of organoids becoming conscious or being employed in commercial applications. This entails transparency regarding the intended use of their cells, any potential commercialization, and any long-term consequences. In the rapidly developing field of brain organoid research, strongly informed consent procedures safeguard donors' rights, promote openness, and preserve ethical principles.

Misuse for any Harmful Purposes: The potential misuse of brain organoids for harmful purposes poses a significant ethical concern, particularly as this technology continues to grow and become more widely accessible. Ideal applications for brain organoids could include enhancing neurotoxicity research for bioweapons or developing cognitive models for hostile AI systems. The multiple uses of brain organoids can lead to misuse, necessitating strict control and regulation. We must develop ethical guidelines, closely monitor research activity, and uphold international agreements to prevent misuse of brain organoid technology and ensure its proper use for beneficial and humanitarian endeavours.

Therefore, the neuroethical issues surrounding brain organoids and organoid intelligence research are complex and multifaceted. A multidisciplinary strategy that respects the developing capabilities of these cutting-edge technologies and ensures the implementation of ethical guidelines and safety measures is necessary to address these issues.

FUTURE IMPLICATIONS

Brain organoids derived from human pluripotent cells offer a useful framework for simulating aspects of both healthy and diseased development. This technology has the potential to change the understanding of brain development and neurological disorders. Previous research focused on using animal models; future research should focus on incorporating newer technology with brain organoids such as "liquid air culture," "flat disc," and "rotating bioreactor" to promote gas and nutrition exchange (Kim & Chang, 2023). Moreover, incorporating various cell types, such as vascular and microglial cells, which are absent in the real world, or assembling multiple brain organoids together could enhance neuronal development and result in a more accurate representation of the human brain. These models can also aid in understanding neurodegenerative illnesses and neurodevelopmental disabilities. In addition to this, future researchers can focus on collaborative research with diverse stakeholders that can provide a new paradigm for brain organoid technology. Encouraging brain organoid researchers and bioethicists to work together at the benchside to jointly identify newly developing ethical challenges in actual time throughout the agreement's lifetime is one example of an integrated strategy. Future improvements in drug testing methods could pave the way for a more ethical approach to drug development. Lastly, it is crucial to develop and implement a clear ethical standard for the manufacture and use of brain organoids (Hyun et al., 2020).

CONCLUSION

The ability to understand the initial development of the human brain using brain organoids is unparalleled, and it has great potential; it holds great promise for comprehending neurological and psychological illnesses and designing effective treatments by offering a platform for in-depth research on the human brain. The integration of artificial intelligence could expedite research by enabling swift analysis of complex organism data, potentially leading to the development of innovative treatment approaches. However, the potential for creating advanced uses of brain organoids requires ethical foresight. We have included the historical and current aspects of brain organoids, their usage, and neuroethical considerations in the present work. It is crucial to establish a sophisticated governance framework through transparent, global discourse that considers all viewpoints.

REFERENCES

Adlakha, Y. K. (2023). Human 3D brain organoids: Steering the demolecularization of brain and neurological diseases. *Cell Death Discovery*, *9*(1), 221. Advance online publication. DOI: 10.1038/s41420-023-01523-w PMID: 37400464

Andrews, M. G., & Nowakowski, T. J. (2019). Human brain development through the lens of cerebral organoid models. *Brain Research*, *1725*, 146470. DOI: 10.1016/j.brainres.2019.146470 PMID: 31542572

Ballav, S., Ranjan, A., Sur, S., & Basu, S. (2024). *Organoid Intelligence: Bridging Artificial Intelligence for Biological Computing and Neurological Insights*. IntechOpen., DOI: 10.5772/intechopen.114304

De Jongh, D., Massey, E. K., Berishvili, E., Fonseca, L. M., Lebreton, F., Bellofatto, K., Bignard, J., Seissler, J., Buerck, L. W., Honarpisheh, M., Zhang, Y., Lei, Y., Pehl, M., Follenzi, A., Olgasi, C., Cucci, A., Borsotti, C., Assanelli, S., & Bunnik, E. M. (2022). Organoids: A systematic review of ethical issues. *Stem Cell Research & Therapy*, *13*(1), 337. Advance online publication. DOI: 10.1186/s13287-022-02950-9 PMID: 35870991

Eichmüller, O. L., & Knoblich, J. A. (2022). Human cerebral organoids — A new tool for clinical neurology research. *Nature Reviews. Neurology*, *18*(11), 661–680. DOI: 10.1038/s41582-022-00723-9 PMID: 36253568

Giandomenico, S. L., Mierau, S. B., Gibbons, G. M., Wenger, L. M., Masullo, L., Sit, T., Sutcliffe, M., Boulanger, J., Tripodi, M., Derivery, E., Paulsen, O., Lakatos, A., & Lancaster, M. A. (2019). Cerebral organoids at the air–liquid interface generate diverse nerve tracts with functional output. *Nature Neuroscience*, *22*(4), 669–679. DOI: 10.1038/s41593-019-0350-2 PMID: 30886407

Greely, H. T. H., & Kreitmair, K. V. (2021). Should Cerebral Organoids be Used for Research if they Have the Capacity for Consciousness?. *Cambridge quarterly of healthcare ethics: CQ: the international journal of healthcare ethics committees*, *30*(4), 575–584. https://doi.org/DOI: 10.1017/S0963180121000050

Hartung, T., Morales Pantoja, I. E., & Smirnova, L. (2024). Brain organoids and organoid intelligence from ethical, legal, and social points of view. *Frontiers in Artificial Intelligence*, *6*, 1307613. Advance online publication. DOI: 10.3389/frai.2023.1307613 PMID: 38249793

Hoppe, M., Habib, A., Desai, R., Edwards, L., Kodavali, C., Sherry Psy, N. S., & Zinn, P. O. (2023). Human brain organoid code of conduct. *Frontiers in Molecular Medicine*, *3*, 1143298. Advance online publication. DOI: 10.3389/fmmed.2023.1143298 PMID: 39086687

Hyun, I., Scharf-Deering, J. C., & Lunshof, J. E. (2020). Ethical issues related to brain organoid research. *Brain Research*, *1732*, 146653. DOI: 10.1016/j.brainres.2020.146653 PMID: 32017900

Kim, S., & Chang, M. (2023). Application of human brain Organoids—Opportunities and challenges in modeling human brain development and neurodevelopmental diseases. *International Journal of Molecular Sciences*, *24*(15), 12528. DOI: 10.3390/ijms241512528 PMID: 37569905

Lancaster, M. A., & Knoblich, J. A. (2014). Generation of cerebral organoids from human pluripotent stem cells. *Nature Protocols*, *9*(10), 2329–2340. DOI: 10.1038/nprot.2014.158 PMID: 25188634

Lancaster, M. A., Renner, M., Martin, C., Wenzel, D., Bicknell, L. S., Hurles, M. E., Homfray, T., Penninger, J. M., Jackson, A. P., & Knoblich, J. A. (2013). Cerebral organoids model human brain development and microcephaly. *Nature*, *501*(7467), 373–379. DOI: 10.1038/nature12517 PMID: 23995685

Lavazza, A., & Pizzetti, F. G. (2020). Human cerebral organoids as a new legal and ethical challenge†. *Journal of Law and the Biosciences*, *7*(1), lsaa005. Advance online publication. DOI: 10.1093/jlb/lsaa005 PMID: 34221418

Lee, T., & Sawai, T. (2023). Global governance of human brain organoid research and applications: A role for the World Health Organization? *Molecular Psychology: Brain. Molecular Psychology*, *2*, 11. DOI: 10.12688/molpsych.17548.1

Little, M. H. (2017). Organoids: A special issue. *Development (Cambridge, England)*, *144*(6), 935–937. DOI: 10.1242/dev.150292 PMID: 28292836

Lu, X., Yang, J., & Xiang, Y. (2022). Modeling human neurodevelopmental diseases with brain organoids. *Cell Regeneration (London, England)*, *11*(1), 1. Advance online publication. DOI: 10.1186/s13619-021-00103-6 PMID: 34982276

Mansour, A. A., Gonçalves, J. T., Bloyd, C. W., Li, H., Fernandes, S., Quang, D., Johnston, S., Parylak, S. L., Jin, X., & Gage, F. H. (2018). An in vivo model of functional and vascularized human brain organoids. *Nature Biotechnology*, *36*(5), 432–441. DOI: 10.1038/nbt.4127 PMID: 29658944

Modafferi, S., Zhong, X., Kleensang, A., Murata, Y., Fagiani, F., Pamies, D., Hogberg, H. T., Calabrese, V., Lachman, H., Hartung, T., & Smirnova, L. (2021). Gene–environment interactions in developmental neurotoxicity: A case study of synergy between Chlorpyrifos and CHD8 knockout in human BrainSpheres. *Environmental Health Perspectives*, *129*(7), 077001. Advance online publication. DOI: 10.1289/EHP8580 PMID: 34259569

Paşca, A. M., Sloan, S. A., Clarke, L. E., Tian, Y., Makinson, C. D., Huber, N., Kim, C. H., Park, J., O'Rourke, N. A., Nguyen, K. D., Smith, S. J., Huguenard, J. R., Geschwind, D. H., Barres, B. A., & Paşca, S. P. (2015). Functional cortical neurons and astrocytes from human pluripotent stem cells in 3D culture. *Nature Methods*, *12*(7), 671–678. DOI: 10.1038/nmeth.3415 PMID: 26005811

Pham, M. T., Pollock, K. M., Rose, M. D., Cary, W. A., Stewart, H. R., Zhou, P., Nolta, J. A., & Waldau, B. (2018). Generation of human vascularized brain organoids. *Neuroreport*, *29*(7), 588–593. DOI: 10.1097/WNR.0000000000001014 PMID: 29570159

Plummer, S., Wallace, S., Ball, G., Lloyd, R., Schiapparelli, P., Quiñones-Hinojosa, A., Hartung, T., & Pamies, D. (2019). A human iPSC-derived 3D platform using primary brain cancer cells to study drug development and personalized medicine. *Scientific Reports*, *9*(1), 1407. Advance online publication. DOI: 10.1038/s41598-018-38130-0 PMID: 30723234

Qian, X., Song, H., & Ming, G. L. (2019). Brain organoids: Advances, applications and challenges. *Development (Cambridge, England)*, *146*(8), dev166074. Advance online publication. DOI: 10.1242/dev.166074 PMID: 30992274

Quadrato, G., Nguyen, T., Macosko, E. Z., Sherwood, J. L., Min Yang, S., Berger, D. R., Maria, N., Scholvin, J., Goldman, M., Kinney, J. P., Boyden, E. S., Lichtman, J. W., Williams, Z. M., McCarroll, S. A., & Arlotta, P. (2017). Cell diversity and network dynamics in photosensitive human brain organoids. *Nature*, *545*(7652), 48–53. DOI: 10.1038/nature22047 PMID: 28445462

Ravn, T., Sørensen, M. P., Capulli, E., Kavouras, P., Pegoraro, R., Picozzi, M., Saugstrup, L. I., Spyrakou, E., & Stavridi, V. (2023). Public perceptions and expectations: Disentangling the hope and hype of organoid research. *Stem Cell Reports*, *18*(4), 841–852. DOI: 10.1016/j.stemcr.2023.03.003 PMID: 37001517

Romito, A., & Cobellis, G. (2015). Pluripotent stem cells: Current understanding and future directions. *Stem Cells International*, *2016*(1), 9451492. Advance online publication. DOI: 10.1155/2016/9451492 PMID: 26798367

Sakaguchi, H., Ozaki, Y., Ashida, T., Matsubara, T., Oishi, N., Kihara, S., & Takahashi, J. (2019). Self-organized synchronous calcium transients in a cultured human neural network derived from cerebral Organoids. *Stem Cell Reports*, *13*(3), 458–478. DOI: 10.1016/j.stemcr.2019.05.029 PMID: 31257131

Sato, T., Vries, R. G., Snippert, H. J., Van de Wetering, M., Barker, N., Stange, D. E., Van Es, J. H., Abo, A., Kujala, P., Peters, P. J., & Clevers, H. (2009). Single Lgr5 stem cells build crypt-villus structures in vitro without a mesenchymal niche. *Nature*, *459*(7244), 262–265. DOI: 10.1038/nature07935 PMID: 19329995

Smirnova, L., Caffo, B. S., Gracias, D. H., Huang, Q., Morales Pantoja, I. E., Tang, B., Zack, D. J., Berlinicke, C. A., Boyd, J. L., Harris, T. D., Johnson, E. C., Kagan, B. J., Kahn, J., Muotri, A. R., Paulhamus, B. L., Schwamborn, J. C., Plotkin, J., Szalay, A. S., Vogelstein, J. T., & Hartung, T. (2023). Organoid intelligence (OI): The new frontier in biocomputing and intelligence-in-a-dish. *Frontiers in Science*, *1*, 1017235. Advance online publication. DOI: 10.3389/fsci.2023.1017235

Trujillo, C. A., Gao, R., Negraes, P. D., Gu, J., Buchanan, J., Preissl, S., Wang, A., Wu, W., Haddad, G. G., Chaim, I. A., Domissy, A., Vandenberghe, M., Devor, A., Yeo, G. W., Voytek, B., & Muotri, A. R. (2019). Complex oscillatory waves emerging from cortical Organoids model early human brain network development. *Cell Stem Cell*, *25*(4), 558–569.e7. DOI: 10.1016/j.stem.2019.08.002 PMID: 31474560

Xue, X., Sun, Y., Resto-Irizarry, A. M., Yuan, Y., Aw Yong, K. M., Zheng, Y., Weng, S., Shao, Y., Chai, Y., Studer, L., & Fu, J. (2018). Mechanics-guided embryonic patterning of neuroectoderm tissue from human pluripotent stem cells. *Nature Materials*, *17*(7), 633–641. DOI: 10.1038/s41563-018-0082-9 PMID: 29784997

Zhong, X., Harris, G., Smirnova, L., Zufferey, V., Sá, R. D., Baldino Russo, F., Baleeiro Beltrao Braga, P. C., Chesnut, M., Zurich, M., Hogberg, H. T., Hartung, T., & Pamies, D. (2020). Antidepressant Paroxetine exerts developmental neurotoxicity in an ipsc-derived 3D human brain model. *Frontiers in Cellular Neuroscience*, *14*, 14. DOI: 10.3389/fncel.2020.00025 PMID: 32153365

Chapter 9
Fostering Green Economy via Catalyst of Sustainability:
Navigating Entrepreneurial Landscape for Eco-Innovation

Revti Rani Roy
https://orcid.org/0009-0009-5354-4996
Chanakya National Law University, India

Ajay Kumar
https://orcid.org/0009-0003-1575-7778
Chanakya National Law University, India

Manju Kaushik
https://orcid.org/0000-0002-0720-1249
Amity University, Jaipur, India

Aashish Goswami
https://orcid.org/0009-0008-0248-5482
Amity University Rajasthan, Jaipur, India

ABSTRACT

There is no shame in acknowledging that profiteering through wealth generation is the most vicious motivator to satiate entrepreneurial aspirations. However, in this globalizing world where resource availability is faced with problem of resource exploitation, whilst economic development bears the cost of climate change, there is a need for a structural change in business behavior to encourage sustainability. The study examines how the state-of-the-art entrepreneurial ventures are insufficient

DOI: 10.4018/979-8-3693-8009-3.ch009

to address the concerns of green economy; and how it can be transformed through simulation of eco-efficient innovation techniques by synergizing determinants of sustainability. The study is an exploration into spectrum of issues surrounding green entrepreneurship and how R&D and regulatory regimes in eco-innovation can contribute to a holistic solution. Using entrepreneurial Responsibility as a variable for sustainable entrepreneurship besides policy interventions- instilling entrepreneurial motivations for firms to go green, will surely stimulate a green economy driven by ethical ecopreneurs.

INTRODUCTION

Man's pursuit towards individual happiness is driven by self-interests! An individual entrepreneur, thus, is always motivated to take higher risks for fulfilment of his desires to earn more profits. Whether he has reached the ends of profiteering by purely capitalist means or slightly sustainable means can be examined by checking the various parameters that act as a checklist for passing the *green test!* This green test includes the best practices and initiatives towards a sustainable environment that acts as a moral compass for entrepreneurs to evaluate their Individual Social Responsibility and for their firms/ businesses to evaluate their Corporate Social Responsibility. Although there has been a significant upsurge in the number of start-ups and "green firms" since the past few years, there still remains several unanswered questions pertaining to ecological preservation and firm's practices thereto.

When one talks about sustainability initiatives in relation to thriving businesses in the current world, *green entrepreneurs, green firms* and *green economy* emerge out as buzz words. The term *"sustainability"*, as simple as it may sound, has far more wider connotations and can be applied in all aspects of life today. In a rather restrictive sense, entrepreneurship might be related to profiteering and wealth maximization, but a wider perspective also attaches social responsibility upon individual entrepreneurs to the society at-large, the environment and the future generation alike, suggesting to an idea of *sustainable entrepreneurship.*

As early as 1997, when sustainable entrepreneurship was a relatively new topic globally, Robert Isaak, in his work *"Globalisation and Green Entrepreneurship"*, while talking about the global trends to be taken into consideration by a 21st-century risk-taking entrepreneur for a viable future of future generations, documented the need to invest in *socially responsible companies that pass a 'green screen'* (Issak 1997). While there is no iota of doubt that sustainability will foster a healthy ecosystem and sustainable practices are the need of the hour, but the same comes in conflict with human needs and aspirations. This debate of the cost-benefit constraints surrounding green practices by businesses and individuals has always been a major concern. While the entrepreneurs and businesses may shield themselves in the argument of less profit and higher green costs for opting green approaches (Schaltegger and Burritt, 2018) towards production and distribution patterns; individual consumers cloak their arguments with problems of higher expenditure towards green consumption patterns considering their income. One response towards this might be that shifting towards greener production alternatives shall go hand-in-hand with the demand of green products by sustainable consumers engraining individual social responsibility. Another response might be that the intervention of State actors is required to regulate production and consumption patterns and to make a green shift! A third alternative response might be to regulate production patterns based on market analysis, using technological tools.

Sustainable Development, Green Economy and Role of Sustainability Entrepreneurs

Entrepreneurship outcomes differ in different social and cultural contexts based on several dimensions and factors distinguishing commercial entrepreneurship with social/ sustainable entrepreneurship (Collins and Kearins, 2010; Lumpkin, et al., 2013). In order to flourish, sustainability necessitates an understanding of planetary boundaries Put another way it is necessary to identify that Earth has a limited carrying capacity for life and perpetual prosperity over time. Regretfully important biophysical limits are currently being crossed by human activity and turning in to sustainability is one major solution to this dilemma. The term "sustainable" evokes ideas of energy efficiency, environmental responsibility and green jobs all of which are undoubtedly inherent in prosperous 21st-century businesses. The modern entrepreneur, according to Lillich seeks to live in the era of *sustainable capitalism*, a framework that aims at maximization long-term economic gain through market reforms, addressing real needs while integrating environmental, societal and regulatory mechanisms throughout their managerial and administrative processes (Lillich, 2012). This idea of sustainable capitalism is applicable to the whole investment value chain and it transcends national boundaries, sectors and classes, and multiple interested parties- from start-ups to big publicly traded corporations, established entrepreneurial firms to institutional investors, staffs to business leaders, activists to legislators!

Research on entrepreneurship reflections show how the fundamental socioeconomic change required to ensure flourishing green economy could be brought about through entrepreneurship (Isaak 1997; Isaak, 2002). We think that the article also highlights the idea of sustainable entrepreneurship and possesses transformative potential (Schaefer et al., 2015), which is something that sustainability scholars should focus on (Walley and Taylor, 2002). Building on the key determinants of green entrepreneurs proposed by Walley and Taylor (2002), the researchers interpret in this chapter, the sustainability principles or financial incentives, that green entrepreneurs are swayed by, for achieving sustainable development (Yu and Gibbs, 2020). The current research also indicates that green jobs, green entrepreneurship and the green economy are means, and not ends for attaining sustainable development, and thus, can be seen as essential to reaching the desirable Sustainable Development Goals (SDGs) (van der Ree, 2019).

In this chapter, the researchers, under various segments, seek an answer to the following research questions:

1. Whether there can be a *one pill suits all* solution for all green entrepreneurial initiatives?
2. Whether prevalent innovation models sufficiently address the green concerns or there is a requirement of further research and development for green entrepreneurship to thrive?
3. What is the role of State actors in promoting sustainable entrepreneurship for a green economy?

Hypotheses: The first hypothesis is that research and development bringing sustainable and green innovation techniques for production, packaging and marketing is weaker and thus does not serve as a motivation for sustainable entrepreneurship.

The second hypothesis is that regulatory policies imposing "green tax", and "corporate social responsibility" mandates are not sufficient for turning an entrepreneur to an ethical ecopreneur; rather, apart from instilling responsibility, government policies and schemes must also provide motivations in form of subsidies or exemptions for entrepreneurs to adopt greener alternatives.

Significance of the Study. The research focuses on the idea of instilling social responsibility and accountability on entrepreneurs towards sustainable development; and addresses the issues surrounding it. The research is a distinct study, classifying the various actors involved in sustainable entrepreneurship, identifying the factors, motivators and limitations for green ventures in contemporary times; and suggests the evolution of eco-innovation in technological frameworks along with a sound State policy to foster the growth of green entrepreneurs in a sustainable economy, not compromising on economic growth.

DYNAMICS OF SUSTAINABLE TRANSITIONAL DEVELOPMENT OF GREEN ENTREPRENEURSHIP

The early researchers in the area of social responsibility towards a sustainable green economy considered the idea of social responsibility as normatively and empirically unsustainable, yet they left a little room for another idea in the form of a capitalist discourse upon producers' and consumers' behaviour promoting consumerism and rapid industrialization- leading to environmental degradation in a functional economy (Jones, 1996; Taeusch, 1935; Wood, 1991). However, the idea of sustainable businesses often ended up merely in piles of trash papers of big enterprises, for the cost of eco-sustainability was always higher (Taneja et al., 2011). The problem was not so prominent for established firms, rather it was for the new start-ups who had to make the cost-benefit analysis for adopting sustainable practices (Schick et al., 2002) for *increasing eco-efficiency, reducing resource expenses and waste.*

Today, the world is transitioning to a safer sustainable place and market is one of the key determinants for a sustainable environment (Gibbs, 2006; Isaak, 2002). For the market to be sustainable, it is pertinent for the key players of market forces to think and act sustainably, and here comes the role of existing and emerging entrepreneurs to come up with such business models, ideas, processes, technologies and organizational structures which prioritize the environment. Now the idea of entrepreneurship is transitioning to a reformed idea of ecopreneurship (Gibbs, 2006; Hessels et. Al., 2008; Schaltegger, 2002; Schaper, 2002). The commercial consciousness of entrepreneurs is thus evolving to a sustainable commercial consciousness, where the entrepreneurs are not merely investing and working towards sustaining the environment, but at the same time, their conscious efforts to preserve the environment does not cost them their profit margins in their commercial ventures. This entire socio-economic-environmental perspective of businesses has emerged as what we understand today as *green entrepreneurship,* and the idea is still in its novice phase.

Nevertheless, the idea of green entrepreneurship is broader than one might perceive, and it includes an entrepreneur's positive efforts towards reducing waste, conserving resources, preserving energy and imperative steps towards environmentally-conscious behaviour (Mrkajic et al., 2019) through constantly upgrading to sustainable technologies and business models. But such conscious efforts might require some actual visible incentive in line with the aspirations of a modern entrepreneur, else there might arise a conflict between the costs and benefits attached to eco-modernizing the businesses to meet the environmental standards.

Eco-modernization, Responsible Entrepreneurship and Economic Growth

Ecological preservation and eco-modernization must go hand-in-hand for the modern entrepreneurs to achieve the twin goals of wealth maximization and sustainable development. Some researchers have remarkably given a bold statement conveying that the real goal of sustainability can be achieved through eco-modernization (Tilley and Young, 2006), whereby the modern entrepreneur incorporates modern innovation techniques to boost productivity and maximize profits but not at the cost of environment. A radical economist propagating the theory of entrepreneurship in light of wealth accumulation, might regard the ultimate motive of entrepreneurship as wealth generation and profit maximization, however, the proponents of modern entrepreneurial theory like Joseph A. Schumpeter himself, takes a broader view to debunk the role of entrepreneurs, integrating their traditional capitalist roles with social and environmental roles. To this extent, Schumpeter regards the modern entrepreneurs as innovators and attributes to them the responsibility of what he called as the *"creative destruction"- the idea wherein an entrepreneur brings about a change in the existing entrepreneurial setup to bring about a constructive change in the society* (Schumpeter, 1976). A modern entrepreneur, is thus headed towards sustainable entrepreneurship, wherein the aspirations of an entrepreneur coincide with the requirements of the society.

Entrepreneurship is not only a catalyst of societal wellness and social health but it is also a key factor for economic growth, as it creates employment, enables innovation and promotes productivity (Global Entrepreneurship Monitor Report, 2023). *Sustainable entrepreneurship* and *socially responsible investment* have emerged as the most strategic idea for businesses to strike a balance between environmental sustainability along with meeting the goals of the business organization, thus fostering a green economy (Figueroa-Armijos and Johnson, 2016). SDG 8 talks about *"inclusive sustainable economic growth and decent work for all"*. Entrepreneurs are seen as crucial to the transition to a global village (Tilley and Young, 2006), and in addition to increasing employment there is a belief that entrepreneurship can boost productivity, economic growth and regeneration.

Studies have shown that sustainable business policies and sustainable investing helps the external as well as internal operations of such businesses- a brand's sustainability not only helps in reduction of carbon footprints, it also helps in creation and building of the brand name (BRASS Program Planning Committee, 2010, Javed et al, 2024). Findings even suggest that the eco-awareness attached with a brand plays a key role in invoking ethical consumption patterns (Filho et al., 2024; Marshall, 2011) and thus, a modern responsible entrepreneur (Walley and Taylor, 2002) who is dedicated towards sustainability would make sure to make environment-friendly products, use eco-modernized tools for production and innovation of processes, and at the same time, he would also ensure that the brand's sustainable purpose reach to the consumers, so as to create an organized ethical consumption pattern of ethical communities (Papaoikonomou, 2012) in a sustainable market. This entire cyclic nature of transactions would lead to creation of a green economy with circular flow of money, leading to sustainable economic growth and prosperity (Figueroa-Armijos and Johnson, 2016).

It is responsible entrepreneurship and sustainable practices which, through environmental performances (Wu et al., 2023) will help achieve the dual target of ecological preservation and economic growth (Hye et al., 2023) at the same time.

The Intersection Between Ecological Modernisation and Green Entrepreneurship

The notion of *"modernizing the modernity"* (Smith and Garza-Rubalcava, 2019) is at the heart of the ecological modernization theory. It is believed that the intended modernization brought about by ecological modernization represents a transformation of the industrialized society that solely prioritizes production- into one that ensures environmental preservation while producing goods (Leonidou et al., 2017). Modern society does not have to give up on technological advancements as a result of the transformation. The goal of ecological modernization theory is analysing how modern industrialized societies respond to environmental crises. Upon careful examination of several determinants, like institutional structure, economic efficiency, productivity, organizational capacity and particularly cultural background and motivations of entrepreneurs, one may question the circumstances under which ecological modernization is likely to occur.

It can be said that that strong environmental consciousness and a strong public commitment (Kaesehage et al., 2019) are two major cultural traits influencing the capacity of a risk-taking entrepreneur (Hessels et al., 2008) to adopt the fundamentals of ecological modernization, which propagates that it is possible to change production and consumption patterns through eco-innovative tools in a way that is environmentally sustainable (Smith and Garza-Rubalcava, 2019). This idea can be backed by Marshall's proposed model of *"international for-profit social entrepreneurs"*, whereby he characterises entrepreneurialism based on certain personality traits of entrepreneurs, such as- proactive mindset, opportunity recognition, social and cultural networks (O'Neill, Jr. et al., 2006) and performance-based outcomes (Marshall, 2011) and if these attributes of an entrepreneur are integrated with the social and environmental attributes, a "social for-profit entrepreneur" is born who knows how to commercialize his environmental strategies towards eco-quality goods(Volery, 2002).

ECO-INNOVATION STRATEGIES AND TECHNOLOGICAL NICHE PROMOTING CIRCULARITY

The early researchers of green entrepreneurship and sustainable firm behaviour tried to outline the business advantages associated with *pro-active environmental behaviour* (Bianchi and Noci, 1998), highlighting the emerging interests of consumers and investors who clearly identify the businesses based on the sustainability and eco-quality of the brand alongside its business performance (Berry and Rondinelli, 1998). However, the early research majorly missed out on the important aspect of circularity in any corporate or entrepreneurial setup.

While much literature is already available in the area of corporate responsibility and business models for corporate performance, researchers have now started the exploration into the sustainable business models for green entrepreneurship, workable for a transformed green economy- pioneering the evolution of a dynamic business model for entrepreneurial ventures, rooted in eco-innovation techniques (Schaltegger et al., 2016). By incorporating eco-innovation strategies promoting material circularity (Liu et al., 2023) in production and distribution patterns, entrepreneurs can substantially impact the consumption patterns. However, navigating the landscape for eco-innovation can be a daunting task in this era of research and development towards transformation of economy into a green economy, as a shift-change from the classical linear model for waste management towards a business model that promotes material circularity during the life cycle of a product is next to impossible without the assistance of comprehensive understanding and assistance of technological advancements in the field of sustainable innovations, along with process and design innovations and requisite changes in organizational frameworks thereto (Cobo and Angel, 2018). The circular economic model being referred to here is the best example of eco-efficient economic production model, which if incorporated with green innovations, may lead to a self-sustained closed loop system having least waste generation and optimal resource utilization for a product's ultimate life cycle (Chioatto et al., 2020).

Sustainability-Oriented Business Model Innovation: Context and Green Innovation Techniques

Global energy use, greenhouse gas emissions, and waste production are disproportionately concentrated in urban areas as *"urbanization is leaving behind large material footprint and inducing weak waste recovery and prevention systems"*, which has put a tremendous strain on natural resources and accelerated climate change. When *Tim Jackson* documented the idea against the trends of consumerism in modern times, he also identified some of the drivers of sustainable consumption. A common man who is morally and ethically driven, has the slightest of idea about sustainable consumption, would be able to identify its major drivers such as *buying of ethical- environment-friendly alternatives, reduction of personal consumption, reusing and recycling existing goods.* Therefore, there is a requirement of sustainable entrepreneurship drives that attracts consumers towards sustainable and ethical consumption. For a modern entrepreneur, in order to prove their brand and business as sustainable, it is important to make eco-efficient choices when it comes to production, distribution and marketing. Therefore, it becomes imperative to talk about sustainability-oriented eco-efficient business models. Eco-efficiency can be achieved if the firm opts for green innovation techniques at several steps including production, designing, manufacturing and several other areas. In context of economic growth and sustainable change, entrepreneurship has revolutionary capabilities and such changes may be induced by responsible innovators, bringing about production process innovation, techno-legal innovation, technological innovation (Dwivedi et al. 2023), product or service innovation, design innovation or even organisational innovation.

Innovation, here has to be interpreted as "eco-innovation" (Rejeba et al., 2022) not a generic development. Sustainable development Goals- SDG9 aims for an economy with *"inclusive and sustainable industrialization"* that *"fosters innovation."* *Eco-innovation*, thus, cannot be given a strict definition, even though attempts have been made to define the term. For instance, the European Commission Report lays down that *"Eco-innovation is the production, assimilation or exploitation of a product, production process, service or management or business method that is novel to the organization (developing or adopting it) and which results, throughout its life cycle, in a reduction of environmental risk, pollution and other negative impacts of resource use (including energy use) compared to relevant alternatives"* (Kemp and Pearson, 2007). Eco-efficient innovation model for sustainable entrepreneurship must include *green innovation techniques.* Eco-efficient models thriving on green innovation techniques may involve *product and service innovation, process innovation, system innovation, product and design innovation, energy innovations, organizational innovation and technological innovation and other innovations contributing to environmental preservation.*

"Product innovation" would mean modifications in the products and material goods and services itself, such that the ultimate product is more eco-friendly, and its disposal leads to the least impact on the environment. Product innovation is thus related to the nature of goods as well as its design and packaging focusing on the product's *"Design for Environment"* (Benabdellah et al., 2021) that encompasses *"product improvement and product redesign"*, (Iyer et al., 2016) considering the complete life cycle of the product- from acquiring raw materials for its production, to manufacturing, waste disposal and recycling. *"Life Cycle Assessment Tool"* is one

such tool that maps all the technical options related to production, designing and disposal of a product, with their environmental impact, thus this tool can be integrated for product innovation techniques (Rejeba et al., 2022) to assess the environmental impact and sustainability options for an *"extended producer's responsibility"* (Kautto, 2010). *"Process eco-innovation"* includes modification of production systems with a view to improving environmental efficiency and reducing environmental cost- such that production process reduces the depletion of natural resources per unit of production of goods on the one hand, while also minimizing polluting outputs as a by-product of production, on the other. *"System eco-innovation"* encompasses sustainable changes in institutional levels, from sustainably obtaining natural resources to developing eco-friendly alternatives, having increased environmental gains and reduced material footprints on the environment. Investment in *technological eco-innovation* may also lead to a sustainable organizational environment. (Vargas-Hernandez et al., 2025). *"Green Technological Innovation"* (Javed et al., 2024) has emerged as the most dynamic forms of eco-innovation techniques, where entrepreneurs and businesses are encouraged to invest in research and development of environmentally sustainable technologies. *"Organizational eco-innovation"* covers broad ambit including- rearrangement of the framework of the organization, creating new decision-making methods, altering management approaches, operational approaches or strategy aimed at obtaining environmental benefits, including implementation of environmental entrepreneurial policies like environmental audits or new approaches of "production and management of supply chain" (Vargas-Hernandez et al., 2024). Organisational eco-innovation may also include innovative strategies aimed at accomplishing the CSR objectives (Williamson et al., 2006) of the firm, including green investments, improvement in eco-equality standards and progress in environmental performance (Wu et al., 2023).

In addition to sustainable business models (Schaltegger et al., 2016; Moliterni, 2017), firms and corporations must also emphasize upon their marketing strategies motivated towards persuading the consumers to invest in greener and sustainable alternatives. Sustainable Marketing strategies focusing on the brand's greener purpose (Akram et al., 2023), where companies present their products by attaching some value standards and eco-quality with them. Studies have been done to find out that a rational consumer would have increased sensitivity to buy a product which has relatively more *eco-quality* and less environmental impact (Amacher et al., 2009) than its other existing alternatives in the market, thus, increasing the relative demand for such goods in the market (Mady et al., 2023). By maintaining the eco-quality of products and linking it with the brand's sustainable purpose firms can present sustainability as a compelling choice for consumers (Fujiwara, 2009), by making it easy for the consumers to differentiate various existing products based on the intensity and level of pollution and environmental impact caused during their production and disposal process.

Material Circularity Fostered by Green Innovation

While it is imperative that a shift to eco-sustainable technologies by entrepreneurial ventures would lead us to a low-carbon economy (Demirel et al., 2019), it is equally vital to understand the significance of a circular economy model, contrary to a liner economic model (Chioatto et al., 2020). The classical linear model is predominantly limited to disposal of raw materials after their extraction, processing it for manufacturing and production of goods for consumption, putting it for consumption or use and ultimately discarding it. Thus, the product's life cycle ends at consumption in a linear model. In contrast, however, the circular model aims at waste management (Albitar et al., 2024), product recreation and optimal utilization of resources up to its economic productivity (Genovese et al., 2017), in a closed-loop circle of product reuse, recycle, remanufacture, thus, contributing to a reduction of material waste and material extraction for production process (van Ewijk and Stegemann, 2023). This circular model, if combined with technologies backed by eco-innovation, can lead to a sustainable future.

Evidently, eco-innovation has emerged as a catalyst for optimizing sustainable business models (Dantas et al., 2022; Wilson and Post, 2013). Although there is a requirement of a clearer framework promoting circular business models- drawing a roadmap for identifying interlinkages between Circular Economy and Eco-Innovation, the concept of a circular business model which has transitioned through various drivers of sustainable entrepreneurship strategies needs to have a co-relation with Eco-Innovation (Abu-Bakar et al., 2024). However, there still remains a gap in adequate innovation strategies related to material circularity in a green economy during corporate life cycle of a product (Cainelli et al., 2020). Cainelli, in his further research advances the argument that environmental degradation can be mitigated majorly through induction of green innovation techniques and this transition might require greening of products and process innovations in line with sustainable business models promoting circularity that are capable of attaining the dual goal of *ecological conservation* and *economic competitiveness* amongst green innovators emerging as sustainable entrepreneurs (De Marchi et al., 2022). The idea is that "*waste should be recirculated to production and consumption processes ensuring that no "waste" would be wasted under the circular economy system.*" (Zaman, 2023).

STATE INITIATIVE TOWARDS GREEN ENTREPRENEURIAL VENTURES

Sustainability-driven entrepreneurs will more often than not establish green entrepreneurial ventures with enhanced eco-quality of products and systems, with the view to contribute to a transformational socioeconomic condition (Stockholm Environment Institute, 2019). In an idealistic setup, however, one must not ethically expect someone to sacrifice their entrepreneurial interests for future generations (Isaak, 1997), rather they must be motivated by other prevailing interests being incentivized at them. The role of State, thus comes into play because without the intervention and coordination of the State with business-oriented entrepreneurs, there cannot be an overhauling of the existing commercialized industry and entrepreneurial setup. The entrepreneurs, for instance, cannot simply opt for eco-efficient, low-carbon energy transitions, without the energy sector providing a cost-effective alternative renewable source of energy! Parrish and Foxon (2011) gave a *co-evolutionary model* interlinking the various actors of a sustainable economy and advanced the idea that an economy can achieve the goal of a low-carbon economy with the catalytic force of sustainable entrepreneurship, with the help of an interplay between the dynamics of changing technologies and government coming forth for a public-private interaction. Further research has shown the importance of intervention of public authorities in understanding the nature of entrepreneurial enterprise and taking their help in co-evolutionary sustainable development (Gasbarro et al., 2017). Ball and Kittler (2019) further the research towards involvement of public authorities and envisage a *sustainable transitional model* whereby the role of entrepreneurs has been outlined to be crucial in transitioning towards a sustainable low-carbon economy, though they recognize that the same would be difficult without the interference of public authorities who have access to the data-set of nature of each enterprise along with the design set of socio-technical regulatory regimes aiding the transition process. While there cannot be a one-pill-fits-all solution, however strategic State intervention may lead to substantial market success involving entrepreneurs to produce eco-efficient solutions in terms of knowledge building, resource sharing, production choices, energy usage and value creation towards environmentally beneficial goods and services.

Regulatory Intervention Towards Sustainable Entrepreneurship

If there are no compelling laws and policies in-place, it is highly unlikely that a firm would opt for a sustainable production process or waste recycling and disposal process that might cut down its profits in lieu of lesser environmental cost. Similarly, it is highly unlikely that consumers will go for sustainable consumption, reduce unnecessary buying pattern and work towards absolute consumption of a product-

utilizing, reusing, redesigning and recycling it to its full life cycle- unless they are compelled either by legal norms or societal norms- moral and ethical standards set by the market in a fair-trade society. Therefore, the dialectic discourse upon role of the triad of all the stakeholders- individual consumers, businesses and policymakers (government) becomes critical.

Research by Mady suggests that there might be a positive nexus between regulatory pressure created by government intervention and demand for green products (Mady et al., 2023). Green entrepreneurs get a "sustainable competitive advantage" where their enterprise is operating on principles of sustainability and fosters green innovation at all stages of the product's life cycle, from extraction of raw materials, to production, to marketing, to distribution, to use, reuse and recycle! The State shall thus come up with a co-extensive eco-effective environmental policy (Bhatia and Jakhar, 2021) for entrepreneurs so as to enhance not only the decrease in carbon cost incurred by the State but also to minimize the overall harm caused to the environment due to business ventures.

There already exists the policy of imposition of "green tax" for the environmental cost and expected material footprints resultant from a particular product in its entire lifecycle- from manufacturing to disposal/ recycling. The other changes in policy regulations may be in form of:

a. *Subsidy to firms and businesses on- "eco-friendly production-costs" incurred by them during production process and investment by firms towards "green innovation techniques"* adopted by companies to upgrade the ecological standards of their products.

b. *Imposition of a "minimum eco-quality standard"* upon businesses, ensuring that enterprises are compelled to manufacture products while maintaining the requisite *"eco-quality content",* whereby the government may impose sanctions through tax upon goods not meeting the environmental quality standards on one hand, and subsidize the manufacturing of superior quality goods that pass the eco-quality standard test on the other (Garella, 2021).

c. *Campaigns stimulating ecological awareness amongst consumers about eco-quality products,* whereby policy regulations could encourage "green consumerism" to grow. Government initiatives to raise consumer awareness of environmental issues might include informational and educational programs that try to bolster private and philanthropic motives that lead customers to place a premium on environmentally friendly products. If a government campaign is successful in raising consumers' awareness of eco-quality products and getting them to pay more for them, it will only benefit the demand for green products and, consequently, the profits of green businesses (Filho et al., 2024).

Therefore, it is imperative that there must be a regulator responsible for regulating the market to ensure responsible entrepreneurship, green production, and green consumerism (Giallonardo and Mulino, 2024). To this end, the role of government may be outlined along the lines of the need for environmental policy instruments promoting sustainable production patterns and responsible consumption.

An environmental policy promoting green entrepreneurship, as suggested by the researchers in this chapter may be developed using the carrot-and-stick method! That is to say that, on the one hand, the State may bring a *Restrictive Regulation*, which may, for instance put a limited restriction on the use of such products or production methods that hamper the environment directly or if their production requires use of scarce natural resources. Such a limited restriction here, may be in the form of increasing the cost of such products and production through imposition of higher green tax on the enterprise. But, environmental policy stringency (Aydin et al., 2024) must be implemented wisely, and since we live in a laizzez-faire economy, some liberty to the entrepreneurs has to be given, so as to stimulate production and economic growth, for entrepreneurs contribute widely to the economic growth of the country! Therefore, such an effective State regulation, based on sustainability may positively encourage the entrepreneurs to switch to eco-friendly alternatives and eco-efficient methods of business if the policy incentivizes the entrepreneurs through subsidies for production cost on green products and on using eco-friendly technologies and eco-innovation for an increase in ecological footprint and decrease in carbon footprints! Further, environmental policy may also include regulations regarding imposition of "emission taxes" or "carbon tax" for encouraging eco-innovation in energy solutions- so that the entrepreneurs switch to eco-efficient energy transitions (Aydin et al., 2024) rather than the traditional sources of energy, that has an adverse impact on the environment. Therefore, in order to effectively target entrepreneurial deficiencies, which may be overlooked, these regulatory mechanisms incorporated in environmental policies can tackle the issues of environmental degradation and contribute to the creation of a green economy- slowly and gradually, as there is a substantially adverse relationship between eco-efficiency and environmental degradation.

Apart from restrictive regulatory policy, there is also a substantial requirement for the state to invest in research and development strengthening and promoting "Green Technology Innovation", an advancement of which would definitely encourage more green choices towards energy efficiency, through production and process innovations. This would ensure a reduction in environmental waste and would repair the ecological damage at the same time. This solution is also supported by the study of Wu et al. (2023), that talks about the positive and visible impacts of "Green Technology Innovation" on "environmental performance" of the firms and business ventures (Wu et al., 2023). Green technology Innovation has shown remarkable developments towards green economic growth in OECD (Mahmood et al., 2022). An exploratory study conducted by Javed (2024), endorsed by several other empirical studies, also suggests towards empirical finidings pointing towards the positive implications of switching towards green innovation techniques on environmental health in various nations. Javed also highlights the importance of investment in Research & Development (R&D) towards innovation of green technologies and calls upon the various actors including the State, the government and private educational institutions, research institutions funded by the state, other interested private parties and private sector businesses to aid to R&D. Mere creation of such policies is not enough, rather the government should also monitor its implementation at all levels. State policies may encourage the knowledge creation and promotion of such green research and development initiatives through funding, recognition and other incentives. This would ensure a green economy running on eco-efficient sources powered by green innovation- meaning efficient use of natural resources and energy conervation on one hand and environmental protection on the other. This means that *"environmental policy might affect environmental quality"* (Javed et al, 2024).

Green Financing and Public Funding for Green Ventures

Gharleghi et al. (2024) define "green finance" as *"investments in green bonds, reduction in CO2 emission, and efficiency of the environmental tax by individual countries towards green growth and transition to a green economy while reducing the negative environmental impacts."* Some recent findings also suggest that green financing has an adverse relationship with factors contributing to environmental degradation, such as "energy consumption, CO2 emissions, greenhouse emissions and natural resources depletion" (Afzal et al., 2022). Moreover, green finance policies have been proven to have a positive and direct relation with Corporate Social Responsibility (CSR) activities being carried out by entrepreneurial ventures (Sheng et al., 2024).

Any risk-taking entrepreneur needs some hard-driven objective and incentive to carry out socially-driven operations. While some study shows that in order to attain sustainable solution from entrepreneurial processes, government must also back them up with green policy incentives backed by environmental, social, and governance (ESG) determinants Gharleghi et al. (2024). The entrepreneurs receiving public funding tend to solve more social welfare problems (Lumpkin et al., 2013). Study by Gliedt and Parker (2007) suggests that government funding towards subsidizing green operations to cut-out extra-costs incurred by non-profit organizations have contributed to more societal ventures being taken up by such enterprises, however at the same time, there is little scope for research and development in green field because the public funding for R&D is not enough (Javed et al., 2024). Therefore, a social entrepreneur is often stuck between the choices of turning the non-profit venture to a for-profit green entrepreneur to enable public funding and financial aids because large-scale funding is required for research and development. However, to tackle the problems associated with paucity of funds and limitation of available resources, it is also suggested that, to offset the cost of environmental hazards public funding for research and developments in the field of eco innovation and green inventiveness can be made subject to demonstration of skillset of aspiring entrepreneurs (Lumpkin et al., 2013). This would engage the entrepreneurs in a competitive innovative experimentation to achieve sustainable solutions. Moreover, collaboration of the government with such social ventures and green entrepreneurs and funding for eco-innovation in the short run will only prove to be beneficial for the economy in the long run towards the fulfilment of its sustainable development goals! Therefore, Green Financing and public funding towards sustainable entrepreneurship are encouraged if the funds have been disbursed to achieve sustainable outcomes for mitigating carbon footprints.

Entrepreneurship Policy Encouraging Eco-Innovation and Competitiveness

The idea of *ecological standards* for processes involved in production of goods that meet the *eco-quality* is not a novel idea, rather it has been talked about, directly or indirectly in various early researches surrounding sustainable and green economy (Isaak, 1997; Berry and Rondinelli, 1998; Isaac, 2002). However, it has only been around a decade that researchers have also started exploring the area of *"Environmental-quality competition"* and research in this field shows how sustainability acts as a key parameter for assessing a firm's growth and proactive customer engagement, which creates an environment for *"sustainable competitive advantage"* (Mady et al. 2023). The idea is to create a positive healthy competitive environment amongst entrepreneurs and assess them based on their willingness and active involvement towards meeting sustainability standards as per the expectations of the various stakeholders involved- (socially conscious ethical consumers, government regulators and policymakers, green service organizations and researchers). This finding can be backed by the research of Schaltegger and Burritt (2018), who focussed upon the idea of *"responsible sustainability management"* in entrepreneurial setups and came up to the finding that responsible entrepreneurship in line with national and international eco-standards may lead to the creation of a self-regulated sustainable business environment, driven by improved ecological standards and performance-based indicators of growth. Research also indicates that sustainability approaches in product life-cycle assessment and organizational improvement show fulfilment of regulatory standards and thus distinguishes entrepreneurs based on sustainability parameters (Schaltegger and Burritt, 2018). Businesses must strive to attain their profitable outcomes along with embracing proactive sustainable strategies in their business models. A company subscribing to higher sustainable standards and eco-compatible products and processes is more likely to have long-term value creation of its brand and a cult following of environmentally conscious ethical consumers, which form a separate consumer base, and thus incorporating sustainability practices and eco-standards would yield the modern entrepreneur *a distinct value* in the society along with profits. Researches have shown that the firms creating their own identity based on proactive sustainable practices also have another coincidental effect on the other businesses (Mady et al. 2023). It has been identified that socially conscious consumers are willing to bear the additional environmental cost attached to a green product, thus creating a buzz demand for eco-friendly alternatives for existing non-green products. This leades to an indirect force for fellow businesses and entrepreneurs to go in the direction of green entrepreneurship (Mondal et al., 2023)- adopting sustainable standards. It might be said that it creates a pull-and-push effect for the customers and entrepreneurs! A green entrepreneur subscribing to all the eco-equality standards, who produces green sustainable products with least damage to environment is sought to attract and pull the consumers towards investing in their products/ services. This in turn pushes the competing businesses to adopt and enhance their eco-standards and come up with equally good or better sustainable alternatives! This is a perfect example of a sustainable capitalist economy, which fosters a healthy competition and gives a sustainable competitive advantage to sustainability pioneers. Demand for eco-friendly products by responsible ethical consumers is sought to change the consumption pattern and the overall market demand. Accordingly, the quality of market supply of alternative goods will also have to be altered by competing businesses in a free-flow economy, because it is those firms that keep up with the eco-quality, ecological standards and sustainable practices, who will have a *sustainable competetive advantage* over the others (Mady et al. 2023).

The above discussion is suggestive towards *green entrepreneurship policy* (Figueroa-Armijos and Johnson, 2016), and not merely entrepreneurship policies driven by motivators of entrepreneurial aspirations (Johnson, 2007). Figueroa-Armijos and Johnson (2016) argue in favour of beneficial tax-credit system for enthusiasts of sustainable entrepreneurship so as to foster the adaptation of green entrepreneurial strategies. However, the research also suggests that apart from the tax credit system, the government may come up with green incentives at structural level, such as local assistance for implementation of eco-innovation tools. Therefore, an entrepreneurship policy that encourages eco-innovation and promotes it is also a welcome step towards the creation of a green economy. Regulatory policies having compliance parameters will, on the one hand, act as a stringent determinant for assessing competitiveness, on the other hand entrepreneurship policies aiding in eco-innovation (Akram et al., 2023) and contributing to firm's green growth (Mahmood et al., 2022) will attract the entrepreneurs towards performance based on the environmental parameters and keep their business operations environmentally conscious (Mady et al. 2023). The same can be supported by the findings of Sánchez and Deza (2015), who talks about "Porter's win-win hypothesis" which lays down that *"properly designed environmental standards can trigger innovation offsets, which can not only lower the net cost of meeting environmental regulations, but can even lead to absolute advantages over firms in foreign countries not subject to similar regulations"*. With the objective to incentivize invention, research and development in eco-innovation (Demirel et al., 2019), it is also suggested that regulators should not simply make a blanket ban on available technologies, rather, in order to promote *inclusivity* of all categories of entrepreneurs, green entrepreneurship policies must also include policy towards incentivizing enttrepreneurs who invest in *"best available technology"* as their pocket allows, so that the cost to benefit ratio is maintained along with their chances to gain the competitive advantage based on sustainable indicators.

DISCUSSION AND CONCLUSIONS

Promoting green and sustainable entrepreneurship necessitates a multifaceted strategy that includes strong regulatory frameworks governing efficient corporate social responsibility procedures. In addition to removing obstacles to sustainable production and emphasizing the advantages of eco-friendly products, businesses must endeavor to comprehend the needs and desires of their target market. Businesses may significantly influence the transition to a sustainable lifestyle by using marketing principles to connect consumers with a brand's sustainable purpose and make sustainability enticing. The trend towards a greener economy will only increase as more businesses embrace sustainable methods and consumers become more conscious of their effects. Above all, sustainability challenges should be framed in public debate in a way that puts them closer to consumers. At this point, sustainability will be seen as more than simply a duty, as it may also be considered to be a wise economic strategy for firms and businesses alike.

Towards the end of this study, it is pertinent to understand that by no far stretch of imagination the study can be said to be an exhaustive study. The area of eco-innovation in relation to green entrepreneurship is a relatively new area which requires a lot of research. It can be said that in an economy that fosters a capitalist mindset for economic growth, individual entrepreneurs often overlook their share in sustainability initiatives for green economy owing to their entrepreneurial aspirations. However, this study addresses certain important concerns surrounding social responsibility and accountability on entrepreneurs towards sustainable development; in a capitalist economy. The study suggests incorporating sustainability into capitalism and is an exploration into the aspects of sustainable capitalism. The present research also distinctly classifies the various actors involved in sustainable entrepreneurship, and it can be said that it is not a one-man job of entrepreneurs alone- rather there has to be a co-extensive framework of environmentally conscious consumers, government, policymakers, regulators, technologists, innovators, environment enthusiasts and any other incidental actors. It has been also identified during this research that there might be several motivators and limitations for green entrepreneurship ventures in contemporary times. Investments in eco-innovation and technological frameworks fostering sustainable ecosystem can be a key motivator on the one hand, while a sound State policy fostering the growth of green entrepreneurs in a sustainable economy, not compromising on economic growth can be a push factor for the entrepreneurs to engage in green business, on the other hand. The paucity of funds distribution, encouraging green production is one of the key limitations, as identified in the study, because Research and Development is highly dependent on huge capital and funds. Therefore, the government and policy makers should strengthen the research and development in the area of eco-innovation in order to achieve the long-term sustainable goal of green entrepreneurship.

In the beginning of the chapter, the researchers had identified certain key research questions sought to be answered during the course of various segments in this chapter. The first research question was: Whether there can be a *one pill suits all* solution for all green entrepreneurial initiatives? To this, the researchers have come to the solution that it is not possible to devise a single solution for all green entrepreneurial initiatives, as the nature of entrepreneurship businesses may vary in spatial and temporal terms, it may also vary in various cultural setups, or it may vary with the variation in typologies of product or even production methods. The possibilities are endless and therefore, there cannot be a one pill suits all solution to all issues surrounding different entrepreneurs headed towards sustainability drive.

The second research question outlined in the beginning of this study was: whether prevalent innovation models sufficiently address the green concerns or there is a requirement of further research and development for green entrepreneurship to thrive? To this, it can be said that there is a lot of scope for improvement and since we are living in the technological era, the present innovation models cannot be said to sufficiently address all the green concerns. Rather, newer Green Innovation Models may be devised for a thriving sustainable economy.

The last research question to be answered during the research was framed as: What is the role of State actors in promoting sustainable entrepreneurship for a green economy? It can be concluded through the above study that it is the State that strives to achieve the goal of a green economy, and therefore, the role of State actors and policy regulators becomes very essential for promoting a sustainable green entrepreneurship. Without State intervention and restrictive mandates regulating entrepreneurial ventures it would be highly unlikely that any business-oriented entrepreneur will take the risk of eco-innovations and maintain the eco-quality standards of their firm that passes the "green screen"! Moreover, it is the State that is the most interested party and the best beneficiary on behalf of its masses, and therefore, the role of State cannot be separated when we talk about green entrepreneurship. Another reason to strengthen this argument is that eco-innovation for fostering green entrepreneurship requires huge investments in Research and Developments, for which best reliance can be placed on State actors.

In light of the above discussions, it is important to test the hypotheses framed during the beginning of this chapter.

Testing of Hypotheses

The first hypothesis was, that "research and development bringing sustainable and green innovation techniques for production, packaging and marketing is weaker and thus does not serve as a motivation for sustainable entrepreneurship." The researchers, throughout this chapter have placed heavy emphasis on the requirement of strengthening of research and development, maintaining that there is a substantial requirement for the state to invest in research and development strengthening and promoting "Green Technology Innovation", and also to encourage green entrepreneurs to engage in green innovation research. The present study also concludes that the existing Innovation methods cannot be said to be pragmatic for all start-ups, though it might be implemented in big entrepreneurial ventures and established firms. Thus, the hypothesis tested out to be true.

The last hypothesis was that "regulatory policies imposing "green tax", and "corporate social responsibility" mandates are not sufficient for turning an entrepreneur to an ethical ecopreneur; rather, apart from instilling responsibility, government policies and schemes must also provide motivations in form of subsidies or exemptions for entrepreneurs to adopt greener alternatives."

This hypothesis also tested out to be true towards the end of the study, because the study in the last segment has addressed the importance of government policies towards subsidies and tax credit policies along with creation of stronger green entrepreneurship policies.

Further Research

Further research may be expanded to cover the aspects concerning use of eco-innovation models in a circular business model, which in itself can contribute to a vast field of literature. Further research into this area may also include an exploration into the possibilities of advancement of the existing eco-innovative models through use of Artificial Intelligence for predictive analysis and data profiling.

REFERENCES

Abu-Bakar, H., Charnley, F., Hopkinson, P., & Morasae, E. K. (2024). Towards a typological framework for circular economy roadmaps: A comprehensive analysis of global adoption strategies. *Journal of Cleaner Production*, *434*, 140066. Advance online publication. DOI: 10.1016/j.jclepro.2023.140066

Afzal, A., Rasoulinezhad, E., & Malik, Z. (2022). Green finance and sustainable development in Europe. *Ekonomska Istrazivanja*, *35*(1), 5150–5163. DOI: 10.1080/1331677X.2021.2024081

Akram, M. W., Yang, S., Hafeez, M., Kaium, M. A., Zahan, I., & Salahodjaev, R. (2023). Eco-innovation and environmental entrepreneurship: Steps towards business growth. *Environmental Science and Pollution Research International*, *30*(23), 63427–63434. DOI: 10.1007/s11356-023-26680-4 PMID: 37022542

Albitar, K., Nasrallah, N., Hussainey, K., & Wang, Y. (2024). Eco-innovation and corporate waste management: The moderating role of ESG performance. *Review of Quantitative Finance and Accounting*, *63*(2), 781–805. DOI: 10.1007/s11156-024-01281-5

Amacher, G. S., Koskela, E., & Ollikainen, M. (2004). Environmental quality competition and eco-labelling. *Journal of Environmental Economics and Management*, *47*(2), 284–306. DOI: 10.1016/S0095-0696(03)00078-0

Aydin, M., Degirmenci, T., Erdem, A., Sogut, Y., & Demirtas, N. (2024). From public policy towards the green energy transition: Do economic freedom, economic globalization, environmental policy stringency, and material productivity matter? *Energy*, *311*, 133404. Advance online publication. DOI: 10.1016/j.energy.2024.133404

Ball, C., & Kittler, M. (2019). Removing environmental market failure through support mechanisms: Insights from green start-ups in the British, French and German energy sectors. *Small Business Economics*, *52*(4), 831–844. https://www.jstor.org/stable/48701962. DOI: 10.1007/s11187-017-9937-8

Benabdellah, A. C., Zekhnini, K., Cherrafi, A., Garza-Reyes, J. A., & Kumar, A. (2021). Design for the environment: An ontology-based knowledge management model for green product development. *Business Strategy and the Environment*, *30*(8), 4037–4053. DOI: 10.1002/bse.2855

Berry, M., & Rondinelli, D. (1998). Proactive corporate environmental management: A new industrial revolution. *The Academy of Management Perspectives*, *12*(2), 38–50. Advance online publication. DOI: 10.5465/ame.1998.650515

Bhatia, M. S., & Jakhar, S. K. (2021). The effect of environmental regulations, top management commitment, and organizational learning on green product innovation: Evidence from automobile industry. *Business Strategy and the Environment*, *30*(8), 3907–3918. DOI: 10.1002/bse.2848

Bianchi, R., & Noci, G. (1998). "Greening" SMEs' Competitiveness. *Small Business Economics*, *11*(3), 269–281. https://www.jstor.org/stable/40228980. DOI: 10.1023/A:1007980420087

BRASS Program Planning Committee. (2010). Clean, Green, and Not So Mean: Can Business Save the World? *Reference and User Services Quarterly*, *50*(2), 135–140. https://www.jstor.org/stable/20865382. DOI: 10.5860/rusq.50n2.135

Cainelli, G., D'Amato, A., & Mazzanti, M. (2020). Resource efficient eco-innovations for a circular economy: Evidence from EU. *Research Policy*, *49*(1), 103827. DOI: 10.1016/j.respol.2019.103827

Chioatto, E., Zecca, E., & D'Amato, A. (2020). Which innovations for Circular Business Models?: A Product Life-Cycle Approach. *Fondazione Eni Enrico Mattei (FEEM)*. https://www.jstor.org/stable/resrep27688

Cobo, S., & Angel, I. A. D.-R. (2018). From linear to circular integrated waste management systems: A review of methodological approaches. *Resources, Conservation and Recycling*, *135*, 279–295. DOI: 10.1016/j.resconrec.2017.08.003

Collins, E. M., & Kearins, K. (2010). Delivering on Sustainability's Global and Local Orientation. *Academy of Management Learning & Education*, *9*(3), 499–506. https://www.jstor.org/stable/25782033

Dantas, R. M., Ilyas, A., Martins, J. M., & Rita, J. X. (2022). Circular Entrepreneurship in Emerging Markets through the Lens of Sustainability. *Journal of Open Innovation*, *8*(4), 211. Advance online publication. DOI: 10.3390/joitmc8040211

De Marchi, V., Cainelli, G., & Grandinetti, R. (2022). Multinational subsidiaries and green innovation. *International Business Review*, *31*(6), 102027. Advance online publication. DOI: 10.1016/j.ibusrev.2022.102027

Demirel, P., Li, Q. C., Rentocchini, F., & Tamvada, J. P. (2019). Born to be green: New insights into the economics and management of green entrepreneurship. *Small Business Economics*, *52*(4), 759–771. https://www.jstor.org/stable/48701958. DOI: 10.1007/s11187-017-9933-z

Dwivedi, A., Sassanelli, C., Agrawal, D., Gonzalez, E. S., & D'Adamo, I. (2023). Technological innovation toward sustainability in manufacturing organizations: A circular economy perspective. *Sustainable Chemistry and Pharmacy*, *35*, 101211. Advance online publication. DOI: 10.1016/j.scp.2023.101211

Figueroa-Armijos, M., & Johnson, T. G. (2016). Entrepreneurship policy and economic growth: Solution or delusion? Evidence from a state initiative. *Small Business Economics*, *47*(4), 1033–1047. https://www.jstor.org/stable/26154684. DOI: 10.1007/s11187-016-9750-9

Filho, M. G., & Gonella, J. (2024, September). dos S. L., Latan, H., Ganga, G. M. D. (2024). Awareness as a catalyst for sustainable behaviors: A theoretical exploration of planned behavior and value-belief-norms in the circular economy. *Journal of Environmental Management*, *368*, 122181. Advance online publication. DOI: 10.1016/j.jenvman.2024.122181

Fujiwara, K. (2009). Environmental policies in a differentiated oligopoly revisited. *Resource and Energy Economics*, *31*(3), 239–247. DOI: 10.1016/j.reseneeco.2009.03.002

Garella, P. G. (2021). The effects of taxes and subsidies on environmental qualities in a differentiated duopoly. *Letters in Spatial and Resource Sciences*, *14*(2), 197–209. DOI: 10.1007/s12076-021-00272-7

Gasbarro, F., Annunziata, E., Rizzi, F., & Frey, M. (2017). The Interplay Between Sustainable Entrepreneurs and Public Authorities: Evidence from Sustainable Energy Transitions. *Organization & Environment*, *30*(3), 226–252. https://www.jstor.org/stable/26408339. DOI: 10.1177/1086026616669211

GEM (Global Entrepreneurship Monitor). (2023). Global Entrepreneurship Monitor 2023/2024 Global Report: 25 Years and Growing. *London: GEM*.https://www.gemconsortium.org/reports/latest-global-report

Genovese, A., Acquaye, A., Figueroa, A., & Lenny Koh, S. C. (2017). Sustainable supply chain management and the transition towards a circular economy: Evidence and some applications. *Omega*, *66-B*, 344–357. DOI: 10.1016/j.omega.2015.05.015

Gharleghi, B., Shafighi, N., and Nawaser, K. (2024). Green finance and its role in sustainability in the EU. *Journal of Economy and Technology*. DOI: 10.1016/j.ject.2024.07.004

Giallonardo, L., & Mulino, M. (2024). Green Consumerism and Firms' Environmental Behaviour Under Monopolistic Competition: A Two-Sector Model. *Italian Economic Journal: Springer.*, *10*(1), 347–376. DOI: 10.1007/s40797-023-00223-9

Gibbs, D. (2006). Sustainability Entrepreneurs, Ecopreneurs and the Development of a Sustainable Economy. *Greener Management International, 55*, 63–78. https://www.jstor.org/stable/greemanainte.55.63

Gliedt, T., & Parker, P. (2007). Green community entrepreneurship: Creative destruction in the social economy. *International Journal of Social Economics*, *34*(8), 538–553. DOI: 10.1108/03068290710763053

Hessels, J., van Gelderen, M., & Thurik, R. (2008). Entrepreneurial aspirations, motivations, and their drivers. *Small Business Economics*, *31*(3), 323–339. https://www.jstor.org/stable/40650947. DOI: 10.1007/s11187-008-9134-x

Hye, Q. M., Ul-Haq, J., Visas, H., & Rehan, R. (2023). The role of eco-innovation, renewable energy consumption, economic risks, globalization, and economic growth in achieving sustainable environment in emerging market economies. *Environmental Science and Pollution Research International*, *30*(40), 92469–92481. DOI: 10.1007/s11356-023-28945-4 PMID: 37491494

Isaak, R. (1997). Globalisation and Green Entrepreneurship. *Greener Management International, 18*, 80–90. https://www.jstor.org/stable/45259405

Isaak, R. (2002). The Making of the Ecopreneur. *Greener Management International, 38*, 81–91. https://www.jstor.org/stable/greemanainte.38.81

Iyer, G., & Soberman, D. A. (2016). Social Responsibility and Product Innovation. *Marketing Science*, *35*(5), 727–742. https://www.jstor.org/stable/44012185. DOI: 10.1287/mksc.2015.0975

Javed, A., Rapposelli, A., Khan, F., Javed, A., & Abid, N. (2024). Do green technology innovation, environmental policy, and the transition to renewable energy matter in times of ecological crises? A step towards ecological sustainability. *Technological Forecasting and Social Change*, *207*, 123638. Advance online publication. DOI: 10.1016/j.techfore.2024.123638

Johnson, T. G. (2007). Measuring the benefits of entrepreneur ship development policy. *ICFAI Journal of Entrepreneurship Development*, *4*(2), 35–44.

Jones, M. T. (1996). Social Responsibility and the "Green" Business Firm. *Industrial & Environmental Crisis Quarterly, 9*(3), 327–345. https://www.jstor.org/stable/26162491

Kaesehage, K., Leyshon, M., Ferns, G., & Leyshon, C. (2019). Seriously Personal: The Reasons that Motivate Entrepreneurs to Address Climate Change. *Journal of Business Ethics*, *157*(4), 1091–1109. https://www.jstor.org/stable/45106464. DOI: 10.1007/s10551-017-3624-1

Kautto, P. (2010). New instruments – old practices? The implications of environmental management systems and extended producer responsibility for design for the environment. *Business Strategy and the Environment, 15*(6), 377–388. DOI: 10.1002/bse.454

Kemp, R., & Pearson, P. (2007). *Final Report MEI Project About Measuring Eco-Innovation.* UM Merit.

Leonidou, L. C., Christodoulides, P., Kyrgidou, L. P., & Palihawadana, D. (2017). Internal Drivers and Performance Consequences of Small Firm Green Business Strategy: The Moderating Role of External Forces. *Journal of Business Ethics, 140*(3), 585–607. https://www.jstor.org/stable/44164312. DOI: 10.1007/s10551-015-2670-9

Lillich, M. (2012). Entrepreneurship, Jobs, Sustainable Businesses = Economic Prosperity. *Perspectives on Work, 16*(1/2), 3–7. https://www.jstor.org/stable/41810199

Liu, Z., Han, S., Yao, M., Gupta, S., & Laguir, I. (2023). Exploring drivers of eco-innovation in manufacturing firms' circular economy transition: An awareness, motivation, capability perspective. *Annals of Operations Research.* Advance online publication. DOI: 10.1007/s10479-023-05473-5

Lumpkin, G. T., Moss, T. W., Gras, D. M., Kato, S., & Amezcua, A. S. (2013). Entrepreneurial processes in social contexts: How are they different, if at all? *Small Business Economics, 40*(3), 761–783. https://www.jstor.org/stable/23360622. DOI: 10.1007/s11187-011-9399-3

Mady, K., Anwar, I., & Abdelkareem, R. S. (2023). Nexus between regulatory pressure, eco-friendly product demand and sustainable competitive advantage of manufacturing small and medium-sized enterprises: The mediating role of eco-innovation. *Environment, Development and Sustainability.* Advance online publication. DOI: 10.1007/s10668-024-05096-1

Mahmood, N., Zhao, Y., Lou, Q., & Geng, J. (2022). Role of environmental regulations and eco-innovation in energy structure transition for green growth: Evidence from OECD. *Technological Forecasting and Social Change, 183*, 121890. Advance online publication. DOI: 10.1016/j.techfore.2022.121890

Marshall, R. S. (2011). Conceptualizing the International For-Profit Social Entrepreneur. *Journal of Business Ethics, 98*(2), 183–198. https://www.jstor.org/stable/41475810. DOI: 10.1007/s10551-010-0545-7

Moliterni, F. (2017). Sustainability-oriented Business Model Innovation: Context and Drivers. *Fondazione Eni Enrico Mattei (FEEM).* https://www.jstor.org/stable/resrep16412

Mondal, S., Singh, S., & Gupta, H. (2023). Assessing enablers of green entrepreneurship in circular economy: An integrated approach. *Journal of Cleaner Production*, *388*, 135999. Advance online publication. DOI: 10.1016/j.jclepro.2023.135999

Mrkajic, B., Murtinu, S., & Scalera, V. G. (2019). Is green the new gold? Venture capital and green entrepreneurship. *Small Business Economics*, *52*(4), 929–950. https://www.jstor.org/stable/48701968. DOI: 10.1007/s11187-017-9943-x

O'Neill, G. D., Jr., Hershauer, J. C., & Golden, J. S. (2006). The Cultural Context of Sustainability Entrepreneurship. *Greener Management International, 55*, 33–46. https://www.jstor.org/stable/greemanainte.55.33

Papaoikonomou, E., Valverde, M., & Ryan, G. (2012). Articulating the Meanings of Collective Experiences of Ethical Consumption. *Journal of Business Ethics*, *110*(1), 15–32. https://www.jstor.org/stable/41684010. DOI: 10.1007/s10551-011-1144-y

Parrish, B. D., & Foxon, T. J. (2006). Sustainability Entrepreneurship and Equitable Transitions to a Low-Carbon Economy. *Greener Management International, 55*, 47–62. https://www.jstor.org/stable/greemanainte.55.47

Rejeba, H. B., Monnierb, E., Rioa, M., Evrarda, D., Tardifb, F., & Zwolinski, P. (2022). From Innovation to Eco-Innovation: Co-Created Training Materials as a Change Driver for Research and Technology Organisations. *29th CIRP Life Cycle Engineering Conference. Science Direct Procedia CIRP, 105.* 98-103. DOI: 10.1016/j.procir.2022.02.017

Sánchez, Á. P., & Deza, X. V. (2015). Environmental Policy Instruments and Eco-innovation: An Overview of Recent Studies. *Innovar: Revista de Ciencias Administrativas y Sociales, 25*(58), 65–80. https://www.jstor.org/stable/innrevcieadmsoc.25.58.65

Schaefer, K., Corner, P. D., & Kearins, K. (2015). Social, Environmental and Sustainable Entrepreneurship Research: What Is Needed for Sustainability-as-Flourishing? *Organization & Environment*, *28*(4), 394–413. https://www.jstor.org/stable/26164745. DOI: 10.1177/1086026615621111

Schaltegger, S. (2002). A Framework for Ecopreneurship: Leading Bioneers and Environmental Managers to Ecopreneurship. *Greener Management International, 38*, 45–58. https://www.jstor.org/stable/greemanainte.38.45

Schaltegger, S., & Burritt, R. (2018). Business Cases and Corporate Engagement with Sustainability: Differentiating Ethical Motivations. *Journal of Business Ethics*, *147*(2), 241–259. https://www.jstor.org/stable/45022375. DOI: 10.1007/s10551-015-2938-0

Schaltegger, S., Lüdeke-Freund, F., & Hansen, E. G. (2016). Business Models for Sustainability: A Co-Evolutionary Analysis of Sustainable Entrepreneurship, Innovation, and Transformation. *Organization & Environment, 29*(3), 264–289. https://www.jstor.org/stable/26164769. DOI: 10.1177/1086026616633272

Schaper, M. (2002). *Introduction:* The Essence of Ecopreneurship. *Greener Management International, 38*, 26–30. https://www.jstor.org/stable/greemanainte.38.26

Schick, H., Marxen, S., & Freimann, J. (2002). Sustainability Issues for Start-up Entrepreneurs. *Greener Management International, 38*, 59–70. https://www.jstor.org/stable/greemanainte.38.59

Schumpeter, J. (1976). The Process of Creative Destruction. In *Capitalism, Socialism and Democracy* (5th ed., pp. 81–86). George Allen & Unwin., DOI: 10.4324/9780203202050

Sheng, Y., Wang, S., & Wang, Y. (2024). Doing good in times of need: Green finance policy and strategic corporate social responsibility. *Economic Analysis and Policy, 84*, 1029–1045. Advance online publication. DOI: 10.1016/j.eap.2024.10.008

Smith, A. V., & Garza-Rubalcava, U. (2019). "Ecological Modernization Theory: Developing a Consensus with the Addition of Green and Sustainable Remediation". Springer Nature (Switzerland W. Leal Filho et al. (eds.)). *Industry, Innovation and Infrastructure, Encyclopedia of the UN Sustainable Development Goals.* DOI: 10.1007/978-3-319-71059-4_39-1

Stockholm Environment Institute. (2019). *Transformational change through a circular economy.* Stockholm Environment Institute. https://www.jstor.org/stable/resrep22978

Swain, S. (2014). From philanthropy to social entrepreneurship. In Damousi, J., Rubenstein, K., & Tomsic, M. (Eds.), *Diversity in Leadership: Australian women, past and present* (pp. 189–206). ANU Press., https://www.jstor.org/stable/j.ctt13wwvj5.13 DOI: 10.22459/DL.11.2014.10

Taeusch, C. F. (1935). The Relation between Legal Ethics and Business Ethics. *California Law Review, 24*(1), 79–95. DOI: 10.2307/3476485

Taneja, S. S., Taneja, P. K., & Gupta, R. K. (2011). Researches in Corporate Social Responsibility: A Review of Shifting Focus, Paradigms, and Methodologies. *Journal of Business Ethics, 101*(3), 343–364. https://www.jstor.org/stable/41475906. DOI: 10.1007/s10551-010-0732-6

Tilley, F., & Young, W. (2006). Sustainability Entrepreneurs: Could They Be the True Wealth Generators of the Future? *Greener Management International, 55*, 79–92. https://www.jstor.org/stable/greemanainte.55.79

van der Ree, K. (2019). Promoting Green Jobs: Decent Work in the Transition to Low-carbon, Green Economies. In C. Gironde & G. Carbonnier (Eds.), *The ILO @ 100: Addressing the Past and Future of Work and Social Protection* (Vol. 11, pp. 248–272). Brill. https://www.jstor.org/stable/10.1163/j.ctvrxk4c6.19

van Ewijk, S., & Stegemann, J. (2023). THE CIRCULAR ECONOMY. In *An Introduction to Waste Management and Circular Economy* (pp. 306–348). UCL Press., DOI: 10.2307/jj.4350575.17

Vargas-Hernandez, J. G., González, F. J., Orozco-Qijano, E. P., & Vargas-Gonzàlez, O. C. (2024). Green Organizational Management and Technological Innovation on Green Sustainable Organizational Performance. In Cepni, E. (Ed.), *Chaos, Complexity, and Sustainability in Management* (pp. 115–140). IGI Global., DOI: 10.4018/979-8-3693-2125-6.ch007

Vargas-Hernandez, J. G., González-Àvila, F. J., Vargas-Gonzàlez, O. C., Castañeda-Burciaga, S., & Guirette-Barbosa, O. A. (2025). Green Technology Innovation and Its Implications in the Sustainable Organizational Environment. In Ullah, A., Pandey, J., & Masengu, R. (Eds.), *Impacts of Technology on Operations Management: Adoption, Adaptation, and Optimization* (pp. 205–234). IGI Global., DOI: 10.4018/979-8-3693-6205-1.ch008

Volery, T. (2002). An Entrepreneur Commercialises Conservation: The Case of Earth Sanctuaries Ltd. *Greener Management International, 38*, 109–116. https://www.jstor.org/stable/greemanainte.38.109

Walley, E. E. (Liz), & Taylor, D. W. (David). (2002). Opportunists, Champions, Mavericks …? A Typology of Green Entrepreneurs. *Greener Management International, 38*, 31–43. https://www.jstor.org/stable/greemanainte.38.31

Williamson, D., Lynch-Wood, G., & Ramsay, J. (2006). Drivers of Environmental Behaviour in Manufacturing SMEs and the Implications for CSR. *Journal of Business Ethics, 67*(3), 317–330. https://www.jstor.org/stable/25123876. DOI: 10.1007/s10551-006-9187-1

Wilson, F., & Post, J. E. (2013). Business models for people, planet (& profits): Exploring the phenomena of social business, a market-based approach to social value creation. *Small Business Economics, 40*(3), 715–737. https://www.jstor.org/stable/23360620. DOI: 10.1007/s11187-011-9401-0

Wood, D. J. (1991). Corporate Social Performance Revisited. *Academy of Management Review, 16*(4), 691–718. DOI: 10.2307/258977

Wu, L., Wang, L., Philipsen, N. J., & Fang, X. (2023). The impact of eco-innovation on environmental performance in different regional settings: New evidence from Chinese cities. *Environment, Development and Sustainability*. Advance online publication. DOI: 10.1007/s10668-023-04280-z

Yu, Z., & Gibbs, D. (2020). Unravelling the role of green entrepreneurs in urban sustainability transitions: A case study of China's Solar City. *Urban Studies (Edinburgh, Scotland)*, *57*(14), 2901–2917. https://www.jstor.org/stable/26959607. DOI: 10.1177/0042098019888144

Zaman, A. (2023). Zero-Waste: A New Sustainability Paradigm for Addressing the Global Waste Problem, in *The Vision Zero Handbook: Theory, Technology and Management for a Zero Casualty Policy*, (Edvardsson, K. et al., (Eds.), *Springer.*) DOI: 10.1007/978-3-030-76505-7_46

Chapter 10
Microlearning:
Its Adaptation, Challenges, and Path Ahead

Monica Kunte

https://orcid.org/0000-0003-3486-0642

Symbiosis Centre for Management and Human Resource Development, Symbiosis International University, India

Joseph Mathew

Symbiosis Centre for Management and Human Resource Development, Symbiosis International University, India

Aman Choudhary

Symbiosis Centre for Management and Human Resource Development, Symbiosis International University, India

ABSTRACT

The research study aims to explore significance of microlearning in modern training initiatives. It seeks to investigate the benefits of microlearning in terms of employee engagement and its effectiveness in achieving concise training targets. This study employs a quantitative as well as qualitative research design to investigate the effectiveness, adaptation and challenges of microlearning in IT sector. Thematic exploration has been utilized to analyse the findings from in depth interviews with Learning and Development Managers and HR managers. The findings suggest that delivering targeted and personalized learning experiences, microlearning can keep employees motivated and continuously updated with relevant skills. The study's originality lies in its in-depth analysis of microlearning's adaptation, challenges, and the potential path ahead in the context of modern organizational training. The research provides managers with valuable insights and practical implications for incorporating microlearning into their organization's learning and development

DOI: 10.4018/979-8-3693-8009-3.ch010

initiatives.

INTRODUCTION

In the face of dynamic business requirements and the continuous need to upskill the workforce, organizations have turned to microlearning as an effective learning strategy (Borgman & Raza, 2020). Microlearning is a small bite-sized learning initiative which is tailor-made to organizational requirements in order to ensure employees learn new concepts within a short span of time, normally within 2-3 minutes per learning session. Focusing on specific learning demands depends on how microlearning addresses the little amount of learning content that is made up of fine-grained, interrelated, and loosely-coupled brief learning activities. In this world of short-video reels, Twitter space, people's attention span is reducing quite significantly, they check their mobile phones every now and then. Moreover, the personalized nature of microlearning empowers learners to control their learning journey, resulting in increased motivation and knowledge retention (Choi & Lee, 2019).

These days, young people don't want to waste their time listening to lectures. They prefer to use or use their newfound knowledge and abilities right away. They need access to learning opportunities outside of traditional classroom settings. Therefore, it's critical to make an effort to understand that contemporary children tend to be more active than their parents generally and that they select skills and information that they can easily see or exhibit as useful. In their compliance training, learners frequently need to grasp a lot of words, information, or guidelines. A platform for continual microlearning disseminates the information gradually in a variety of ways. That knowledge becomes ingrained in the learner's long-term memory with repeated exposure. Through ongoing microlearning, employees will interact with the material throughout the year in little bursts. The relevant material will be presented to them repeatedly but dispersedly. Each session is absolutely unique, which makes it interesting. By automating the process and using spaced repetition, training material that they already know will appear less frequently while practice material will be more frequently accessible. And it works well. Cramming doesn't work, according to a study. A three-hour eLearning course doesn't either. Not if you want to retain your knowledge. Spaced repetition, or repeated exposure to brief, manageable pieces of information, is necessary for long-term memory retention.

Learners resent being obliged to do lengthy online training, such as traditional 60- or 90-minute eLearning courses, because they perceive such courses to be burdensome, dull, or unhelpful for their employment. They put things off and avoid doing them. The integration of gamification and interactive elements further enriches the learning experience, fostering higher engagement and knowledge application (Lee & Lin, 2020).Additionally, busy workers dislike having to plan training and being interrupted from their everyday tasks or work to attend training sessions or access eLearning courses. This signifies the need and importance of microlearning in current employee-engagement scenario. The path ahead for microlearning involves a deeper understanding of learners' preferences and needs, enabling organizations to create tailored and impactful learning interventions (Chen & Hsieh, 2021).

Through our research we are trying to identify the gaps related to microlearning such as the Microlearning program's lack of alignment with learning objectives, lack of interactivity and many other challenges and also assessing the effectiveness of microlearning in reinforcing organizational performance.

LITERATURE REVIEW

Microlearning refers to short forms of learning and consists of short, fine-grained, inter- connected and loosely coupled learning activities with micro content (Lindner, 2006; Schmidt, 2007).In the Microlearning Guide to Microlearning, the author Carla Torgerson, a pioneer in the field defines it as "A piece of learning content that can be digested in little more than five minutes" (Torgerson 2016).Learning expert Will Thalheimer opines that a 20-minute or less interaction known as a microlearning initiative can be made up of any combination of content presentation, review, practise, reflection, behavioural prompting, performance support, goal- reminder messaging, persuasive messaging, task assignments, social interaction, diagnosis, coaching, management interaction, or other learning-related methodologies. (Thalheimer 2017).

The concept of microlearning can be traced back to the 1960s when educational psychologist George A. Miller introduced the idea of "chunking" in his paper on human cognitive capacity (Miller, 1956). Miller posited that the human brain can effectively process information by breaking it into smaller, manageable units. This insight laid the foundation for the idea that shorter, focused learning modules could lead to better knowledge retention and understanding. In the 1980s and 1990s, computer-based training (CBT) and multimedia learning emerged as innovative approaches to education and training (Barron, 1998; Clark, 1994). These technologies provided opportunities to deliver interactive and engaging content to learners. The advent of the internet and advancements in mobile technology revolutionized the learning landscape, giving rise to mobile learning (m-learning) (Sharples, Taylor, & Vavoula, 2010). As organisations recognised the benefits of microlearning, it found widespread application in corporate training programs. Companies began to adopt microlearning to cater to the diverse learning preferences of their employees, enhance knowledge retention, and address the challenges posed by traditional training methods (Kapp, 2012). Josh Bersin company (2022) has picturised the evolution of corporate learning: From 1998-2002 corporates provided users with self-study online learning, which blended e-learning with offline learning. The format of learning was based on a course catalogue through an online university. The second stage is talent-driven learning, where the learning was career focused and the employees had comprehensive learning opportunities. The period ranged from 2005 to 2012. During this period LMS is used as a talent platform. The third stage is digital learning, where there was embedded learning based on demand, rather than focusing on every content. The format was digital learning. In this stage, LMS was used as an experience platform. Now the current trend is learning in the flow of work, with the focus on everyone, all the time and everywhere. Microlearning, creator economy and intelligent skills system is prioritised so there are proper results in corporate talent management strategies.

Figure 1. Number of articles on microlearning by authors associated with SCOPUS journal.

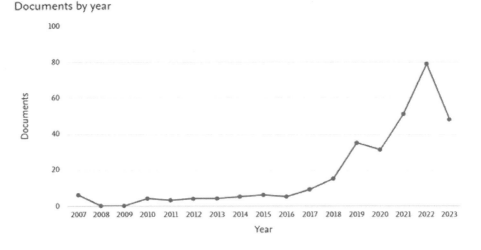

Figure 1. depicts the trend of research articles related to microlearning from 2007 to 2023. Before 2017, there were limited research articles dedicated to microlearning. However, starting in 2017, there was a substantial increase in research activities, leading to a peak in 2022 with 80 research articles solely focused on microlearning. This shows the importance of microlearning among learning and development professionals.

Access to smartphones and high-speed internet became widespread globally, learners sought on-demand and bite-sized content. This led to the development of microlearning, which delivers learning content in small, focused chunks, making it easily consumable and accessible across various devices. Research by Anders Berggren (2017) highlights the effectiveness of microlearning in improving knowledge retention and learner engagement. Berggren's study conducted in multinational corporations across Europe and Asia demonstrated that employees preferred microlearning modules for their concise and targeted nature, enabling them to quickly acquire new skills while maintaining productivity. The international adoption of microlearning has been accelerated by the dynamic and fast-paced nature of the business environment. Organisations operating globally face the challenge of providing consistent and relevant training to a diverse workforce. Microlearning's adaptability to different cultures, languages, and learning preferences has made it a preferred choice for multinational corporations (Redwood & Adams, 2019).

Several studies have examined the benefits of microlearning in specific regions. For instance, a study by Wu and Chen (2020) in the Asia-Pacific region revealed that microlearning improved learners' self-directed learning abilities, leading to better knowledge application and problem-solving skills. In Latin America, research by Gomez and Rivera (2018) indicated that microlearning contributed to higher learner motivation and engagement, resulting in increased completion rates of training programs. While the evolution of microlearning has been positive, challenges persist in its implementation. Cross-cultural considerations, translation and localization of content, and technical infrastructure limitations are among the challenges faced by organizations in international contexts (Gagnon & Heather, 2018). Additionally, cultural norms related to learning styles and preferences may impact the effectiveness of microlearning initiatives (Curtis & Rossetti, 2019). Despite challenges, the path ahead for microlearning in the international context appears promising. Organisations are investing in microlearning platforms and creating customised content to cater to diverse learners globally. Technological advancements, such as augmented reality and artificial intelligence, are further enhancing the effectiveness and personalization of microlearning experiences (Briggs & Hanson, 2021).

One study by Sağsan and Özden (2021) provides a historical overview of microlearning, tracing its evolution from the early 2000s to the present day. The study highlights the key milestones and developments in microlearning, including the introduction of mobile devices and the growth of social media platforms. The study also notes the increasing importance of personalized and adaptive learning in the evolution of microlearning. Another study by Rodriguez-Aflecht and Fuentealba (2021) explores the role of technology in the evolution of microlearning. The study notes that the development of new technologies, such as artificial intelligence and virtual reality, has the potential to transform microlearning by providing more immersive and interactive learning experiences. Similarly, a review of the literature by Thorne (2017) highlights the importance of mobile technology in the evolution of microlearning. The review notes that the rise of smartphones and tablets has enabled learners to access microlearning content anytime and anywhere, making it a more convenient and flexible learning approach. In addition to these studies, there is also evidence to suggest that the evolution of microlearning is being driven by changes in learner preferences and needs. A study by Yang and Lin (2019) found that learners are increasingly looking for personalized and adaptive learning experiences that are tailored to their individual needs and preferences.

Overall, the literature suggests that microlearning has evolved significantly over the past two decades, driven by changes in technology, learner preferences, and the increasing demand for personalized and adaptive learning. However, more research is needed to determine the optimal design and delivery of microlearning interventions for different learning outcomes and contexts.According to Christa Elliott(2016), shorter content may be better when it comes to executing workforce training. Although it isn't the best method for every learning topic, micro- learning can be a useful tool for companies trying to engage with contemporary corporate learners. Deep learning strategies are more appropriate for difficult or precise subjects.

Few other benefits and roles which disrupt the learning and development segment are:

The learning path can include bite-sized information, giving students the freedom to decide what they want to study and on what sort of device they want to learn it. Numerous different learning styles are addressed by microlearning. Through a study conducted by Nicolae (2017) it has been established that microlearning enables learners to more readily assimilate and remember the knowledge presented through the course's manageable and digestible sessions and exercises. Josh Cadoz points out that microlearning is well aligned with the demands of the modern learner since it often uses a rich-media format and is shorter in duration. Because knowledge is divided up into smaller chunks, microlearning may be utilised in a number of settings (such as personalised learning routes) and can even be reused depending on the situation. Microlearning is mobile-friendly and even available in offline versions because it is built for multi-device delivery. Microlearning is a particularly helpful material format when combined with the "pull-factor" of its design. A study by Muntean (2011) examined the benefits of microlearning in the context of mobile learning. The study found that microlearning was associated with increased engagement, motivation, and satisfaction among learners. The study also noted that microlearning was well-suited to mobile devices due to its short, easily digestible format.Just in time learning: If they see that the content is pertinent to their present developmental requirements, employees are more likely to consume it. Christa Elliot(2016) believes that, content should be distributed in response to specific difficulties an employee is having at work or in response to events like onboarding or the introduction of new policies. Right now microlearning would be the most suitable learning and development initiative for this requirement.

According to a research study conducted by Josh Cardoz (2017) companies are 30% more likely to increase customer satisfaction levels year over year if they use microlearning for employee development. Using microlearning as a reinforcement approach in a continuous learning environment to extend the shelf life of your training programmes. This is a fantastic technique to provide learners with "pick-me-up" material in the event of knowledge gaps or declining retention. Additionally, this will prevent expensive retraining and reduce lost production. According to a Delloite survey, employees only have to devote 1% of a regular workweek to training and development. That is just 25 minutes in a 40 hour work week. With the short span of learning intention only microlearning suits the learning initiatives, implemented whenever the management feels employees are in the prime of concentration and learning.

Similarly, a review of the literature by Sitzmann et al. (2010) highlighted the benefits of microlearning for knowledge acquisition and retention. The review found that microlearning was associated with improved retention of factual information, as well as better application of knowledge to real-world scenarios. In addition to these benefits, microlearning has also been shown to be a cost-effective approach to learning. A study by Perez-Sanagustin et al. (2013) found that microlearning was associated with lower costs compared to traditional classroom- based learning. Overall, the literature suggests that microlearning has numerous benefits for learners, including improved learning outcomes, increased engagement and motivation, and cost-effectiveness. However, more research is needed to determine the optimal design and delivery of microlearning interventions for different learning outcomes and contexts. One study by Nandan and Nandan (2020) found that microlearning was an effective method for teaching coding skills to beginners. The study used a randomized controlled trial to compare the effectiveness of microlearning with traditional classroom-based learning. The results showed that the microlearning group had a significantly higher mean score on the coding assessment compared to the traditional learning group. Another study by de Jong et al. (2015) investigated the effectiveness of microlearning for improving information security awareness among employees. The study used a quasi-experimental design and found that the microlearning intervention was effective in improving employees' knowledge and behaviour related to information security.

Similarly, a study by Chen and Wang (2020) found that microlearning was effective in improving employees' technical skills in the IT sector. The study used a quasi-experimental design and found that the microlearning intervention was associated with significant improvements in employees' technical knowledge and skills. In addition to these studies, there is also evidence to suggest that microlearning can improve knowledge retention and application. A review of the literature by Spitzer and Lippoldt (2019) found that microlearning was effective in improving retention and transfer of knowledge, particularly when combined with other learning interventions. One study by Chen and Kao (2021) investigated the effectiveness of microlearning in the workplace. The study found that microlearning was associated with improved learning outcomes, including increased knowledge retention, better transfer of learning to the job, and improved job performance

Overall, the literature suggests that microlearning is an effective approach to learning in the IT sector. Studies have shown that it can be used to teach technical skills, improve information security awareness, and improve knowledge retention and application. However, more research is needed to determine the optimal design and delivery of microlearning interventions for different learning outcomes in the IT sector. One of the main challenges in microlearning is the risk of information overload. Because microlearning modules are short and focused, there is a tendency to include too much information, which can overwhelm learners and impede their ability to retain and apply what they have learned (Barron & Iyer, 2019). To address this challenge, educators must carefully curate the content of microlearning modules, ensuring that each piece of information is relevant and necessary for achieving the learning objectives (Shen & Chen, 2019).

Another challenge in microlearning is maintaining learner engagement. Because microlearning modules are brief, learners may quickly lose interest or become distracted (Mayer & Johnson, 2020). To address this challenge, educators must design microlearning modules that are interactive and engaging, incorporating multimedia elements such as videos, animations, and quizzes to keep learners motivated and focused (Barron & Iyer, 2019). A third challenge in microlearning is the difficulty of assessing learning outcomes. Traditional methods of assessing learning, such as exams and papers, may not be suitable for microlearning modules, which may focus on discrete skills or concepts (Shen & Chen, 2019). To address this challenge, educators must develop alternative assessment strategies, such as performance-based assessments or portfolios, that allow learners to demonstrate their mastery of the material in a meaningful way (Mayer & Johnson, 2020).

Microlessons often just target short-term objectives; they only cover one learning objective that you want your staff to accomplish shortly. Jeanellie Avelino(2021) states an instance of adopting daily or weekly micro-courses might assist your staff to perform better if your objective is to consistently give exceptional customer service. To maximise their potential and assist them in acquiring new skills and abilities, you must invest in more thorough training and development if your long-term objective is to attain 100% customer satisfaction at the end of the year. Finally, a challenge in microlearning is the lack of social interaction and collaboration that is typically present in traditional classroom settings. Because microlearning is often delivered through digital platforms, learners may feel isolated and unsupported (Barron & Iyer, 2019). To address this challenge, educators must incorporate opportunities for peer-to-peer interaction and collaboration, such as discussion forums or group projects, into their microlearning designs (Shen & Chen, 2019). One study by Sangwan and Sangwan (2019) discussed the potential of microlearning to support lifelong learning. The study suggested that microlearning could provide learners with continuous access to learning materials and support ongoing skill development throughout their careers. Another study by Davis and Wong (2020) highlighted the potential of microlearning to support personalized learning. The study noted that microlearning could be tailored to individual learners' needs, preferences, and learning styles, allowing for a more personalized and effective learning experience.

In addition to these benefits, the literature suggests that the path ahead for microlearning includes the use of emerging technologies such as artificial intelligence (AI) and virtual reality (VR) to enhance the learning experience. A study by Papacharissi and Mendelson (2021) discussed the potential of AI to support personalized, adaptive microlearning experiences that can be tailored to individual learners' needs and preferences. Similarly, a study by Tuncay et al. (2020) highlighted the potential of VR to create immersive microlearning experiences that can improve engagement and knowledge retention. Finally, the literature suggests that the path ahead for microlearning includes the integration of microlearning into blended learning approaches that combine multiple learning modalities. A study by Tsaousoglou et al. (2021) discussed the potential of blended learning approaches that combine microlearning with traditional classroom-based learning and online learning to support a more comprehensive and effective learning experience.

Overall, the literature suggests that the path ahead for microlearning includes the use of emerging technologies, the integration of microlearning into blended learning approaches, and the continued exploration of microlearning's potential to support lifelong learning and personalized learning. However, more research is needed to determine the most effective ways to design and deliver microlearning interventions in these contexts. According to a research study conducted by Nicolae Balcessu, there is relatively little interaction between the information and the learner when lengthy content is presented all at once to students. In this approach, the learning environment may become unpredictable since the content may be more than what the working memory can hold. Knowing what to anticipate when new information is provided is essential for improving retention, which is another aspect of the issue. Therefore, it has been established that microlearning enables students/learners to more readily absorb and retain the information offered by the course's manageable and digestible lessons and exercises. Some of the emerging tools and platforms used in microlearning are:

- Grovo Platform
- Coursmos Platform
- Yammer Platform

As microlearning gains popularity as an effective learning and development strategy, it is essential to understand the challenges that organizations may encounter during its implementation and adoption. This section aims to review the existing research on challenges in microlearning and identify potential areas for further exploration. Integrating microlearning into existing learning management systems (LMS) or training platforms can be a complex task. Compatibility issues, data migration, and user authentication may present challenges during the integration process (DeRouin et al., 2017). Understanding how microlearning can seamlessly complement and enhance the overall learning ecosystem is a potential area for further investigation. While microlearning is designed to be engaging, maintaining learner motivation over time can be a challenge. Learners may become disengaged if the content lacks interactivity, personalization or fails to cater to diverse learning preferences (Giancarlo, 2019). Exploring gamification, social learning elements, and other strategies to sustain learner engagement could be an area for future research. As organizations scale up their microlearning initiatives, ensuring the sustainability and cost-effectiveness of the program becomes critical. The challenge lies in managing and updating a growing library of microlearning content while accommodating the learning needs of an expanding workforce (Koutropoulos, 2017). Investigating scalable content creation models and long-term cost-benefit analyses may offer insights into maintaining microlearning's impact over time.

RESEARCH METHODOLOGY

This study employs a quantitative as well as qualitative research design to investigate the effectiveness, adaptation and challenges of microlearning in IT sector. A cross-sectional survey approach was used to collect data from a sample of employees working in IT companies as core employees and interview is taken from Learning and development professional in IT companies as a sources of data collection.

A quantitative research design was chosen as it allows for the use of statistical analysis to examine the relationship between variables. A cross-sectional survey approach was used to collect data at a single point in time from a diverse sample of employees working in different IT companies. The survey instrument used in this study was developed based on a review of the literature on implementation of microlearning and reinforcement of it post implementation, as well as feedback from experts in the field. The sample for this study was drawn from employees working in several IT companies in India. The IT companies were selected based on their size and reputation, and employees were recruited through their respective HR departments. A total of 100 employees were invited to participate in the study, with 40 completing the survey for a response rate of 40%. Data was collected through an online survey administered via a secure online platform. The survey instrument consisted of 32 items, divided into three sections: demographic information, effectiveness of microlearning implementing, and organizational performance post implementation. The demographic section included questions related to age, gender, work experience, and organization. The effectiveness section included questions related to duration, skills covered, and other platforms through microlearning. The post implementation section included questions related to knowledge retention, reskilling, and challenges in microlearning.

Data collected from the survey was analyzed using descriptive and inferential statistics. Descriptive statistics were used to summarize the characteristics of the sample, while inferential statistics were used to test the research hypotheses. Specifically, multiple regression analysis was used to examine the relationship between employee engagement and organizational performance, while controlling for demographic variables. Participants were informed of the purpose of the study and their right to withdraw at any time without penalty. Informed consent was obtained from all participants prior to completing the survey, and all data was kept confidential and anonymous. One limitation of this study is the use of a cross-sectional design, which limits the ability to establish causal relationships between variables. Another limitation is the use of self-reported data, which may be subject to bias or social desirability effects. Finally, the sample size may limit the generalizability of the findings to other industries or regions.

RESULTS AND DISCUSSION

Qualitative Analysis

This qualitative study focuses on understanding the developments, requirements, benefits, and challenges of Microlearning in the IT industry. Interviews were conducted with HR managers from various IT organisations to gain insights into different aspects of the system. By coding and analysing the responses, this research aims to provide valuable insights into the topic.

"Microlearning has proven effective in achieving learning outcomes and enhancing engagement at Wipro. Employees have provided positive feedback, citing its impact on knowledge retention and skill development..."(R1)

Employees across these organisations have expressed positive feedback, indicating the effectiveness of microlearning in achieving learning outcomes. Microlearning is widely recognised for its ability to provide flexible and personalised learning experiences, leading to increased engagement and improved knowledge retention. These findings highlight the value of microlearning as a viable approach in contemporary learning and development strategies.

"Microlearning fills significant gaps left by traditional learning methods. Its flexibility and accessibility allow learners to engage with targeted content at their convenience, which is crucial in a rapidly changing work environment. The bite-sized format of microlearning enhances knowledge retention and promotes the immediate application of learning in the workplace."(R2)

IT organisations emphasise the delivery of micro-learning content in bite-sized modules that range from 5 minutes to approximately 15 minutes. This approach allows learners to access targeted content in manageable increments, promoting efficient learning. The frequency of microlearning intake is tailored based on factors such as topic complexity and learner availability, striking a balance between regular content access and the time needed for content absorption and application. The interviews with HR managers underscore the ability of microlearning to fill gaps present in traditional learning methods. Microlearning offers flexibility and accessibility, enabling learners to engage with content at their convenience, which is crucial in a rapidly changing work environment. The bite-sized format of micro-learning enhances knowledge retention and promotes the immediate application of learning in the workplace. The incorporation of interactive elements such as quizzes, simulations, and gamification promotes active engagement and addresses gaps associated with passive learning approaches.

Strategies employed by IT organisations to overcome resistance include raising awareness through workshops, webinars, and awareness sessions. The organisations showcased real-life examples and success stories to demonstrate the benefits of microlearning. Emphasizing the self-paced and personalized nature of microlearning helped employees understand its value, leading to increased acceptance over time. These findings highlight the importance of change management strategies when introducing new learning approaches. *"We take a data-driven approach to prioritize microlearning topics. We align them strategically with the organization's goals, leveraging inputs from subject matter experts and emerging industry trends."(R3)*

The HR managers emphasize a data-driven approach that considers factors such as business needs, emerging trends, skill gaps, and learner feedback. Collaboration with subject matter experts and business leaders ensures the alignment of microlearning content with organizational learning objectives. Various modes and formats, including videos, quizzes, gamified elements, and virtual training, are utilized to deliver microlearning content. The inclusion of interactive elements and real-life scenarios enhances learner engagement and promotes effective knowledge retention. The HR managers highlighted the significance of measuring the reinforcement and impact of microlearning. Tracking completion rates, gathering learner feedback through surveys and assessments, conducting performance evaluations, and analyzing business metrics are employed as means of evaluation. These approaches provide organizations with valuable insights into the effectiveness of microlearning interventions, the application of learning outcomes in real work scenarios, and areas for improvement. Measuring the impact of microlearning aligns with organizational learning goals and facilitates the continuous enhancement of learning and development strategies.

These insights contribute to the existing body of knowledge and inform organizations and researchers seeking to implement or explore microlearning as an effective training method in contemporary learning and development practices.

Managerial implication

The research on "Microlearning: Its Adaptation, Challenges, and Path Ahead" provides valuable managerial implications for organizations looking to incorporate microlearning in their learning and development strategies.

Firstly, the research highlights the effectiveness of microlearning in achieving learning outcomes and enhancing knowledge retention. Managers can leverage microlearning as a viable approach to improve employee engagement and performance. By delivering targeted and personalized learning experiences, microlearning can keep employees motivated and continuously updated with relevant skills.

Secondly, the study emphasizes the importance of change management when introducing microlearning. Managers should be prepared to address resistance from employees unfamiliar with this approach. Through awareness sessions and showcasing success stories, managers can create a positive outlook on microlearning, leading to greater acceptance and participation.

Thirdly, the data-driven approach to content design and delivery is crucial for successful microlearning implementation. Managers should utilize data analytics to identify skill gaps and align microlearning content with specific learning objectives and business needs. This ensures that the microlearning modules are relevant and impactful.

Furthermore, microlearning can foster a continuous learning culture within the organization. Managers can use microlearning to provide employees with just-in-time support and ongoing professional development opportunities. This will contribute to a skilled and adaptable workforce that can thrive in a rapidly changing business environment.

Lastly, managers need to establish a robust measurement and evaluation system for microlearning initiatives. Tracking completion rates, gathering learner feedback, and analyzing performance metrics will provide valuable insights into the effectiveness of microlearning interventions. This data-driven approach allows managers to make informed decisions to continually enhance their learning and development strategies.

"To overcome resistance, we conducted awareness sessions and engaged in open discussions with our workforce. We emphasized the benefits of microlearning, highlighting its personalized nature and flexibility" The combined approach of awareness sessions and open discussions has succeeded in promoting the adoption of microlearning as a viable and effective training method within the organization. The increased acceptance, understanding of benefits, and positive feedback from employees signify the potential of microlearning to become an integral part of the organization's learning and development strategy. By addressing the resistance through these means, the organization can ensure a smoother transition to micro-learning and leverage its advantages for continuous improvement and professional growth of its workforce.

Overall, the research on microlearning provides managers with valuable insights and practical implications for incorporating microlearning into their organization's learning and development initiatives.

CONCLUSION

This research paper has provided a comprehensive exploration of the topic of microlearning adaptation, challenges, and the path ahead. Through qualitative analyses and interviews with Learning and Development Managers in various organizations, valuable insights have been gained. The research reveals that microlearning has been successfully adopted as a training method in the IT industry, offering flexible and personalized learning experiences. It fills gaps left by traditional learning approaches, promoting active engagement, knowledge retention, and immediate application of learning in the workplace. Challenges in adopting microlearning, such as resistance and content curation, have been identified, and effective strategies for overcoming these challenges have been highlighted. Additionally, the research emphasizes the significance of data-driven content design, various delivery modes, and measurement of impact in ensuring the effectiveness of microlearning initiatives. Looking ahead, the path for microlearning lies in embracing emerging technologies and incorporating interactive elements to enhance engagement and impact. The adaptability of microlearning will continue to play a pivotal role in meeting the evolving demands of the modern workforce. Overall, this research contributes valuable insights to the field of learning and development, showcasing the potential of microlearning as a powerful and efficient approach for continuous skill development in the rapidly changing IT industry. As organisations seek to stay competitive and support employee growth, microlearning offers a promising path ahead for creating agile and future-ready learning experiences.

REFERENCES

Barron, A. E. (1998). Integrating computer-based multimedia training into micro-computer software training. *Computers in Human Behavior*, *14*(2), 193–206.

Barron, A. E., & Iyer, R. (2019). Microlearning: Challenges, opportunities, and best practices. In Graziano, K. (Ed.), *Proceedings of Society for Information Technology &*.

Bersin, J. (2022), https://joshbersin.com/wp-content/uploads/2021/12/WT-21_12 - HR-Predictions-for-2022-Report.pdf

Borgman, H., & Raza, S. (2020). Leveraging microlearning for just-in-time performance support. *Journal of Workplace Learning*, *32*(5), 313–328.

Chen, H., & Hsieh, C. (2021). Understanding the factors influencing users' acceptance of microlearning. *Computers in Human Behavior*, *114*, 106571.

Chen, H. L., & Wang, H. C. (2020). Evaluating the Effectiveness of Microlearning in Enhancing Employees' Technical Competencies: An Empirical Study. *Journal of Information & Knowledge Management*, *19*(02), 2050012.

Choi, Y., & Lee, S. (2019). The effects of microlearning on motivation and learning satisfaction in a mobile learning environment. *Sustainability*, *11*(21), 6003.

de Jong, M., Albers, C. J., & Koster, R. (2015). Effects of microlearning, game-based learning, and gamification on learners' motivation and learning performance. *Journal of Educational Psychology*, *107*(4), 1083–1099.

Elliott, C. (2016), https://www.hrotoday.com/news/engaged- workforce/learning/bite-sized-learning-2/

Gabrielli, S., Kimani, S., & Catarci, T. 2006.The Design of MicroLearning Experiences: A Research Agenda.https://www.researchgate.net/publication/253150976_The_Design_of_Micro Learning_Experiences_A_Research_Agenda

Hug, T. (2005). Microlearning: A New Pedagogical Challenge (Introductory Note). In Microlearning: Emerging Concepts, Practices and Technologies after eLearning. Proceeding of Microlearning 2005, Learning & Working in New Media. Book Editors: Theo Hug, Martin Lindberg, Peter A. Bruck, Innsbruck university press, 7–1

Lee, M., & Lin, Y. (2020). Game-based microlearning for improving university students' physics learning. *Computers & Education*, *154*, 103933.

Mayer, R. E., & Johnson, C. I. (2020). Microlearning for engineering education: A review. *Journal of Engineering Education*, *109*(1), 1–11.

Miller, G. A. (1956). The magical number seven, plus or minus two: Some limits on our capacity for processing information. *Psychological Review*, *63*(2), 81–97. DOI: 10.1037/h0043158 PMID: 13310704

Nandan, S., & Nandan, S. (2020). Microlearning: A Boon for Learning Coding Skills for Beginners. *Journal of Educational Technology & Society*, *23*(2), 70–82.

Rodriguez-Aflecht, G., & Fuentealba, R. (2021). The Evolution of Microlearning: A Review. *Sustainability*, *13*(2), 669.

Sağsan, M., & Özden, Y. (2021). An Overview of Microlearning: A Historical Development Perspective. [iJET]. *International Journal of Emerging Technologies in Learning*, *16*(7), 188–202.

Schmidt, A. (2007). Microlearning and the Knowledge Maturing Process:Micromedia and Corporate Learning. *Proceedings of the 3rd International Microlearning 2007*, Innsbruck, Austria, June 2007, Innsbruck University Press, 99-105.

Sharples, M., Taylor, J., & Vavoula, G. (2010). A theory of learning for the mobile age. In *Medienbildung in neuen Kulturräumen* (pp. 87–99). VS Verlag für Sozial-wissenschaften. DOI: 10.1007/978-3-531-92133-4_6

Shen, D., & Chen, W. (2019). Challenges and opportunities of microlearning: A review of the literature. *Journal of Educational Technology & Society*, *22*(3), 62–76.

Spitzer, K. L., & Lippoldt, D. (2019). Microlearning: A Review of the Literature. *Journal of Educational Technology Systems*, *47*(2), 151–172.

Teacher Education International Conference (pp. 2457-2463). Association for the Advancement of Computing in Education (AACE).

Thorne, S. L. (2017). The Evolution of Mobile Microlearning: From Personalization to Adaptive Learning. *Journal of Learning Analytics*, *4*(3), 114–122.

Torgerson, C. (2016). *The Microlearning Guide to Microlearning*. Torgerson Consulting.

Yang, S. J. H., & Lin, Y. C. (2019). A Personalized Microlearning Approach for EFL Learners: A Study of Learning Effectiveness and Learner Satisfaction. *Journal of Educational Technology & Society*, *22*(1), 103–115.

Chapter 11
Strategies for Cost–Effective Implementation of AI in Project Management:
Maximizing ROI for Entrepreneurs

Rachit Agarwal
https://orcid.org/0000-0003-0164-5292
Chandigarh University, Mohali, India

Tanya Kumar
https://orcid.org/0000-0001-7608-9820
Chandigarh Business School of Administration, Mohali, India

Ramneek Ahluwalia
https://orcid.org/0009-0003-1441-6082
Chandigarh University, Mohali. India

ABSTRACT

In the evolving landscape of project management, integrating Artificial Intelligence (AI) offers entrepreneurs unprecedented opportunities to enhance operational efficiency, mitigate risks, and achieve strategic goals. This book chapter explores key strategies to optimise return on investment (ROI) through the cost-effective implementation of AI in project management. The case study approach represents how entrepreneurs can achieve flexibility and cost savings by adopting cloud-based AI services for project planning, resource management, and performance analysis. Furthermore, the chapter explores the role of predictive analytics and machine learning algorithms in optimizing project outcomes and minimizing operational

DOI: 10.4018/979-8-3693-8009-3.ch011

expenses. Moreover, the chapter explores emerging trends and future directions in AI-driven project management, including advancements in natural language processing and autonomous project management systems. In conclusion, this book chapter provides entrepreneurs with practical strategies for achieving cost-effective implementation of AI in project management.

INTRODUCTION

In this age of lightning-fast technical development, Artificial Intelligence (AI) has emerged as a game-changer for companies of all stripes. Artificial intelligence (AI) has the potential to revolutionize project management by increasing accuracy, productivity, and flexibility. Artificial intelligence's revolutionary potential in project management allows firms to optimize decision-making, automate repetitive operations, and decrease operational inefficiencies, all while improving project results and adapting to changing market dynamics (Feng et al., 2022). Particularly at the front of this change are entrepreneurs, who are always looking for new ways to use AI to boost performance and keep costs in check. But the real test for businesses is finding the sweet spot between initial investment expenses and making the most of available resources when applying AI technologies. Artificial intelligence (AI) is becoming more important in project management as companies face more complicated settings and multidimensional issues, including limited time, resources, and budgets. With fewer resources at their disposal, entrepreneurs must make calculated bets to maximize the return on investment (ROI) from each artificial intelligence (AI) investment. In contrast to huge organizations, who have the means to test out a wide range of AI tools, entrepreneurs need to be frugal and focus on innovative but financially responsible techniques that don't break the bank (Toledano et al., 2024). Exploring best practices, techniques, and frameworks targeted at improving ROI while integrating AI technology smoothly into project management processes may be done at this confluence of AI and entrepreneurship. By giving predictive insights, optimizing resource allocation, boosting decision-making, and enhancing automation, artificial intelligence is changing the face of project management (Anomah et al., 2024). Automating mundane operations like scheduling and resource monitoring is just one of the many ways that AI-powered solutions are revolutionizing the way projects are managed. These developments improve project results by increasing productivity and freeing up project managers to concentrate on high-value activities (Kureljusic & Karger, 2024). When it comes to sectors that prioritize speed-to-market, operational efficiency, and cost reduction, entrepreneurs may gain a competitive edge by implementing AI technology (Al et al., 2024).

REVIEW OF LITERATURE

AI can help in handling accounting forecast mechanism by providing detailed insights towards solving accounting problems (Kureljusic Karger, 2024). AI leads to digital transformation in carrying out day-to-day business tasks, collaboration among resources and transforming the overall decision-making processes (Tursunbayeva & Gal, 2024). (Lehner & Knoll, 2022) discussed regarding the application of AI and Big Data in the field of accounting and auditing as what can be the opportunities and risk factors in managing the human machine collaborations in respective domain. Project cost handling is one of the key responsibilities of project managers. These days managers are attempting to utilise AI for cost minimisation by doing cash flow predictions in advance (Cheng & Roy, 2011). (Zema et al., 2022) discussed upon deep and machine learning that can enable the forecasting of opportunity costs having deviations in actual values. (Shete et al., 2024) researched upon the cybersecurity threats of Meesho company as well as cryptocurrency trends that can lead to cyber fraud. (Vegar & Mijač, 2024) stated the utilisation of AI in project management in terms of aiding its decision-making process. (Thirumagal et al., 2024) stated the usage of AI in handling the risk management in banking operations and how it can result in fraud detection by usage of AI as innovative approach of cost saving. (Lai et al., 2024) explored the usage of AI in terms of sustainability of investment risk management. (Ifeanyichukwu, 2024) stated the usage of AI in different sectors of the economy and businesses are adopting them in the form of different techniques like VR, AR, IoT, DaaS and so on. (Salleh et al., 2024) provided their views on the concept of augmented project management by facilitating managerial decisions based on information technology. AI provides cost effective tools in handling manufacturing projects and project managers can track the cost consecutively (Sahli et al., 2023). AI enables the managers to think critically as well as effective communication in fostering qualitative inputs in projects (Gaines & Balac, 2000). (Bayraktar et al., 2011) provided necessary factors that are to be taken care of in taking project related decisions in minimising the project cost and schedule. (Ayub et al., 2016) discussed the concept of cost contingency in project execution for managing the risk factors for future managerial performance. With the advancement in AI, the workforce of the project can also be handled in cost management process (Rathod & Sonawane, 2022). With the usage of AI, management mechanism can be effectively handled for achieving business goals as comparatively with that in context of human intelligence (Kobbacy, 2012). (Wang, 2023) combined the usage of AI on project management decision making. Authors also utilised data driven models for solving project related problems as per the quality of decision making in project. (Bakhtawar et al., 2019) developed a model for risk mitigation for sustainable development in fostering in-novation for handling large infrastructure or projects in risk assessment. (Dikmen et al., 2006) provided their insight on integrated mechanism of risk management used as a tool for decision support resulting in cost estimation. Authors employed Monte Carlo Simulation technique to evaluate the risk scenario by improving their performance. (Xue et al., 2009) suggested upon product development cost result in lead time and reliability in product development support in facilitating the decision-making process. (Gibson et al., 2005) addressed the current mechanism of business intelligence and its usage in IT for decision making support and how these can be utilised for large scale operational development. (Chaidarun et al., 2015) stated that how portfolio management skills can generate the simulated option and stock prices for handling the trading platforms. (Prasad et al., 2024) stated the complexity of handling the public private partnership projects and its contribution towards infra-structural development by making effective usage of different resources. (Serrano, 2022) discussed various challenges faced in handling the physical infrastructure of a project as how it can lead to agile project management based upon artificial intelligence. Distributed ledger technology is helps to maintain the data of project and its users. (Robertson, 2022) stressed upon predicting the outcomes of projects by taking into consideration the professional review as how they forecast and accurately handle the business-related problems.

TRANSFORMATIVE IMPACT OF AI ON PROJECT MANAGEMENT PRACTICES

Businesses are rethinking their methods of planning, executing, and making decisions in light of the revolutionary effects of AI on project management. Artificial intelligence (AI) offers strong tools that improve efficiency, simplify procedures, and promote wiser decision-making, which is crucial for enterprises dealing with the growing complexity of project management (Kotowska & Sikorska, 2023). It revolutionizes the field of project management by freeing up managers to concentrate on strategic endeavors through the automation of mundane duties, provision of predictive insights, and optimization of resource allocation. Decision-making is further improved by AI, which helps project managers make better choices by delivering data-driven insights. Machine learning techniques allow AI to sift through mountains of data, spot trends, and make suggestions that improve budget forecasts, work prioritization, and risk assessment (Rahim & Chishti, 2024). Predictive analytics solutions powered by AI, for instance, may examine past data from comparable projects to spot possible hazards or delays. Project managers can now stay on top of issues before they get out of hand, which keeps projects on schedule and within budget (Hu, 2022). Additionally, AI greatly enhances resource management, which is essential for the effective completion of projects. Artificial intelligence (AI) can improve resource allocation by forecasting demand, finding bottlenecks, and making real-time adjustments (Secinaro et al., 2024). This is especially important when dealing with human, financial, or material resources. In order to maximize efficiency, machine learning models may evaluate team members' workloads, monitor their performance, and recommend the best methods to allocate resources. Entrepreneurs and small enterprises, who typically face major challenges due to limited resources, can greatly benefit from this optimization. Artificial intelligence (AI) helps keep projects on track, under budget, and free of team member fatigue by facilitating more efficient resource allocation (Heji et al., 2023). AI is already revolutionizing project management with its predictive insights and ability to foretell outcomes. One of AI's strongest suits is predictive analytics, which can tell you when a project will finish, how much it will cost, and how to avoid problems. With this kind of planning ahead, project managers may better distribute resources, anticipate and prepare for potential problems, and modify timetables to minimize delays. Project results are therefore more predictable, and teams are able to attain greater success rates through proactive risk mitigation (Crawford & Nilsson, 2023). Tasks like scheduling, reporting, and communication may be automated with the help of AI. While administrative work is essential, project managers frequently waste a lot of time on it since it does not contribute to the project's overall strategy (Boffa, 2023). Chatbots and automated scheduling systems, both driven by AI, can take care of these menial jobs, allowing managers more time to concentrate on strategic planning and managing stakeholders (Kanakov & Prokhorov, 2022). There is a marked decrease in human error and an increase in total project efficiency as a result of this automation. The capacity of AI to sift through massive datasets in search of patterns and make predictions is one of its most valuable contributions to the field of project management (Radhi et al., 2024). In order to foretell possible dangers, project delays, or budget overruns, predictive analytics driven by AI may examine past project data (Mishra et al., 2024). Entrepreneurs greatly benefit from this talent since it allows them to anticipate problems and take proactive measures to resolve them before they escalate into expensive problems (Baiod & Hussain, 2024). Efficient risk mitigation, appropriate resource allocation, and deadline ad-herence are all guaranteed by accurate forecasting. The key to a successful project is making the most of all available resources, whether they be human, monetary, or material. Algorithms powered by AI can examine patterns of resource distribution and make suggestions for making the most of them. For instance, AI can find team members who aren't getting enough work or offer suggestions for job redistribution to eliminate bottlenecks. By adjusting resources on the fly, entrepreneurs may keep projects on schedule and under budget while still meeting the evolving demands of their ventures (Rodgers et al., 2022). Data analysis, evaluation of various factors, and outcome prediction are commonplace in project management decision-making. By offering insights and suggestions based on data, AI helps in this process. Machine learning models can analyze various project scenarios, predict their consequences, and recommend the best course of action. As a result, project managers are able to make better judgments, which boosts success rates and decreases the chances of expensive mistakes (Chaturvedi & Raja Mohammed, 2024).

CHALLENGES IN AI IMPLEMENTATION FOR ENTREPRENEURS

When it comes to using AI for project management, entrepreneurs encounter a number of obstacles, the most common of which are related to expenses, scalability, skill shortages, and system integration. A lot of money, time, and effort goes into building or buying AI systems, as well as the necessary hardware, software, and training. Entrepreneurs, particularly those operating SMEs, may find these upfront expenses to be too high, which hinders their capacity to properly integrate AI (Gonçalves et al., 2022). A company's AI systems need to be scalable so they can expand with the company. But conventional on-premise AI solutions frequently can't change with the demands of businesses without spending a fortune more. For their businesses to grow, entrepreneurs want solutions that can scale up or down according to their needs. Another major obstacle that businesses face is the skills gap. Data scientists, ML experts, and AI system administrators are essential for any AI project. Smaller companies often have challenges when it comes to finding or affording the right people to properly implement and manage AI technology (Qasim et al., 2021). The usefulness of artificial intelligence (AI) tools might be diminished if the right people aren't on staff to make the most of them. Lastly, there is always the matter of integrating with current systems. Unfortunately, many companies are still using antiquated project management tools that aren't always compatible with AI. Making the necessary changes to current procedures to incorporate AI into these systems may be a lengthy and expensive process (Yang et al., 2019. To make sure AI improves rather than interrupts their project management procedures, entrepreneurs must cautiously traverse these integration hurdles. Taken as a whole, these obstacles show how important it is to allocate resources strategically and prepare ahead for AI adoption if we want it to be effective and affordable. AI technology, especially when built and implemented in-house, might demand a hefty sum up front (Hoque, 2017). It might be challenging for entrepreneurs to justify big investments in AI without first establishing a clear picture of the possible return on investment (ROI). For instance, it may be rather costly, particularly for smaller organizations, to construct AI systems that need extensive infrastructure, data processing skills, and continuous maintenance (Kachroo et al., 2020). In addition, there is an initial cost associated with AI adoption due to the investment needed to teach people on how to utilize these systems efficiently. Another important issue for entrepreneurs is scalability (Maione & Leoni, 2021). Artificial intelligence solutions should be able to adapt to the changing demands of project management as companies expand. It may be rather costly to react to these shifting needs with traditional on-premises AI systems since they lack the scalability necessary. Artificial intelligence solutions should be scalable and cost-effective for entrepreneurs. Small and medium-sized businesses (SMEs) sometimes lack the specialized technical skills necessary to effectively adopt artificial intelligence (AI). Finding competent people to design, build, and oversee AI-powered project management solutions could be challenging for entrepreneurs. Underutilized technology and poor ROI can occur when firms fail to attract and retain the necessary workforce, which makes it difficult to fully exploit AI's possibilities (Stancheva & Bogdanova, 2021). AI solutions with current project management tools and procedures should be a top priority (Ashley, 2020). Legacy systems that were not built to be AI compatible can make it even more difficult to integrate modern AI technology into them. In order to keep project workflows running smoothly and make sure that the new AI technologies enhance current processes instead of making them more complicated, entrepreneurs need to meticulously prepare for these integrations (Xiaohua et al., 2021).

STRATEGIES FOR COST-EFFECTIVE AI IMPLEMENTATION

When it comes to using AI for project management, entrepreneurs encounter a number of obstacles, the most common of which are related to expenses, scalability, skill shortages, and system integration. A lot of money, time, and effort goes into building or buying AI systems, as well as the necessary hardware, software, and training. Entrepreneurs, particularly those operating SMEs, may find these upfront expenses to be too high, which hinders their capacity to properly integrate AI. A company's AI systems need to be scalable so they can expand with the company (Brands & Elam, 2015). But conventional on-premise AI solutions frequently can't change with the demands of businesses without spending a fortune more. For their businesses to grow, entrepreneurs want solutions that can scale up or down according to their needs. Another major obstacle that businesses face is the skills gap (Yang et al., 2021). Data scientists, ML experts, and AI system administrators are essential for any AI project. Smaller companies often have challenges when it comes to finding or affording the right people to properly implement and manage AI technology. The usefulness of artificial intelligence (AI) tools might be diminished if the right people aren't on staff to make the most of them. Lastly, there is always the matter of integrating with current systems (Robertson, 2022). Unfortunately, many companies are still using antiquated project management tools that aren't always compatible with AI. Making the necessary changes to current procedures to incorporate AI into these systems may be a lengthy and expensive process. To make sure AI improves rather than interrupts their project management procedures, entrepreneurs must cautiously traverse these integration hurdles. Taken as a whole, these obstacles show how important it is to allocate resources strategically and prepare ahead for AI adoption if we want it to be effective and affordable (Serrano, 2022). AI technology, especially when built and implemented in-house, might demand a hefty sum up front. It might be challenging for entrepreneurs to justify big investments in AI without first establishing a clear picture of the possible return on investment (ROI). For instance, it may be rather costly, particularly for smaller organizations, to construct AI systems that need extensive infrastructure, data processing skills, and continuous maintenance. In addition, there is an initial cost associated with AI adoption due to the investment needed to teach people on how to utilize these systems efficiently. Another important issue for entrepreneurs is scalability. Artificial intelligence solutions should be able to adapt to the changing demands of project management as companies expand. It may be rather costly to react to these shifting needs with traditional on-premises AI systems since they lack the scalability necessary. Artificial intelligence solutions should be scalable and cost-effective for entrepreneurs. Small and medium-sized businesses (SMEs) sometimes lack the specialized technical skills necessary to effectively adopt artificial intelligence (AI). Finding competent people to design, build, and oversee AI-powered project management solutions could be challenging for entrepreneurs (Prasad et al., 2024). Underutilized technology and poor ROI can occur when firms fail to attract and retain the necessary workforce, which makes it difficult to fully exploit AI's possibilities. How to Integrate with Current Systems: Integrating AI solutions with current project management tools and procedures should be a top priority. Legacy systems that were not built to be AI compatible can make it even more difficult to integrate modern AI technology into them (Chaidarun et al., 2015). In order to keep project workflows running smoothly and make sure that the new AI technologies enhance current processes instead of making them more complicated, entrepreneurs need to meticulously prepare for these integrations.

ORGANIZATIONAL READINESS AND CHANGE MANAGEMENT

Entrepreneurs need to make sure their companies are prepared to use AI if they want to integrate it into project management. Checking if the company has the resources, personnel, and backing from upper management to effectively deploy AI is what we mean when we talk about organizational readiness. To make sure workers are on board, educated, and ready to use AI-driven solutions, change management must also be a priority. In order to effectively manage change, it is important to match stakeholder expectations, encourage a culture of innovation, and provide thorough training to all team members so that they can use AI technologies efficiently. Successful AI adoption and increased return on investment (ROI) are more probable for entrepreneurs who put these factors first. How prepared an organization is to use artificial intelligence (AI) depends on how well it can integrate and use AI technology into its daily operations (Gibson et al., 2005). Organizations should consider their readiness to embrace AI-related changes in terms of infrastructure, skills, leadership buy-in, and cultural receptivity before making the leap. If the organization wants to incorporate AI technologies without causing major disruptions to its IT systems, it must first determine if its present technical infrastructure is strong and adaptable enough to do so (Salleh et al., 2024). For example, in order to meet the requirements of AI-driven processes, it may be necessary to modify or replace existing systems. In addition, businesses should assess their preparedness for the workforce. Data scientists, machine learning experts, and AI system managers are typically needed for AI implementation, although they might not be easily available on the current team. To fill this knowledge vacuum, it is necessary to engage in training programs, recruit qualified candidates, or work with outside AI specialists (Vidmar et al., 2023). The support of upper-level management is also essential. The best way to get everyone on board with AI projects is for upper management to not just provide their stamp of approval, but to really push for them. Appropriate resources should be set aside for the adoption of AI, and the required organizational reforms should be driven by strong leadership. The degree to which a company is culturally prepared to accept and even welcome change and innovation is just as critical. Unless a culture of innovation and constant learning is fostered, people may be resistant to the changes brought forth by AI, which disrupt old workflows (Xue et al., 2009) (Ifeanyichukwu, 2024). Businesses may help employees adjust by promoting an attitude that sees AI more as an enhancing tool than a replacement. Thus, in order for an organization to be prepared to implement AI, it must first guarantee that it has the required resources, personnel, support from upper management, and flexibility to adapt to new ways of doing things. The capacity of the firm to effectively integrate AI technology and reap their advantages is determined by all of these elements taken together. The preparedness of a business to embrace new technology should be evaluated by entrepreneurs prior to deploying AI (Geraldi et al., 2024). The technical infrastructure, the capabilities of the staff, and the buy-in of leadership are crucial aspects to assess. The technological infrastructure refers to the existing information technology systems and how well they work with AI solutions. Team members' AI knowledge and the necessity of outside talent or training programs are examples of workforce skills. Leadership Buy-In guarantees that top-level executives are prepared to fund and back AI-driven projects. For artificial intelligence (AI) projects to be a success, change management tactics are essential. Stakeholder involvement should be a primary goal for entrepreneurs. They should include stakeholders early on to align expectations and eliminate opposition (Sánchez et al., 2024). To guarantee a seamless integration, training programs involve offering AI-specific training to team members and project managers (Dikmen et al., 2006). Teams are encouraged to experiment with AI-driven tools and processes via cultural adaption, which leads to a culture of innovation and continual development (Tursunbayeva & Gal, 2024).

CASE STUDIES

The following case studies will provide the insights towards various companies who have applied cloud-based AI service to carry on their operations and what are the specific outcomes attained from its application are listed in detail in the Table 1.

Table 1. Case studies

Company Name	Entrepreneur	Cloud-Based AI Service Used	Application	Benefits Achieved	Outcome
Zoom Video Communications	Eric Yuan, Founder	Google Cloud AI	Resource management and demand forecasting	AI-driven analytics to optimize server capacity during high traffic	Cost savings on infrastructure by using flexible cloud services, enhanced performance
Lyft	Logan Green, Co-founder	Google Cloud Machine Learning	Project planning for fleet management and predictive maintenance	AI predictive analytics for real-time fleet maintenance needs	Cost reduction in operations & optimized fleet deployment through preventive maintenance
Grammarly	Alex Shevchenko, Founder	Microsoft Azure AI	Performance analysis for product development	AI-driven feedback on user interaction & usage patterns	Flexibility in product improvements, cost savings through accurate market insights

continued on following page

Table 1. Continued

Company Name	Entrepreneur	Cloud-Based AI Service Used	Application	Benefits Achieved	Outcome
HubSpot	Brian Halligan & Dharmesh Shah, Founders	AWS AI & Amazon SageMaker	Customer resource management (CRM) system automation	AI to streamline customer data management & lead generation processes	Enhanced resource allocation, cost savings on data management and marketing
Netflix	Reed Hastings, Co-founder	AWS AI & Amazon S3	Content recommendation and resource optimization	AI-driven content analytics for user behavior and preference analysis	Reduced content delivery costs, improved user engagement through personalized recommendations

(Source: Authors' Compilation)

The corresponding companies and their entrepreneurs as such as Zoom video communications, Lyft, Grammarly, HubSpot and Netflix. Zoom leveraged AI-driven analytics to optimize server capacity during high-traffic periods, such as the surge in demand experienced during the global pandemic. Zoom avoided overcommitting resources by dynamically scaling its infrastructure to meet demand utilizing cloud-based AI technologies. This case study demonstrates how cloud-based AI is essential for effective infrastructure management, particularly for businesses dealing with significant growth or unpredictable demand. Optimized performance at a manageable cost was achieved by Zoom through its scalability enabled by AI-driven resource management. To keep tabs on its fleet and foresee when repairs will be necessary, Lyft deployed predictive analytics driven by artificial intelligence. By taking this preventative measure, we were able to save operating expenses while increasing the efficiency of fleet deployment and decreasing vehicle downtime. Cloud-based machine learning may have a direct influence on operational efficiency, as demonstrated by Lyft's use of AI for predictive maintenance (Kusmin et al., 2024). Lyft improved their fleet management and cut maintenance expenses by anticipating impending problems and allocating resources accordingly. In order to enhance its products on an ongoing basis, Grammarly used AI-driven input on user interactions and use trends. Grammarly was able to make data-driven judgments about product improvements and new additions by studying user interactions with the platform. The case study of Grammarly highlights the power of cloud-based AI to analyze user behavior, leading to product advancements and enhanced customer satisfaction (Lai et al., 2024). Additionally, this exemplifies the adaptability and cost-control that AI gives to product creation. AI was used by HubSpot to automate lead creation and customer data management. AI solutions made consumer data processing quicker and more precise, allowing humans to focus on more strategic work. One example of how cloud-based AI may improve marketing and sales productivity, especially in situations with limited resources, is HubSpot's use of AI in CRM. With the help of automation, the firm was able to boost performance and save a ton of money on client acquisition. Netflix optimized their content recommendation engine by analyzing user preferences and behavior using AI-driven content analytics (Kureljusic Karger, 2024). As a result, Netflix was able to increase user engagement and retention through the delivery of highly tailored content. The effectiveness of cloud-based AI in improving the user experience and decreasing operational expenses is illustrated by Netflix's usage of AI for content customisation. Netflix improved engagement and return on investment by allocating resources more effectively for content delivery.

ETHICAL CONSIDERATIONS IN AI IMPLEMENTATION

As AI technologies increasingly permeate project management, ethical issues like data protection, algorithmic transparency, and bias reduction must be prioritized in AI adoption methods. Entrepreneurs must guarantee that AI systems adhere to pertinent data protection rules, such as the General Data Protection Regulation (GDPR), to prevent legal issues. Moreover, openness in the decision-making processes of AI systems is essential for fostering confidence among stakeholders and guaranteeing equitable, impartial results. Entrepreneurs must verify that AI systems adhere to data privacy standards (Dikmen et al., 2006). AI systems frequently handle substantial volumes of sensitive data, necessitating the implementation of stringent data protection protocols (Roberts & Candi, 2024). AI algorithms must exhibit transparency and be devoid of biases that may result in immoral decision-making. Ensuring transparency necessitates the documentation of decision-making processes in AI systems and the implementation of bias mitigation methods, including the auditing of algorithms for fairness (Yi & Luo, 2024). An effective governance structure is essential to guarantee that AI-driven project management processes adhere to legal and ethical requirements. Entrepreneurs must establish governance frameworks to regulate AI utilization, ensure compliance, and tackle ethical dilemmas as they emerge (Bakhtawar et al., 2019).

EMERGING TRENDS AND FUTURE DIRECTIONS

The integration of Artificial Intelligence (AI) into project management has rapidly transformed the way businesses operate, enabling more efficient, data-driven, and automated processes. As AI technologies continue to evolve, several emerging trends are reshaping project management practices, offering new opportunities for innovation, efficiency, and improved outcomes (Cheng & Roy, 2011) (Müller et al., 2024). One of the most significant trends in AI-driven project management is the increasing use of Natural Language Processing (NLP) for communication and reporting automation. NLP enables AI systems to understand and interpret human language, making it possible for project management tools to process large amounts of unstructured data from emails, reports, and meeting notes (Wang, 2023). AI-powered chatbots and virtual assistants are now able to handle routine inquiries, send updates, and even assist in managing team communications (Gaines & Balac, 2000). This automation reduces the time project managers spend on administrative tasks, enabling them to focus on higher-level strategic work (Hamid et al., 2020). Additionally, NLP-driven tools are enhancing real-time collaboration, improving the speed and clarity of communication between stakeholders. Another emerging trend is the development of autonomous project management systems that can handle end-to-end project workflows with minimal human intervention. These AI-driven systems use machine learning algorithms to manage project timelines, allocate resources, and monitor progress autonomously. Autonomous systems continuously learn from past project data, allowing them to make real-time adjustments to schedules, budgets, and resources based on current project conditions. This dynamic decision-making helps prevent delays and budget overruns, leading to more successful project outcomes (Kobbacy, 2012). As these systems become more sophisticated, they are expected to take on more complex tasks, further reducing the need for manual project management. Predictive analytics is also playing a growing role in AI-driven project management (Pishdad & Onungwa, 2024). Predictive models analyze historical project data to forecast future risks, delays, and performance issues (Bayraktar et al., 2011). By identifying potential problems early, project managers can proactively address issues before they impact project success. For entrepreneurs, predictive analytics can significantly improve decision-making by providing accurate forecasts on project timelines, resource requirements, and financial outcomes. This trend enables businesses to make data-driven decisions that enhance efficiency and reduce costs, ultimately increasing their return on investment (ROI) (Sahli et al., 2023). In addition to these advancements, AI integration with the Internet of Things (IoT) is emerging as a powerful tool for project management. IoT devices collect real-time data from various sources such as construction sites, manufacturing plants, or remote project locations (Ahuja et al., 2024). AI algorithms can analyze this data to monitor equipment usage, environmental conditions, and employee performance, providing project managers with actionable insights (Thirumagal et al., 2024). This integration allows for real-time adjustments to project plans based on the actual conditions in the field, improving both project accuracy and resource allocation. Another trend to watch is the rise of AI-enhanced project risk management (Rathod & Sonawane, 2022). AI systems are becoming increasingly capable of identifying and mitigating risks through continuous monitoring and real-time analysis of project data (Felemban et al., 2024). Machine learning algorithms can detect patterns that may indicate potential risks, such as scope creep, resource constraints, or schedule delays (Vegar & Mijač, 2024). This early detection allows project managers to implement risk mitigation strategies before issues escalate, thus safeguarding project success (Ayub et al., 2016) (Hentati & Boulila, 2024). Finally, ethical AI implementation is gaining attention as businesses become more aware of the need to ensure transparency, fairness, and accountability in AI-driven project management. With concerns about data privacy, algorithmic bias, and the fairness of AI decision-making processes, businesses are prioritizing the development of ethical guidelines and governance frameworks for AI usage (Aggrawal & Thomas, 2024). Ensuring that AI systems are transparent and free from bias is essential to building trust among stakeholders and ensuring responsible project management.

CONCLUSION

The integration of AI in project management provides entrepreneurs substantial opportunity to improve operational efficiency and optimize ROI. By using economical tactics like utilizing cloud-based AI platforms and employing predictive analytics, firms may save expenses while gaining advantages from AI. A robust emphasis on organizational preparedness, ethical implications, and future trends will facilitate sustainable and responsible AI implementation. Entrepreneurs that adeptly use these methods will enhance their ROI and position their enterprises for sustained development and innovation in a progressively AI-driven environment (Qiao & Wang, 2012). The economic integration of AI in project management offers entrepreneurs a distinctive potential to augment productivity, optimize procedures, and boost ROI. Nonetheless, this necessitates meticulous evaluation of the obstacles, tactics, and ethical ramifications associated with AI implementation. Entrepreneurs may effectively incorporate AI into their project management methods by utilizing cloud-based solutions, predictive analytics, and emphasizing organizational preparedness, all while ensuring cost-efficiency. The combination of AI with IoT, risk management, and ethical AI is transforming the field of project management. These technologies provide organizations, particularly entrepreneurs, the capacity to enhance their operations, minimize expenses, and elevate project results. As artificial intelligence progresses, its influence on project management will expand, fostering more creativity and efficiency across all sectors. This economical strategy guarantees that enterprises not only fulfil their immediate project management objectives but also prepare for sustained success in a progressively AI-oriented market.

REFERENCES

Aggrawal, S., & Thomas, P. J. (2024). Investigating the Industry Perceptions and Use of AI Tools in Project Management: Implications for Educating Future Engineers. In *2024 ASEE Annual Conference & Exposition*. DOI: 10.18260/1-2--47700

Ahuja, L., Thakur, A., Seth, A., & Seth, K. (2023). Integrating Cloud, Blockchain and AI Technologies—Challenges and Scope. In *International Conference on Entrepreneurship, Innovation, and Leadership* (pp. 377-386). Singapore: Springer Nature Singapore.

Al Wael, H., Abdallah, W., Ghura, H., & Buallay, A. (2023). Factors influencing artificial intelligence adoption in the accounting profession: The case of public sector in Kuwait. *Competitiveness Review*, *34*(1), 3–27. DOI: 10.1108/CR-09-2022-0137

Anomah, S., Ayeboafo, B., Owusu, A., & Aduamoah, M. (2024). Adapting to AI: exploring the implications of AI integration in shaping the accounting and auditing profession for developing economies. *EDPACS*, 1-25.

Ashley, K. D. (2020). Accounting for legal values. In *Computational Legal Studies* (pp. 190–214). Edward Elgar Publishing. DOI: 10.4337/9781788977456.00014

Ayub, B., Thaheem, M. J., & Din, Z. (2016). Dynamic management of cost contingency: Impact of KPIs and risk perception. *Procedia Engineering*, *145*, 82–87. DOI: 10.1016/j.proeng.2016.04.021

Baiod, W., & Hussain, M. M. (2024). The impact and adoption of emerging technologies on accounting: perceptions of Canadian companies. *International Journal of Accounting & Information Management*.

Bakhtawar, B., Thaheem, M. J., & Arshad, H. (2019). Integrating sustainability into project risk management; an application in PPP projects. In *Proceedings* (Vol. 2019). Annual Conference-Canadian Society for Civil Engineering.

Bayraktar, M. E., Hastak, M., Gokhale, S., & Safi, B. (2011). Decision tool for selecting the optimal techniques for cost and schedule reduction in capital projects. *Journal of Construction Engineering and Management*, *137*(9), 645–655. DOI: 10.1061/(ASCE)CO.1943-7862.0000345

Boffa, J. (2023). *AI Assisted Business Analytics: Techniques for Reshaping Competitiveness*. Springer Nature. DOI: 10.1007/978-3-031-40821-2

Brands, K., & Elam, D. (2015). Identifying quality enablers for online graduate accounting courses using an appreciative inquiry case study. *International Journal of Human Resources Development and Management*, *15*(2-4), 128–141. DOI: 10.1504/IJHRDM.2015.071164

Chaidarun, N., Tepsuporn, S., Hayes, R., Beling, P., Scherer, W., & Grazioli, S. (2014, December). Computational Intelligence in Financial Engineering Trading Competition: A system for project-based learning. [IEEE.]. *Proceedings of the ... Winter Simulation Conference. Winter Simulation Conference*, *2014*, 3552–3560. DOI: 10.1109/WSC.2014.7020185

Chaturvedi, V., & Raja Mohammed, K. B. N. (2023). Dynamic Inventory Management Using AI: A Case on Datarobot. In *International Conference on Artificial Intelligence and Knowledge Processing* (pp. 3-14). Cham: Springer Nature Switzerland.

Cheng, M. Y., & Roy, A. F. (2011). Evolutionary fuzzy decision model for cash flow prediction using time-dependent support vector machines. *International Journal of Project Management*, *29*(1), 56–65. DOI: 10.1016/j.ijproman.2010.01.004

Crawford, J., & Nilsson, F. (2023). Integrating ESG risks into control and reporting: Evidence from practice in Sweden. In *Handbook of Big Data and Analytics in Accounting and Auditing* (pp. 255–277). Springer Nature Singapore. DOI: 10.1007/978-981-19-4460-4_12

Cristofoli, C., & Clemmensen, T. (2023). Underlying Factors of Technology Acceptance and User Experience of Machine Learning Functions in Accounting Software: A Qualitative Content Analysis. In *International Conference on Human-Computer Interaction* (pp. 413-433). Cham: Springer Nature Switzerland. DOI: 10.1007/978-3-031-48060-7_31

Dikmen, I., Birgonul, M. T., & Arikan, A. E. (2006). Application of An Integrated Risk Management System (IRMS) to An International Construction Project. *Management*, *153*, 163.

Felemban, H., Sohail, M., & Ruikar, K. (2024). Exploring the Readiness of Organisations to Adopt Artificial Intelligence. *Buildings (Basel, Switzerland)*, *14*(8), 2460. DOI: 10.3390/buildings14082460

Feng, X., Conrad, M., & Hussein, K. (2022). NHS big data intelligence on Blockchain applications. In *Big Data Intelligence for Smart Applications* (pp. 191–208). Springer International Publishing. DOI: 10.1007/978-3-030-87954-9_8

Gaines, D., & Balac, N. (2000). Using mobile robots to teach artificial intelligence research skills. In *Proceedings of the 2000 ASEE Annual Conference*.

Geraldi, J., Locatelli, G., Dei, G., Söderlund, J., & Clegg, S. (2024). AI for Management and Organization Research: Examples and Reflections from Project Studies. *Project Management Journal*, *55*(4), 339–351. DOI: 10.1177/87569728241266938

Gibson, M. C., & Arnott, D. R. (2005). The evaluation of business intelligence: A case study in a major financial institution. In *Australasian Conference on Information Systems 2005* (pp. 1-12). Australasian Chapter of the Association for Information Systems.

Gonçalves, M. J. A., da Silva, A. C. F., & Ferreira, C. G. (2022). The future of accounting: How will digital transformation impact the sector? [). MDPI.]. *Informatics (MDPI)*, *9*(1), 19. DOI: 10.3390/informatics9010019

Hamid, M., Zeshan, F., Ahmad, A., Munawar, S., Aimeur, E., Ahmed, S., Abu Elsoud, M., & Yousif, M. (2020). An intelligent decision support system for effective handling of IT projects. *Journal of Intelligent & Fuzzy Systems*, *38*(3), 2635–2647. DOI: 10.3233/JIFS-179550

Heji, A. E., Alansari, O. E., & Al-Sartawi, A. (2023). Artificial intelligence and its impact on accounting systems. In *Artificial Intelligence, Internet of Things, and Society 5.0* (pp. 363–376). Springer Nature Switzerland. DOI: 10.1007/978-3-031-43300-9_30

Hentati, H., & Boulila, N. (2023). Digital maturity index for accounting firms. *Journal of Accounting & Organizational Change*, (ahead-of-print).

Hoque, Z. (2017). Appreciative inquiry for accounting research. In *The Routledge Companion to Qualitative Accounting Research Methods* (pp. 129–144). Routledge. DOI: 10.4324/9781315674797

Ifeanyichukwu, E. E. (2024). Technological Implementation in the Service Sector: A Case Study. In *Artificial Intelligence for Smart Technology in the Hospitality and Tourism Industry* (pp. 305-336). Apple Academic Press. DOI: 10.1201/9781003432951-18

Kachroo, P., Saiewitz, A., Raschke, R., Agarwal, S., & Huang, A. J. (2020). A New Language and Input–Output Hidden Markov Model for Automated Audit Inquiry. *IEEE Intelligent Systems*, *35*(6), 39–49. DOI: 10.1109/MIS.2019.2963653

Kanakov, F., & Prokhorov, I. (2022). Analysis and applicability of artificial intelligence technologies in the field of RPA software robots for automating business processes. *Procedia Computer Science*, *213*, 296–300. DOI: 10.1016/j.procs.2022.11.070

Kobbacy, K. A. (2012). Application of artificial intelligence in maintenance modelling and management. *IFAC Proceedings Volumes, 45*(31), 54-59.

Kotowska, B., & Sikorska, M. (2023). Digital transformation of a Polish accounting firm: Tools, impediments, business performance benefits and implications–case study. *Procedia Computer Science*, *225*, 327–336. DOI: 10.1016/j.procs.2023.10.017

Kureljusic, M., & Karger, E. (2023). Forecasting in financial accounting with artificial intelligence–A systematic literature review and future research agenda. *Journal of Applied Accounting Research*, (ahead-of-print).

Kureljusic, M., & Karger, E. (2023). Forecasting in financial accounting with artificial intelligence–A systematic literature review and future research agenda. *Journal of Applied Accounting Research*, (ahead-of-print).

Kusmin, K. L., Normak, P., & Ley, T. (2024). A Methodology for Planning, Implementation and Evaluation of Skills Intelligence Management-Results of a Design Science Project in Technology Organisations. *Frontiers in Artificial Intelligence*, *7*, 1424924. DOI: 10.3389/frai.2024.1424924 PMID: 39169913

Lai, S., Zhang, S., Hassan, A., & Mushtaq, R. T. (2024). Analyzing Role of Artificial Intelligence in Project Management and Investment Risk: A CiteSpace Insight. In *Proceedings of the 2024 16th International Conference on Machine Learning and Computing* (pp. 713-719). DOI: 10.1145/3651671.3651776

Lehner, O. M., & Knoll, C. (2022). *Artificial Intelligence in Accounting*. Routledge. DOI: 10.4324/9781003198123

Maione, G., & Leoni, G. (2021). Artificial intelligence and the public sector: the case of accounting. In *Artificial Intelligence and Its Contexts: Security, Business and Governance* (pp. 131–143). Springer International Publishing. DOI: 10.1007/978-3-030-88972-2_9

Mishra, R. K., Srivastava, S., Singh, S., & Upadhyay, M. K. (2024). Exploring the Opportunities of AI Integral with DL and ML Models in Financial and Accounting Systems. In *2024 4th International Conference on Advance Computing and Innovative Technologies in Engineering (ICACITE)* (pp. 999-1003). IEEE. DOI: 10.1109/ICACITE60783.2024.10616847

Müller, R., Locatelli, G., Holzmann, V., Nilsson, M., & Sagay, T. (2024). Artificial intelligence and project management: Empirical overview, state of the art, and guidelines for future research. *Project Management Journal*, *55*(1), 9–15. DOI: 10.1177/87569728231225198

Pishdad, P., & Onungwa, I. O. (2024). ANALYSIS OF 5D BIM FOR COST ESTIMATION, COST CONTROL, AND PAYMENTS. [ITcon]. *Journal of Information Technology in Construction*, *29*(24), 525–548. DOI: 10.36680/j.itcon.2024.024

Prasad, K. R., Karanam, S. R., Ganesh, D., Liyakat, K. K. S., Talasila, V., & Purushotham, P. (2024). AI in public-private partnership for IT infrastructure development. *The Journal of High Technology Management Research*, *35*(1), 100496. DOI: 10.1016/j.hitech.2024.100496

Qasim, A., El Refae, G. A., Issa, H., & Eletter, S. (2021). The impact of drone technology on the accounting profession: the case of revenue recognition in long-term construction contracts. In *2021 22nd International Arab Conference on Information Technology (ACIT)* (pp. 1-4). IEEE. DOI: 10.1109/ACIT53391.2021.9677226

Qiao, C., & Wang, H. (2012). The Further Research on the Application of ABC to the Optimization and Control of Project. *Engineering Management Research*, *1*(2), 96. DOI: 10.5539/emr.v1n2p96

Radhi, W. A., Hamdan, A., & Binsaddig, R. (2024). Assessing the Role of Artificial Intelligence (AI) on Tax Fraud Detection. In *Business Development via AI and Digitalization* (Vol. 1, pp. 359–364). Springer Nature Switzerland. DOI: 10.1007/978-3-031-62102-4_30

Rahim, R., & Chishti, M. A. (2024). Artificial Intelligence Applications in Accounting and Finance. In *2024 ASU International Conference in Emerging Technologies for Sustainability and Intelligent Systems (ICETSIS)* (pp. 1782-1786). IEEE. DOI: 10.1109/ICETSIS61505.2024.10459526

Rathod, K., & Sonawane, A. (2022). Application of Artificial Intelligence in Project Planning to Solve Late and Over-Budgeted Construction Projects. In *2022 International Conference on Sustainable Computing and Data Communication Systems (ICSCDS)* (pp. 424-431). IEEE. DOI: 10.1109/ICSCDS53736.2022.9761027

Roberts, D. L., & Candi, M. (2024). Artificial intelligence and innovation management: Charting the evolving landscape. *Technovation*, *136*, 103081. DOI: 10.1016/j.technovation.2024.103081

Robertson, A. (2022). Predicting Project Outcomes with the Association of Project Management. In *Abu Dhabi International Petroleum Exhibition and Conference* (p. D031S081R003). SPE. DOI: 10.2118/210795-MS

Rodgers, W., Degbey, W. Y., Söderbom, A., & Leijon, S. (2022). Leveraging international R&D teams of portfolio entrepreneurs and management controllers to innovate: Implications of algorithmic decision-making. *Journal of Business Research*, *140*, 232–244. DOI: 10.1016/j.jbusres.2021.10.053

Sahli, A., Pei, E., & Evans, R. (2023). A Conceptual Framework for Applying Artificial Intelligence to Manufacturing Projects. In *IFIP International Conference on Advances in Production Management Systems* (pp. 650-661). Cham: Springer Nature Switzerland. DOI: 10.1007/978-3-031-43666-6_44

Salleh, M. H. B., & Aziz, K. A. (2024). Enhancing project management with artificial intelligence: A framework for use case development. In *AIP Conference Proceedings* (Vol. 3153, No. 1). AIP Publishing. DOI: 10.1063/5.0218846

Sánchez, O., Castañeda, K., Vidal-Méndez, S., Carrasco-Beltrán, D., & Lozano-Ramírez, N. E. (2024). Exploring the influence of linear infrastructure projects 4.0 technologies to promote sustainable development in smart cities. *Results in Engineering*, *23*, 102824. DOI: 10.1016/j.rineng.2024.102824

Secinaro, S., Calandra, D., Lanzalonga, F., & Biancone, P. (2024). The Role of Artificial Intelligence in Management Accounting: An Exploratory Case Study. In *Digital Transformation in Accounting and Auditing: Navigating Technological Advances for the Future* (pp. 207–236). Springer International Publishing. DOI: 10.1007/978-3-031-46209-2_8

Serrano, W. (2022). Verification and Validation for data marketplaces via a blockchain and smart contracts. *Blockchain: Research and Applications*, *3*(4), 100100.

Shete, N. L., Maddel, M., & Shaikh, Z. (2024). A Comparative Analysis of Cybersecurity Scams: Unveiling the Evolution from Past to Present. In *2024 IEEE 9th International Conference for Convergence in Technology (I2CT)* (pp. 1-8). IEEE.

Stancheva-Todorova, E., & Bogdanova, B. (2021). Enhancing investors' decision-making–An interdisciplinary AI-based case study for accounting students. In *AIP Conference Proceedings* (Vol. 2333, No. 1). AIP Publishing.

Thirumagal, P. G., Vaddepalli, S., Das, T., Das, S., Madem, S., & Immaculate, P. S. (2024). AI-Enhanced IoT Data Analytics for Risk Management in Banking Operations. In *2024 5th International Conference on Recent Trends in Computer Science and Technology (ICRTCST)* (pp. 177-181). IEEE. DOI: 10.1109/ICRTCST61793.2024.10578533

Toledano, D. S., Toledano, J. I. S. S., & Jiménez, I. M. Á. (2024). Implementation of Analytical Accounting Models and Management Indicators and Development of Business Intelligence and Business Analytics Tools in Urban and Metropolitan Collective Transport Operators: A Case Study. *Revista de Gestão Social e Ambiental*, *18*(9), e06070–e06070. DOI: 10.24857/rgsa.v18n9-023

Tursunbayeva, A., & Gal, H. C. B. (2024). Adoption of artificial intelligence: A TOP framework-based checklist for digital leaders. *Business Horizons*, *67*(4), 357–368. DOI: 10.1016/j.bushor.2024.04.006

Vegar, B., & Mijač, T. (2024). Artificial Intelligence in Project Management: Insights from Croatia. In *2024 47th MIPRO ICT and Electronics Convention (MIPRO)* (pp. 1766-1771). IEEE.

Vidmar, M., Fleck, J., & Williams, R. (2023). AI and Data in Engineering and Innovation: Towards a Sustainable Future? In *2023 IEEE International Conference on Engineering, Technology and Innovation (ICE/ITMC)* (pp. 1-2). IEEE. DOI: 10.1109/ICE/ITMC58018.2023.10332336

Wang, J. (2023). Intelligent Decision Support System for Building Project Management Based on Artificial Intelligence. [). IOP Publishing.]. *Journal of Physics: Conference Series*, *2665*(1), 012022. DOI: 10.1088/1742-6596/2665/1/012022

Xue, F., Sanderson, A. C., & Graves, R. J. (2009). Multiobjective evolutionary decision support for design–supplier–manufacturing planning. *IEEE Transactions on Systems, Man, and Cybernetics. Part A, Systems and Humans*, *39*(2), 309–320. DOI: 10.1109/TSMCA.2008.2010791

Yang, S. N., Chang, L. C., & Chang, F. J. (2019). AI-based design of urban stormwater detention facilities accounting for carryover storage. *Journal of Hydrology (Amsterdam)*, *575*, 1111–1122. DOI: 10.1016/j.jhydrol.2019.06.009

Yi, Z., & Luo, X. (2024). Construction cost estimation model and dynamic management control analysis based on artificial intelligence. *Civil Engineering (Shiraz)*, *48*(1), 577–588. DOI: 10.1007/s40996-023-01173-z

Zema, T., Kozina, A., Sulich, A., Römer, I., & Schieck, M. (2022). Deep learning and forecasting in practice: An alternative costs case. *Procedia Computer Science*, *207*, 2958–2967. DOI: 10.1016/j.procs.2022.09.354

Chapter 12
The Role of Corporate Social Responsibility in Driving Economic Development:
A Systematic Review

Antima Sharma

https://orcid.org/0000-0003-0377-9315

Manipal University Jaipur, India

Vertika Goswami

https://orcid.org/0009-0005-0573-7217

Manipal University Jaipur, India

Durgesh Batra

https://orcid.org/0009-0001-4178-2315

Manipal University Jaipur, India

Preeti Nagar

Manipal University Jaipur, India

Arpita Agarwal

https://orcid.org/0000-0002-3649-4386

Manipal University Jaipur, India

Anadi Trikha

https://orcid.org/0000-0003-0792-0195

Manipal University Jaipur, India

ABSTRACT

Corporate Social Responsibility (CSR) has become crucial in modern business due to its potential to drive economic development. This systematic review analyzes the relationship between CSR initiatives and economic development by reviewing empirical studies, theoretical frameworks, and case analyses. It explores how CSR practices, such as addressing climate change and social inequality, contribute to economic growth by enhancing reputation, improving stakeholder relations, and attracting investment. The review identifies emerging trends in CSR, where companies adopt proactive approaches to social and environmental issues, and underscores the role of CSR in fostering sustainable economic development. Through this, the

DOI: 10.4018/979-8-3693-8009-3.ch012

study provides insights into how businesses can leverage CSR to achieve both social responsibility and economic success.

INTRODUCTION

Corporate Social Responsibility (CSR) has become a major global influence on economic development strategies in recent years, especially when tackling urgent social issues. CSR has developed into a strategic imperative with broad ramifications for several sectors, including healthcare, the environment, and education, as firms increasingly recognise their responsibility in promoting sustainable growth and societal well-being (Wirba, A. V. 2023). The goal of this systematic analysis is to thoroughly investigate how corporate social responsibility (CSR) propels economic progress, with an emphasis on how it affects these vital areas.

Healthcare, the environment, and education are essential cornerstones of socio-economic development, each influencing the productivity and well-being of people and communities in a special and invaluable way. Nevertheless, several obstacles, such as a lack of infrastructure, resource limitations, and social inequality, prevent these industries from realising their full potential as engines of economic expansion (Thirumalesh et al., 2023). In light of this, CSR activities have great potential to address these issues and spark constructive change.

CSR projects in the healthcare industry cover a wide range of interventions targeted at boosting public health awareness, expanding access to healthcare services, and upgrading healthcare infrastructure. CSR initiatives help improve health outcomes, lower the disease burden, and increase productivity, all propelling economic development. Examples of these initiatives include the construction of hospitals and clinics in underserved areas, implementing preventive health programs, and providing necessary medical supplies (Kim et al., 2017; Rahman et al., 2012). For example, corporations have funded mobile health clinics that serve isolated areas by providing short-term medical care and long-term health education. These programs enhance personal health, lower absenteeism, and raise worker productivity, boosting the economy.

Similar to this, CSR is essential to the environmental sector's efforts to advance climate resilience, mitigate environmental degradation, and support sustainable business practices. Corporate investments in conservation, waste management, carbon offsetting, and renewable energy projects support environmental sustainability and open up new commercial prospects for companies in developing green industries, which in turn spurs economic growth. For instance, businesses that invest in wind and solar energy not only lessen their carbon footprint but also develop new markets and employment prospects in the renewable energy sector. Initiatives like reforestation programs and sustainable agricultural methods can boost local communities' livelihoods while enhancing biodiversity and restoring ecosystems.

Moreover, CSR efforts in the education sector prioritise developing human capital, improving educational facilities, and increasing access to high-quality education. Corporate investments in digital literacy efforts, teacher training programs, school construction, scholarships, and vocational training support economic growth by improving educational outcomes, boosting labour participation, and increasing productivity. For example, technology corporations may assist in closing the digital divide and better prepare kids for the workforce by giving underprivileged schools internet access and digital learning resources. Furthermore, educational collaborations and scholarship programs facilitate the pursuit of professional professions and further education by students from underprivileged backgrounds, augmenting social mobility and economic production.

Although corporate social responsibility (CSR) projects have great promise in healthcare, the environment, and education, empirical data regarding their efficacy, scalability, and long-term effects is still lacking (Shoryaditya, 2023). The extant research frequently exhibits inconsistencies in methodology, scope, and measurement frameworks, impeding our comprehension of how corporate social responsibility impacts economic results in these domains. Due to this fragmentation, it isn't easy to come to firm judgements regarding the best practices and most effective CSR tactics in various circumstances.

This work aims to fill these information gaps and offer a thorough understanding of the function of CSR in promoting economic development within the healthcare, environment, and education sectors by systematically evaluating peer-reviewed articles, reports, and case studies. It attempts to find patterns, trends, and best practices that might guide future study, policy, and practice in the area of CSR and economic development by combining empirical data from many contexts and locations (Hossain, M. S., et al., 2020). In addition, this analysis will examine the difficulties and constraints encountered by CSR initiatives—such as those pertaining to accountability, sustainability, and measurement—and offer solutions to improve their efficacy and influence. CSR programs in the education, healthcare, and environmental sectors have enormous potential to promote inclusive growth and drive economic development. Corporations have the potential to significantly impact societal issues and further wider goals of economic development by utilising their resources, knowledge, and power. The goal of this systematic research is to improve our knowledge of the function of corporate social responsibility (CSR) in various industries and offer stakeholders who are dedicated to promoting positive change through ethical business practices useful insights.

Examining the precise mechanisms by which CSR initiatives function is crucial to comprehending the role that CSR plays in economic development. For instance, in the healthcare industry, CSR initiatives frequently entail collaborations with regional administrations, non-governmental organisations, and community groups to guarantee that health interventions suit cultural norms and successfully meet local needs. The longevity of health programs depends on these collaborations' potential to increase community buy-in and make more effective use of available resources. Additionally, CSR in healthcare frequently entails initiatives to upskill regional healthcare workers, enhancing the overall effectiveness of health systems and promoting long-term health gains.

In the environmental domain, corporate social responsibility (CSR) endeavours often centre on mitigating the ecological footprint of commercial activities and endorsing sustainable methodologies throughout supplier chains. The circular economy's guiding principles—which prioritise recycling, reuse, and waste reduction—are being adopted by businesses more and more. These procedures reduce the negative effects on the environment while simultaneously generating new revenue streams and cost savings. For instance, businesses can lower operating expenses and increase their competitiveness by investing in energy efficiency and renewable energy sources. Additionally, lobbying and policy engagement are common components of environmental CSR projects, which aim to persuade governments and other stakeholders to embrace more sustainable practices and policies.

CSR helps the education sector in a number of ways, such as through monetary contributions, in-kind gifts, and staff volunteer initiatives. Businesses frequently sponsor educational programs that help students get ready for professions in technology and engineering, such as STEM (Science, Technology, Engineering, and Mathematics) education initiatives. Businesses that invest in education contribute to the development of a trained labour force that can spur economic expansion and innovation. Additionally, by addressing systemic problems like gender inequality and the lack of access to education for underprivileged populations, CSR projects in education can promote inclusive and equitable economic growth.

Notwithstanding the benefits of corporate social responsibility, these programs are not without difficulties and detractors. Making sure that CSR initiatives have a real impact rather than being purely symbolic or surface-level is a significant problem. Some businesses run the risk of "greenwashing," which is the practice of marketing themselves as environmentally friendly without actually changing how they operate. Similar to this, CSR programs in the healthcare and education sectors need to be well planned and carried out to make sure they deal with the underlying causes of problems rather than just masking their symptoms. For corporate social responsibility (CSR) to be effective, beneficiaries and other stakeholders must be actively involved in order to guarantee that initiatives are pertinent and long-lasting.

Accountability and measurement are two more crucial CSR concerns. Strong systems must be in place to assess the results of CSR projects and make sure the desired benefits are being realised. This entails establishing precise goals, selecting relevant indicators, and carrying out frequent evaluations to monitor development. In order to demonstrate the effectiveness of CSR initiatives and to gain the trust of stakeholders, transparency and reporting are essential. Businesses must be open to learning from and changing their tactics in response to criticism and data, as well as sharing both their achievements and difficulties.

By tackling important issues in the healthcare, environmental, and educational sectors, corporate social responsibility (CSR) has the potential to contribute to economic development significantly. Corporations may accomplish their commercial goals and promote sustainable growth and societal well-being at the same time by making the most of their resources, knowledge, and influence. In addition to identifying best practices and providing stakeholders dedicated to ethical business practices with practical insights, the goals of this systematic review are to give a thorough understanding of the impact of CSR in various industries. Our understanding of CSR's contribution to economic growth can help build a more equitable and sustainable future for all.

Beginning in the latter part of the 20th century, there has been an ongoing discussion on the concept of corporate social responsibility (CSR). With the passage of time, the word "corporate social responsibility" (CSR) has replaced the previous term "social responsibility of business." In addition, this discipline has seen substantial expansion, and as a result, it now encompasses a vast array of ideas, methodologies, and terminologies. A number of terminologies, including but not limited to: society and business, social problems management, public policy and business, stakeholder management, and corporate accountability, are used in order to identify the phenomena that are associated with the concept of corporate responsibility in society. Recently, there has been a resurgence in interest in corporate social obligations, and new alternative conceptions have been presented. These notions include corporate citizenship and corporate sustainability.(Garriga, E., & Melé, D. 2004). expand in paragraph

The shift towards these alternative conceptions underscores a growing awareness of the interconnectedness of business practices with broader societal and environmental issues, suggesting a more holistic approach to corporate responsibility that integrates ethical considerations into all aspects of business operations. Throughout the course of history, terminology such as "society and business," "social problems management," "public policy and business," "stakeholder management," and "corporate accountability" have been used to define various aspects of corporate responsibility. These ideas contributed to the formation of an understanding of how corporations interact with and have an effect on society. They included a wide range of topics, including the management of social concerns, engagement with stakeholders, and commitment to responsibility.

India's approach to Corporate Social Responsibility (CSR) has evolved significantly from traditional models, as evidenced by academic literature examining the shift from voluntary to mandatory CSR frameworks. This development signifies a new understanding of CSR that combines aspects of voluntary and mandatory regulation, demonstrating a more comprehensive approach to corporate accountability. The Indian Companies Act of 2013 (Section 135) exemplifies this changing paradigm. Before the implementation of this revised legislation, CSR practices in India were guided by the Companies Act of 1956, which did not require companies to participate in social or environmental initiatives. (Gatti, L.,2019)

The 2013 amendment represented a significant departure from this previous approach by imposing mandatory CSR obligations on certain categories of companies. Specifically, it mandates that companies meeting specific criteria—such as having a net worth above a certain threshold or generating substantial annual revenue—must allocate at least 2% of their average net profit over the preceding three years to CSR activities.

There is a growing realisation that firms need to manage a complicated terrain of social expectations and environmental repercussions, and this broadened perspective of corporate social responsibility underscores this recognition. The shift towards these alternative conceptions underscores a growing awareness of the interconnectedness of business practices with broader societal and environmental issues, suggesting a more holistic approach to corporate responsibility that integrates ethical considerations into all aspects of business operations.

THEORETICAL FRAMEWORK

Corporate Social Responsibility (CSR) is a business ethos that entails companies proactively addressing social and environmental issues in their day-to-day operations and in their interactions with stakeholders. The theoretical foundation of CSR is rooted in a diverse range of theories and models that emphasize the moral and societal obligations of businesses to contribute positively to society and the environment, and to conduct business in an ethical and sustainable manner. A few theories demonstrated in literature and Indian companies follows are:

Stakeholder Theory

Stakeholder theory, formulated by R. Edward Freeman, suggests that organisations should consider the concerns of all their stakeholders, not only shareholders, when making decisions. This extends to employees, customers, suppliers, members of the community, and others impacted by the company's operations. The concept emphasises that businesses should generate value for all stakeholders to guarantee long-term sustainability. When it comes to making choices, organisations should take into consideration the interests of all of its stakeholders, not just shareholders, according to the stakeholder theory, which was developed by R. Edward Freeman. In addition to workers, customers, and suppliers, this also includes members of the community and other individuals whose lives are influenced by the activities of the firm. The notion places an emphasis on the fact that in order to ensure long-term sustainability, firms should provide value for all aspects of their stakeholders.

Tata Group

Tata Group exemplifies stakeholder theory in action in India. The Tata Trusts, which own a considerable portion of the Tata Group's stock, channel their profits into social initiatives, particularly in the areas of education, health, and rural development. Moreover, Tata's corporate social responsibility efforts encompass community development programs by Tata Steel and digital literacy enhancement programs by Tata Consultancy Services (TCS). The Tata Trusts, a majority stakeholder in Tata Sons, are focusing on improving education and healthcare in India through initiatives like the Tata Institute of Social Sciences and Tata Medical Center. Tata Steel supports sustainable livelihoods through skill development programs and small and medium enterprises.

Triple Bottom Line Theory (TBL)

John Elkington introduced the concept of the Triple Bottom Line (TBL) as a framework for businesses to evaluate their performance in a more comprehensive manner. The TBL approach advocates for businesses to assess their impact on not just financial profits (economic) but also on the environment (environmental) and society (social). By considering these three dimensions, companies are encouraged to adopt sustainable practices that benefit both their stakeholders and the wider community.

ITC Limited

The TBL methodology is used by ITC Limited, which is comprised of a diverse conglomerate in India. The company has been recognised for its efforts in sustainability, which include programs such as its Social and Farm Forestry Program. This program not only improves environmental performance but also helps farmers maintain their livelihoods. The e-Choupal program of the Indian Trade Corporation (ITC) gives farmers more control by supplying them with information on market pricing and best practices. ITC's Social and Farm Forestry initiative has created sustainable livelihoods for farmers, generating 9 million employment days annually. e-Choupal empowers rural farmers with real-time market information, enhancing productivity. ITC's Integrated Watershed Development Program promotes water conservation, rejuvenating over 1 million acres.

Carroll's CSR Pyramid

Archie Carroll came up with the concept of Carroll's CSR Pyramid, which describes four different degrees of corporate responsibility. These levels are as follows: economic (be profitable), legal (respect the law), ethical (do what is right), and philanthropic (be a good corporate citizen). According to the pyramid, firms should first guarantee that they are meeting their economic and legal duties before going on to ensuring that they are meeting their ethical and charitable contributions.

Infosys

The CSR Pyramid is exemplified by Infosys, a prominent information technology services firm in India. In addition to concentrating on achieving financial success, the organisation is also committed to upholding ethical standards and legal regulations. In order to demonstrate the company's dedication to ethical and philanthropic duties, the Infosys Foundation, which is the philanthropic arm of Infosys, works in fields such as education, healthcare, and rural development. Infosys Foundation is a philanthropic organisation that supports social initiatives in education, healthcare, and rural development. It also promotes digital literacy and STEM education through partnerships with educational institutions. Infosys is committed to sustainability by investing in renewable energy and energy-efficient buildings.

Social Contract Theory

In the context of corporate social responsibility (CSR), the Social Contract Theory is a philosophical notion that proposes that an implicit social contract with society binds firms. This contract suggests that businesses have duties to conduct their operations ethically and to make beneficial contributions to the community, in addition to the need to generate profits. Classical social contract theorists like as Thomas Hobbes, John Locke, and Jean-Jacques Rousseau are the originators of this idea. These thinkers maintained that people give their permission, either implicitly or openly, to join societies and adhere to the laws of society in return for protection and advantages.

Mahindra Group

By participating in a wide variety of corporate social responsibility (CSR) initiatives under the brand name "Rise for Good," Mahindra Group adheres to a robust social compact with society. The Mahindra Rise program is dedicated to bringing about good change in the areas in which the company does its business. This includes encouraging education, healthcare, and sustainability via a variety of projects, such as Project Nanhi Kali, which focusses on the education of girls in India who come from disadvantaged backgrounds. Mahindra Group supports underprivileged girls' education, aiming to improve academic performance and reduce dropout rates. They commit to becoming carbon neutral by 2040 and adopt renewable energy, water conservation, and waste management practices. Mahindra's CSR efforts include healthcare, community development, and mobile clinics in rural areas.

The term "Corporate Social Responsibility" (often abbreviated as "CSR") refers to an all-encompassing concept that combines ethical corporate operations with concerns for society and the environment. A framework that allows businesses to strike a balance between generating a profit, adhering to ethical standards, and making contributions to society is provided by a variety of theoretical underpinnings. Companies in India such as the Tata Group, ITC Limited, Infosys, Reliance Industries, and Mahindra Group have shown a significant commitment to corporate social responsibility (CSR) by aligning their business strategy with various theories of CSR. Not only do these businesses employ corporate social responsibility (CSR) to improve their corporate brand, but they also use it to make a good contribution to the communities that they serve and to secure their long-term viability.

OBJECTIVE OF THE STUDY

To identify the implications of CSR on economic development

Research Questions of the Study

o   Who are the main authors with the highest citations?
o   Which are the top investment areas?
o   Which leading companies are investing in CSR?

RESEARCH METHODOLOGY

This research utilizes a combination of qualitative and quantitative research methods to carry out a thorough examination of the literature on Corporate Social Responsibility (CSR) spanning from 2019 to 2024. The aim is to investigate patterns in CSR expenditures and evaluate the influence of CSR on the long-term viability of organizations over a six-year timeframe. The study is based on a total of 96 publications obtained from the Dimensions database, a prominent source of academic research information.

Data Collection

Through the use of the Dimensions database, the first step consisted of conducting a comprehensive search of the CSR-related literature from 2019 to 2024. The parameters for the search were developed with the intention of capturing a broad variety of studies that investigate various elements of corporate social responsibility (CSR), such as its strategy, execution, and effect on diverse sectors and geographies.

The preliminary criteria for inclusion in the study were satisfied by a total of 96 articles, which were found to be eligible for inclusion. This collection of articles covers a wide range of subjects pertaining to corporate social responsibility (CSR), including theoretical underpinnings and conceptual frameworks, empirical investigations, and case analyses, among other things.

Selection of Studies

A comprehensive quantitative analysis was performed on twenty research articles, which were chosen from among the 96 publications examined. According to a number of factors, the decision was made. These criteria included the significance of the study to corporate social responsibility (CSR), the quality of the research methodology, the influence of the citations, and the availability of quantitative data for analysis.

A bibliographic framework and citation analysis were used in order to make the identification of these twenty publications. During the citation analysis, the emphasis was on the frequency and effect of citations in order to determine the level of influence that each research had within the academic community. The bibliographic framework consisted of evaluating the contributions that each publication made to the CSR literature.

Quantitative Analysis with VOS Viewer

Using VOSviewer software, we were able to discover trends and patterns in corporate social responsibility (CSR) spending as well as the influence that it has on the sustainability of the organisation. VOSviewer is a program that can be used to generate and visualise bibliometric networks. These networks include co-authorship, co-citation, and keyword co-occurrence networks, all of which are very helpful in comprehending the structure and dynamics of academic research topics.

Using VOSviewer, the researchers were able to visualise the citation networks of the papers that were chosen for the study. This allowed them to discover prominent publications and prevalent topics in the field of corporate social responsibility research. Through this, we were able to acquire a better grasp of the interrelationships between various studies, the study themes that are gaining significance, and the ways in which CSR practices are developing over time.

A trend analysis was also performed using the program, which allowed for the discovery of phrases and themes that commonly appear together in the papers that were chosen. This made it easier to identify developing patterns in corporate social responsibility (CSR) investment and to understand how businesses are incorporating CSR into their business plans in order to improve sustainability.

Quantitative Analysis with VOSviewer

VOSviewer Software: VOSviewer software was used to find trends and patterns in CSR investment and its effect on organisational sustainability. The program VOSviewer is used to create and display bibliometric networks, which are useful for comprehending the dynamics and structure of academic research domains. Examples of these networks include co-authorship, co-citation, and keyword co-occurrence networks.

Network Visualisation: To determine important works and recurring topics in CSR research, the study used VOSviewer to visualise the citation networks of the chosen publications. This made it easier to comprehend the relationships between various studies, the subjects that are becoming more and more popular for study, and the changes that CSR practices are going through.

Trend Analysis: The program further made it possible to find themes and phrases that recur often in the chosen research. This made it easier to identify new trends in CSR spending and the ways that businesses are incorporating CSR into their operations to improve sustainability.

Data Management with Microsoft Excel

Data Organisation: The data gathered from the Dimensions database and VOSviewer analysis was managed and organised using Microsoft Excel. The chosen papers were catalogued, their citation counts were recorded, and the main ideas and conclusions from each publication were categorised using Excel spreadsheets.

Quantitative Data Analysis: Additional descriptive statistical analyses were carried out using Excel, including mean citation counts, frequency analysis of certain CSR topics, and quantification of the effect of CSR initiatives on organisational sustainability. The CSR literature was quantitatively summarised by these studies, which also aided in the derivation of pertinent conclusions on the field's development during the six-year span.

Figure 1. Research framework of the study

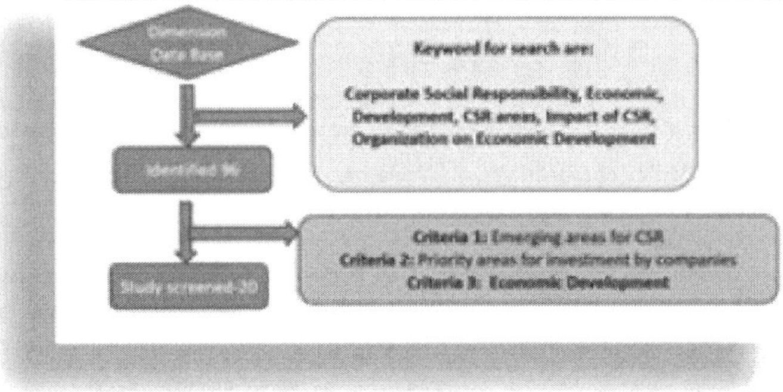

RESULTS AND IMPLICATIONS

Out of the 96 papers identified in this study concerning Corporate Social Responsibility (CSR) and sustainability, several key themes and insights emerged. These papers collectively shed light on the multifaceted impact of CSR initiatives on various stakeholders, including employees, consumers, and the broader community. The research underscores the growing trend of businesses integrating CSR into their core strategies, not merely as a compliance measure but as a fundamental aspect of their operational ethos. 20 papers were selected for further analysis.

Table 1. List of highest citations: Citation-based research documents with authors details

S.No	Title	Year	Citation	Authors
1	Addressing the SDGs in sustainability reports: The relationship with institutional factors	2019	371	Rosati, Francesco; Faria, Lourenço G.D.
2	The effect of green human resources management on corporate social responsibility, green psychological climate and employees' green behavior	2021	74	Sabokro, Mehdi; Masud, Muhammad Mehedi; Kayedian, Azin
3	Translating stakeholders' pressure into environmental practices – The mediating role of knowledge management	2020	59	Shahzad, Mohsin; Qu, Ying; Zafar, Abaid Ullah; Ding, Xiangan; Rehman, Saif Ur
4	Interplay between corporate social responsibility and organizational green culture and their role in employees responsible behavior towards the environment and society	2022	55	Pan, Changjian; Abbas, Jawad; Ãlvarez-Otero, Susana; Khan, Hina; Cai, Cheng
5	Sustainability assurance practices: a systematic review and future research agenda	2021	48	Hazaea, Saddam A.; Zhu, Jinyu; Khatib, Saleh F. A.; Bazhair, Ayman Hassan, Elamer, Ahmed A.
6	Employee happiness and corporate social responsibility: the role of organizational culture	2020	40	Espasandan-Bustelo, Francisco; Ganaza-Vargas, Juan; Diaz-Carrion, Rosalia
7	Online Platforms and the Circular Economy	2019	31	Konietzko, Jan; Bocken, Nancy; Hultink, Erik Jan
8	Criteria for assessing a sustainable hotel business	2020	31	dos Santos, Rodrigo Amado; MÃ©xas, Mirian Picinini; Meiriato, Marcelo Jasmim; Sampaio, Michelle Cristina; Costa, Helder Gomes
9	How Does Green Innovation Strategy Influence Corporate Financing? Corporate Social Responsibility and Gender Diversity Play a Moderating Role	2022	30	Javeed, Sohail Ahmad; Teh, Boon Heng; San Ong, Tze; Chong, Lee Lee; Bin Abd Rahim, Mohd Fairuz; Latief, Rashid
10	Does green finance facilitate firms in achieving corporate social responsibility goals?	2022	28	Wang, Zhuo; Shahid, Muhammad Sadiq; An, Nguyen Binh; Shahzad, Mohsin; Abdul-Samad, Zulkiflee

Table 2 shows the research paper with the highest number of citations indexed in six years from 2019 to 2024. A research paper published in the year 2019 "Addressing the SDGs in sustainability reports: The relationship with institutional factors" got the highest number of publications. So above table shows 10 research papers with the highest citations.

Table 2. A review of central studies focused on CSR and trends in CSR

S.No	Title of the study	Authors Name	Findings
1	How Does Green Innovation Strategy Influence Corporate Financing? Corporate Social Responsibility and Gender Diversity Play a Moderating Role	Javeed et.al,2022	The study's findings reveal a nuanced interplay between corporate social responsibility (CSR), green innovative strategies, and corporate financing. Specifically, CSR appears to weaken the direct relationship between green innovative strategies and the availability of corporate financing. This attenuation suggests that CSR initiatives provide companies with a certain degree of latitude, allowing them to pursue environmentally sustainable innovations without being as immediately constrained by financial considerations. In addition, the research underscores the positive impact of gender diversity on corporate financing. Companies with more gender-diverse leadership teams tend to attract greater financial resources, indicating that investors might view such teams as more capable or effective in navigating financial opportunities. Moreover, the study highlights that gender diversity not only benefits corporate financing but also enhances the effectiveness of green innovative strategies in securing investment. This moderating effect implies that diverse teams are better equipped to leverage green innovations to attract financing, potentially due to their varied perspectives and improved communication strategies. Overall, these findings suggest that while CSR can moderate the link between green innovation and financing, gender diversity plays a crucial role in optimizing both the financial and strategic benefits of green initiatives.

continued on following page

Table 2. Continued

S.No	Title of the study	Authors Name	Findings
2	Corporate environmental investment and sustainable development: based on the perspective of Marxist ecological civilization	Ye et.al., 2023	The research shows a strong correlation between investments made in the environment and businesses' ability to grow sustainably. The term "environmental investment" describes the assets and labour that businesses commit to enhancing their ecological footprint reduction and environmental performance. This investment helps enterprises in a concrete way while also promoting ecological well-being. Businesses may gain social capital, such as improved reputation, closer connections with stakeholders, and more community trust, by contributing to environmental efforts. By promoting improved access to resources and collaborations, this social capital may help relieve the pressure on limited resources. Furthermore, by appealing to customers who are becoming more environmentally sensitive and setting the business apart from rivals, environmental investments may improve market success. Consequently, all of these benefits contribute to the enterprise's long-term viability. Investing in environmental practices essentially helps businesses integrate sustainable development ideas into their core operations, laying the groundwork for future growth, resilience, and success.
3	Corporate environmental investment and sustainable development: based on the perspective of Marxist ecological civilization	Hu., et.al, 2023	The strategic benefit of adopting a green strategy is being more recognised by businesses, particularly in light of the increased interest in sustainable development that is occurring all over the globe. Taking into consideration these results, it is suggested that small and medium-sized businesses (SMEs) modify their strategy and operations in order to have a good influence on the economy, culture, and environment. Within the context of GSCM, this study highlights the importance of a circular economy as a means of promoting sustainability. This transition towards a circular economy will be vital for the success and expansion of small and medium-sized enterprises (SMEs) over the long run.

continued on following page

Table 2. Continued

S.No	Title of the study	Authors Name	Findings
4	Role of Corporate Social Responsibility in Achieving Sustainable Development Goals	Menaga, A.; Vasantha, S.	Studies have shown that the implementation of Corporate Social Responsibility (CSR) may lead to improvements in a number of organisational functions and the influence on society, especially with regard to energy consumption, labour standards, and employment generation. Businesses may enhance their energy consumption habits and increase the sustainability and efficiency of their operations by incorporating corporate social responsibility (CSR) activities into their fundamental strategy. For instance, companies may choose to use renewable energy sources or energy-efficient technology, which not only lessen their carbon footprint but also help achieve more general environmental objectives. Furthermore, enhanced labour practices— such as equitable pay, secure working environments, and respect for employees' rights—are often the result of CSR initiatives. By fostering a healthy work environment, organisations may increase employee happiness and productivity in addition to adhering to ethical norms.
5	Contribution of CSR Towards Development - The Indian Perspective	Ramakrishnan, 2017	Business and industry play a crucial role in the social and economic development of a nation. In the last fifty years, focus of the business has resulted in stakeholder management rather than stockholder management. CSR movement recognizes that Corporations are important and powerful players in today's world. Some of the corporations are more powerful than sovereign states. Development is a complex issue. Business must take full account of the societal expectations. More than classes, it is the masses that nurture business.

continued on following page

Table 2. Continued

S.No	Title of the study	Authors Name	Findings
6	An Analysis of CSR and its Expenditure in India in the Terms of Trends, Impact and Challenges	Shoryaditya, Shoryaditya, 2023	Environmental sustainability, malnutrition, hunger and poverty, livelihood enhancement projects, central government funds including PMNRF, sanitation, art & culture, and vocational skills received more than Rs. 2000 crores each and together contributed to another 29.4% of the expenditure. Safe drinking water, women empowerment, natural resources conservation, gender equality, animal welfare, orphanages, armed forces (veterans, war widows, etc.), special education, Swachch Bharath Kosh, sports, technology incubators, senior citizens welfare, agroforestry, slum area development, and socio-economic inequalities are the other sectors in which CSR amount is used. These sectors accounted for <9% of the CSR expenditure. For the remaining expenditure (~3%), companies have not disclosed the details of the projects.
7	How Corporate Social Responsibility and Sustainable Development Practices Affect the Economic Performance of Indian Businesses: A Critical Analysis	Kujur, Oscar, 2023	This research reviews the corporate social responsibility (CSR) actions of the top ten contributing companies that are aligned with sustainable development (SD) and assesses because of the impact that these operations have on their overall financial success. Embracing corporate social responsibility may result in higher revenue production, decreased capital expenses, and an improved reputation for the company.
8	An Analysis of CSR Expenditure by Indian Companies	Uma, R., & Uma, K. 2021)	The findings indicate that firms do not place a high importance on spending money on environmental protection and pollution control operations. Despite the fact that the effect of corporate social responsibility (CSR) operations cannot always be measured in terms of financial allocations, doing so is still one of the most significant measures of how engaged businesses are with society. Therefore, on the basis of the results, it is possible to draw the conclusion that the incorporation of this provision is a suitable step that the regulators should take in order to make corporations more socially responsible.

The literature pertaining to Corporate Social Responsibility (CSR) accentuates its profound influence on the perpetuity of organizations, elucidating the collaborative roles of the public and private sectors in fostering advancement and ingenuity through CSR initiatives. Research underscores the pivotal role of CSR initiatives in promoting energy efficiency and environmental stewardship. By advocating the uptake of green technologies and sustainable practices, CSR assists organizations in curtailing their carbon footprints and alleviating environmental impact. Notably, companies investing in renewable energy sources, waste reduction, and sustainable resource management contribute to overarching environmental objectives while potentially streamlining operational expenditure.

Furthermore, CSR significantly heightens labor practices by ameliorating working conditions, ensuring equitable remuneration, and upholding employee rights. These enhancements cultivate a more equitable and compassionate work environment, yielding tangible benefits for organizations. Enhanced working conditions may yield heightened employee morale, reduced turnover rates, and amplified productivity, thereby fostering a more engaged and motivated workforce. Companies prioritizing fair labor practices are frequently perceived as more appealing employers, augmenting their capacity to attract and retain premier talent. In addition, CSR endeavors contrive to foster job creation and economic progression through bolstering local development projects and educational initiatives.

By allocating resources to community programs, vocational training, and educational opportunities, businesses can invigorate employment growth and fortify economic stability within their operational domains. Such initiatives aid in cultivating a more adept workforce, bolstering local enterprises, and propelling economic development, thereby fostering a virtuous cycle of growth and opportunity. This comprehensive outlook emphasizes the shared responsibilities of the public and private sectors in shaping and executing CSR strategies. While public sector regulations and incentives are pivotal in configuring the framework for CSR policies, the direct involvement of the private sector in enacting these strategies precipitates significant advancements in sustainability. Subsequent research is anticipated to explore key facets, including the formulation of more sophisticated methodologies for gauging the impacts of CSR, the assimilation of CSR practices into broader organizational strategies, and the scrutiny of variances in CSR practices among diverse industries and regions. Such research endeavors promise deeper insights into effectual leverage of CSR to realize both organizational pursuits and societal advantages, thereby further propelling the domain of corporate sustainability.

Table 3. Keyword occurrence abstract

Clusters	Keywords
Cluster 1	Accountability, brand, business community, commitment, corporate social responsibility, critical role, economic, education, sustainable, employee, employment, environmental, impact etc.
Cluster 2	Climate change, corporate law, democracy, economic era, justice, key role etc.

Figure 2. Abstract keywords occurrence

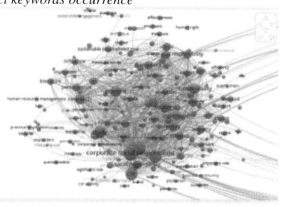

Table 4. CO -authorship occurrence

Clusters	C0-Authorship
Cluster 1	Grayson, David, nelson, Jane
Cluster 2	Antanoras, alexander
Cluster 3	Robe, jane-philippe
Cluster 4	Sheik, salim

Table 4 depicts the name Authors, their occurrence found studies centralizing Corporate Social Responsibility as a research area together with below top cited authors:

i)   Amran, Azlan; Lee, Shiau Ping; Devi, S. Susela
ii) Rosati, Francesco; Faria, LourenÃ§o G.D.
iii) Seta, Dolors
iv) Ramesh, Kumar; Saha, Raiswa; Goswami, Susoban; Sekar; Dahiya, Richa

DISCUSSION AND ANALYSIS

Theme 1: Emerging Trends

The landscape of Corporate Social Responsibility (CSR) is evolving as businesses increasingly recognize the importance of integrating social and environmental considerations into their strategies. The CSR Outlook Report by CSRBOX (2023) highlights emerging trends that reflect this shift, with a notable emphasis on health-care, water, sanitation, and hygiene (WASH), as well as environmental sustainability and education. The report reveals that 26% of businesses prioritize healthcare and WASH initiatives, underscoring a growing corporate commitment to addressing fundamental health and hygiene needs within communities. This focus reflects a broader trend where companies are not only aiming to enhance their social impact but also align their CSR efforts with pressing global challenges. By addressing these critical areas, businesses are contributing to overall societal well-being and sustainability, recognizing that improved health and environmental conditions are integral to fostering resilient and thriving communities.

Due to the fact that public health problems are becoming more generally acknowl-edged, especially in the aftermath of worldwide disasters like as the COVID-19 pandemic, healthcare and WASH have become more of a priority. Some companies have come to the realisation that investing money in sanitary facilities and health-care infrastructure is beneficial not just to public health but also to the confidence that stakeholders have in those businesses. During the course of the outbreak, for instance, a great number of companies raised their corporate social responsibility (CSR) activities by supporting vaccination programs, creating healthcare facilities, and delivering medical supplies. Through the implementation of such measures, communities were able to effectively manage health crises that lasted for a short period of time while also establishing the foundation for long-term health resilience.

Just behind WASH and healthcare, 23% of businesses prioritise environmental sustainability and education promotion. According to the report of CSR Box, 33% of businesses have made environment and sustainability their second top priority, demonstrating a growing understanding of the vital need of environmental stew-ardship. This increased awareness is a reflection of the realisation that sustainable business practices are essential to maintaining both the long-term profitability of firms and the preservation of natural resources. Businesses are embracing more and more sustainable practices, such cutting their carbon footprints, improving energy efficiency, and supporting the ideas of the circular economy. The realisation that sustainable practices might result in cost savings and new markets, along with customer demand for eco-friendly products and regulatory pressures, frequently drives these efforts.

The findings of Ye et al. (2023), which emphasise the important significance of environmental investment in fostering the sustainable development of firms, are consistent with the emphasis on environmental sustainability. Investments in the environment act as a stimulant to increase market performance, reduce resource restrictions, and build social capital. The prioritisation of sustainability and the environment by businesses, which shows a rising understanding of the significance of incorporating environmental factors into corporate plans, further demonstrates this alignment. Businesses are becoming more and more conscious of the fact that ignoring environmental issues can lead to financial losses, legal ramifications, and harm to their reputation. As a result, they are spending money on innovations and methods that lessen their influence on the environment, like waste management programs, renewable energy sources, and sustainable sourcing techniques.

Another noteworthy goal is the promotion of education, which receives attention from about 25% of businesses. This focus on education shows a dedication to funding educational programs and skill development that promote the prosperity and progress of local communities. Businesses understand that innovation and economic growth depend heavily on having a workforce with a good education. In addition to promoting individual empowerment, corporations that fund educational initiatives, scholarships, and career training also create a talent pool of highly qualified individuals who can fuel future expansion. Furthermore, as education increases people's capacity to engage fully in the social and economic spheres, it promotes societal cohesiveness and stability. This makes investing in education consistent with larger sustainability goals.

Furthermore, this dedication to sustainability and education is in line with research by Javeed et al. (2022) and Kujur and Oscar (2023), which highlight the significance of CSR initiatives in reaching Sustainable Development (SD) objectives. The advantages of adopting CSR practices—such as environmental stewardship—for promoting sustainable development are highlighted by these studies. Businesses that put a high priority on CSR initiatives can reap observable advantages like higher revenue production, lower capital expenditures, and improved brand recognition. Enterprises can generate mutually beneficial shared value by incorporating social and environmental factors into their business plans.

These studies' convergence emphasises the connections between corporate social responsibility (CSR), environmental investment, and sustainable growth within businesses. Businesses can improve their financial performance and contribute to long-term sustainability and societal well-being by embracing corporate social responsibility (CSR) and environmental sustainability. This thorough methodology offers insightful information about how CSR and environmental investments affect business performance and sustainability results. A strong commitment to corporate social responsibility (CSR) and sustainable practices is expected to become an increasingly essential differentiation in the marketplace as businesses continue to negotiate the difficulties of the modern business environment.

Theme 2: Leading Companies' Adherence to CSR Requirements

Leading businesses committed significantly to Corporate Social Responsibility (CSR) in the fiscal year 2022–2023 by exceeding their allotted CSR allocations. This proactive approach to corporate social responsibility (CSR) represents a larger trend among leading companies to give social and environmental activities more weight than is required by law. Tata Consultancy Services Ltd. (TCS) said that its expenditure on corporate social responsibility (CSR) exceeded its budget of 457 crore rupees. Similarly, ONGC Ltd. reported spending 476 crore rupees on CSR activities, more than the 452 crore rupees allocated to CSR.

TCS has demonstrated a significant commitment to social responsibility and community development through its substantial CSR expenditures. TCS displays proactive engagement in solving societal problems and investing in sustainable initiatives by allocating resources to various CSR initiatives. TCS's varied CSR portfolio, which includes initiatives pertaining to talent development, healthcare, education, and environmental sustainability, demonstrates this dedication. For instance, TCS has participated in healthcare camps, digital literacy programs, and campaigns to encourage STEM education in poor schools. These initiatives not only support the business's corporate citizenship but also strengthen its standing as a socially conscious one.

In a similar vein, ONGC's real spending, which exceeded the allocated budget, highlights the organization's commitment to CSR initiatives. The three main focusses of ONGC's CSR programs are healthcare, environmental sustainability, and community welfare. The company's efforts include supplying clean drinking water, aiding in the development of renewable energy sources, and encouraging good health and hygiene in rural communities. By tackling important social and environmental issues, ONGC demonstrates its dedication to stakeholder involvement and sustainable development. In addition to improving the company's reputation, this proactive strategy fosters goodwill and confidence among all of its constituents, including local communities, consumers, and staff.

Smaller businesses, such as Sobha Ltd., on the other hand, spend comparatively less on corporate social responsibility (CSR). This could be because of divergent goals or limited resources. This variation in CSR spending demonstrates the range of CSR strategies businesses use across various industries and sizes. While larger enterprises frequently possess the financial ability and resources to allocate major amounts to corporate social responsibility programs, smaller businesses may encounter difficulties in doing so. This does not, however, always reflect a lack of dedication to CSR. Smaller businesses can nevertheless have a significant impact by concentrating on particular problems or collaborating with nonprofit groups to increase their reach.

The financial information from corporations like TCS and ONGC offers a glaring illustration of how top companies meet and even surpass CSR regulations. The a for mentioned trend signifies a wider movement towards stakeholder-centric methods and corporate sustainability as companies realise the importance of incorporating social and environmental factors into their daily operations. Rosati, Francesco, and Faria (2023) have observed that corporate social responsibility (CSR) policies are essential in reducing the correlation between green and creative initiatives and company finance. Their results highlight the significance of integrating social and environmental considerations into corporate goals by indicating that CSR initiatives help mitigate this link. By doing this, businesses can lower the financial risks connected to social and environmental challenges, improve their market positioning, and draw in investors who share their values.

Furthermore, Hu et al. have highlighted the strategic benefit of using a green approach, especially in the context of sustainable development (2023). The advice given to small and medium-sized businesses (SMEs) to modify their business plans in order to have beneficial effects on the environment, culture, and economy highlights the increasing awareness of the necessity for companies to adopt sustainable practices. Adopting sustainable and green practices can help SMEs save money, build their brand's reputation, and reach new markets. SMEs are further encouraged to incorporate sustainability into their business models by government incentives and consumer demand for sustainable goods and services.

The changing role of corporations in social and economic growth is also highlighted by Ramakrishnan (2017), who emphasises the importance of CSR in upholding societal expectations and the move towards stakeholder management. The Sobha Ltd. instance presents significant queries regarding the variables impacting CSR expenditure decisions because of its comparatively lower CSR spending in comparison to larger firms. Stakeholder expectations, industry, financial performance, and company size are a few examples of these variables. Businesses in sectors like manufacturing and energy, for example, that have a significant environmental impact can be under more pressure to fund environmental sustainability projects. On the other hand, businesses in less resource-intensive sectors might put more of an emphasis on charitable causes like healthcare and education.

The summarisation of these results highlights how CSR is changing and how companies are realising how crucial it is from a strategic standpoint to include social and environmental concerns into their daily operations. Companies like TCS and ONGC, who are actively involved in CSR, are prime examples of a wider trend towards stakeholder-centric methods and corporate sustainability. This change is being pushed by a number of factors, including customer expectations, legal requirements, and the realisation that corporate social responsibility (CSR) may benefit society and the business in the long run. The disparity in CSR spending between businesses, however, emphasises how difficult it is to make decisions about CSR and how careful study is required to fully grasp its underlying causes and consequences for sustainable development.

In summary, the evolving landscape of Corporate Social Responsibility (CSR) reflects a growing commitment by businesses to go beyond traditional profit-making activities and engage in practices that have a positive impact on society and the environment. Today, there is an increased emphasis on stakeholder engagement, with companies actively involving communities, employees, customers, and investors in their CSR efforts. This shift towards a more inclusive approach ensures that CSR initiatives are not only beneficial to the company but also resonate with the values and needs of various stakeholders. Additionally, areas such as healthcare, education, and environmental sustainability have become central to CSR strategies, as companies recognize their role in addressing critical global challenges. For example, many businesses are investing in healthcare infrastructure, supporting educational programs, and implementing sustainable practices that reduce environmental footprints. Leading corporations are moving beyond the minimum legal requirements for CSR and are allocating substantial resources towards initiatives that align with their long-term strategic objectives and public expectations. This proactive approach is driven by the understanding that sustainable business practices lead to improved risk management, enhanced reputation, and increased customer loyalty.

As global awareness of sustainability and social responsibility grows, companies of all sizes are compelled to adapt and innovate their CSR strategies to remain relevant and effective. The CSR Outlook Report by CSRBOX (2023) and recent academic research provide valuable insights into the diverse ways CSR initiatives can enhance business performance and contribute to sustainability goals. These resources outline best practices and emerging trends, offering a roadmap for companies seeking to optimize their CSR efforts. By prioritizing CSR, companies can achieve a dual benefit: improving their bottom line through increased efficiency and market differentiation while simultaneously fostering a positive brand image and social impact. Furthermore, effective CSR can drive innovation, attract top talent, and create partnerships that enhance competitive advantage. In essence, by embedding CSR into their core business practices, companies not only fulfill their ethical obligations but also position themselves as leaders in promoting a sustainable and equitable future. This comprehensive approach to CSR enables businesses to respond proactively to the increasing demand for responsible corporate behavior, ensuring long-term success and resilience in an ever-evolving global market.

CONCLUSION

In summary, this systematic review offers a thorough analysis of how corporate social responsibility (CSR) propels economic growth, with an emphasis on the environment, education, and healthcare in particular. According to the findings, CSR activities have a great deal of potential to address important societal issues and promote socioeconomic well-being and sustainable growth. CSR helps the healthcare industry by strengthening the infrastructure of healthcare facilities, increasing public awareness of health issues, and improving access to healthcare services. Better health outcomes, a decrease in the burden of disease, and a rise in productivity are the results of CSR projects like building hospitals and clinics in underprivileged areas, putting preventative health programs into place, and giving basic medical supplies. By fostering a healthier and more cohesive community, these advancements not only benefit the well-being of the individual and the community but also stimulate the economy.

CSR is essential to the environment because it helps businesses adopt sustainable practices and reduces environmental damage. Environmental sustainability is facilitated by corporate expenditures in carbon offsetting programs, waste management initiatives, conservation measures, and renewable energy projects. These programs create new economic prospects for companies in the developing green industries while also aiding in the preservation of natural resources. Businesses that embrace sustainability can lower operating expenses, improve their standing in the marketplace, and turn a profit over the long run, all of which contribute to economic expansion. Another important pillar of socioeconomic development, the education sector, gains a great deal from corporate social responsibility. Enhancing educational outcomes and promoting human capital development are the results of corporate investments in digital literacy efforts, teacher training programs, educational infrastructure, scholarships, and vocational training. More people having access to high-quality education results in a workforce that is more knowledgeable and talented, which is necessary for innovation and economic growth. Businesses can ensure a consistent supply of competent individuals to support their operations and contribute to societal progress by investing in education. The empirical data on the efficacy, scalability, and long-term effects of corporate social responsibility (CSR) efforts is still inconsistent, despite the clear advantages. According to this assessment, CSR research needs more standardised measurement frameworks, wide scopes, and consistent methodology. Filling in these gaps will improve our knowledge of the ways that corporate social responsibility (CSR) affects financial results in the healthcare, environmental, and educational sectors.

The study's conclusions highlight the significance of CSR as a business strategy requirement. Prominent corporations, including ONGC Ltd. and Tata Consultancy Services Ltd. (TCS), demonstrate a proactive approach to corporate social responsibility (CSR) by investing in a variety of programs and surpassing their budgetary allotments. These businesses show that corporate social responsibility (CSR) is an essential part of corporate strategy, helping to achieve sustainable development and positive social impact, rather than just a legal requirement. Nonetheless, the disparity in corporate social responsibility spending between businesses—especially between larger enterprises and smaller enterprises like Sobha Ltd.—highlights the difficulties in making CSR decisions. CSR spending is influenced by a number of factors, including the industry, financial performance, corporate size, and stakeholder expectations. It is essential to comprehend these elements in order to create CSR strategies that effectively support company objectives. A growing corporate emphasis on healthcare, water, sanitation, and hygiene (WASH), environmental sustainability, and education is indicated by emerging trends in corporate social responsibility (CSR). This change is indicative of a growing understanding of the relationship between corporate success and societal well-being. Companies are becoming more conscious of the fact that making investments in these fields improves their own sustainability and competitiveness in addition to addressing pressing societal concerns.

In conclusion, Corporate Social Responsibility (CSR) holds immense potential to drive inclusive growth and foster economic development by addressing crucial societal issues through strategic initiatives. By leveraging their resources, expertise, and influence, companies can contribute meaningfully to broader economic objectives, such as reducing inequalities, improving public health, and enhancing environmental sustainability. This comprehensive review provides valuable insights for stakeholders—ranging from business leaders and policymakers to community organizations—who are committed to advancing positive change through ethical and socially responsible business practices. As businesses increasingly integrate CSR into their core strategies, they are not only positioned to achieve greater economic success but also to promote a more sustainable and equitable world. This dual benefit reinforces the importance of CSR as a vital component of modern business strategy. Furthermore, this research serves as a foundational basis for further investigation into the motivations, strategies, and impacts of CSR initiatives, particularly in areas like healthcare, WASH, environmental sustainability, and education, guiding future efforts to enhance the effectiveness and reach of corporate contributions to societal well-being.

Limitations of the Study

The study is primarily based on secondary data sources, including published research articles and a limited number of industry reports. This reliance means that the study might not capture the most recent industry practices or emerging trends that haven't yet been documented in academic or industry literature. The study focuses exclusively on the practices of large-scale companies. As a result, it does not consider the strategies, challenges, and opportunities unique to medium-sized, small, and startup companies. These organizations often operate under different constraints and may innovate differently compared to their larger counterparts. Excluding them limits the study's applicability across the full spectrum of company sizes. The study may not include longitudinal data, which would provide insights into how practices and strategies have evolved. Understanding these changes over time could offer a more dynamic view of the industry and better predict future trends.

Suggestion for Further Research

There is a substantial possibility of doing research based on the gathering of primary data, despite the fact that the study relied on secondary data throughout its execution. Future research might make use of methods such as surveys, interviews, focus groups, and direct observations in order to get first-hand knowledge from experts working in respective industries. Taking this strategy would make it possible to have a more in-depth comprehension of the prevalent practices, issues, and solutions that are being used by businesses in the present moment. Compared to secondary data, primary data may provide a more comprehensive and nuanced perspective on the behaviours and decision-making processes of an organisation, which secondary data may not be able to completely capture. The present study would make it possible to conduct an in-depth investigation of strategies that have shown to be successful, as well as success stories and lessons acquired, which may be beneficial for both academic research and practical application. Furthermore, it has the potential to assist in the formation of future plans and decision-making procedures. The fact that the present research focusses exclusively on large-scale businesses restricts the extent to which it can be generalised. It is possible that in the future, study may concentrate primarily on medium-sized, small, and startup businesses in order to get an understanding of their viewpoints, behaviours, and patterns in relation to a variety of activities.

REFERENCES

Dos Santos, R. A., Méxas, M. P., Meirino, M. J., Sampaio, M. C., & Costa, H. G. (2020). Criteria for assessing a sustainable hotel business. *Journal of Cleaner Production, 262*, 121347. DOI: 10.1016/j.jclepro.2020.121347

Espasandín-Bustelo, F., Ganaza-Vargas, J., & Diaz-Carrion, R. (2021). Employee happiness and corporate social responsibility: The role of organizational culture. *Employee Relations, 43*(3), 609–629. DOI: 10.1108/ER-07-2020-0343

Evan, W. M., & Freeman, R. E. (1988). A Stakeholder Theory of the Modern Corporation: Kantian Capi talism. In Beauchamp, T., & Bowie, N. (Eds.), *Ethical Theory and Business* (pp. 75–93). Prentice Hall.

Garg, P. (2016). CSR and corporate performance: Evidence from India. *Decision (Washington, D.C.), 43*, 333–349.

Garriga, E., & Melé, D. (2004). Corporate social responsibility theories: Mapping the territory. *Journal of Business Ethics, 53*(1), 51–71. DOI: 10.1023/B:BU-SI.0000039399.90587.34

Gatti, L., Vishwanath, B., Seele, P., & Cottier, B. (2019). Are we moving beyond voluntary CSR? Exploring theoretical and managerial implications of mandatory CSR resulting from the new Indian companies act. *Journal of Business Ethics, 160*(4), 961–972. DOI: 10.1007/s10551-018-3783-8

Gouda, S., Khan, A. G., & Hiremath, S. L. (2017). *Corporate social responsibility in India. Trends, issues and strategies*. Anchor Academic Publishing.

Hazaea, S. A., Zhu, J., Khatib, S. F., Bazhair, A. H., & Elamer, A. A. (2022). Sustainability assurance practices: A systematic review and future research agenda. *Environmental Science and Pollution Research International, 29*(4), 4843–4864. DOI: 10.1007/s11356-021-17359-9 PMID: 34787810

Hossain, M. S., Yahya, S. B., & Khan, M. J. (2020). The effect of corporate social responsibility (CSR) health-care services on patients' satisfaction and loyalty–a case of Bangladesh. *Social Responsibility Journal, 16*(2), 145–158. DOI: 10.1108/SRJ-01-2018-0016

Javeed, S. A., Teh, B. H., Ong, T. S., Chong, L. L., Abd Rahim, M. F. B., & Latief, R. (2022). How does green innovation strategy influence corporate financing? Corporate social responsibility and gender diversity play a moderating role. *International Journal of Environmental Research and Public Health, 19*(14), 8724. DOI: 10.3390/ijerph19148724 PMID: 35886576

Konietzko, J., Bocken, N., & Hultink, E. J. (2019). Online platforms and the circular economy. *Innovation for sustainability: Business transformations towards a better world*, 435-450.

Kujur, O. (2023). How Corporate Social Responsibility and Sustainable Development Practices Affect the Economic Performance of Indian Businesses: A Critical Analysis.

Menaga, A., & Vasantha, S. Role of Corporate Social Responsibility in Achieving Sustainable Development Goals. In *Interdisciplinary Perspectives on Sustainable Development* (pp. 188–191). CRC Press. DOI: 10.1201/9781003457619-39

O'Riordan, L., & Fairbrass, J. (2008). Corporate Social Responsibility (CSR): Models and Theories in Stakeholder Dialogue. *Journal of Business Ethics*, *83*(4), 745–758. DOI: 10.1007/s10551-008-9662-y

Pan, C., Abbas, J., Álvarez-Otero, S., Khan, H., & Cai, C. (2022). Interplay between corporate social responsibility and organizational green culture and their role in employees' responsible behaviour towards the environment and society. *Journal of Cleaner Production*, *366*, 132878. DOI: 10.1016/j.jclepro.2022.132878

Phillips, R. A. (1997). Stakeholder Theory and a Principle of Fairness. *Business Ethics Quarterly*, *7*(1), 51–66. DOI: 10.2307/3857232

Ramakrishnan, D. (2017). Contribution of CSR towards Development-the Indian perspective. *Available at SSRN* 3059833. DOI: 10.2139/ssrn.3059833

Rosati, F., & Faria, L. G. (2019). Addressing the SDGs in sustainability reports: The relationship with institutional factors. *Journal of Cleaner Production*, *215*, 1312–1326. DOI: 10.1016/j.jclepro.2018.12.107

Sabokro, M., Masud, M. M., & Kayedian, A. (2021). The effect of green human resources management on corporate social responsibility, green psychological climate and employees' green behavior. *Journal of Cleaner Production*, *313*, 127963. DOI: 10.1016/j.jclepro.2021.127963

Shahzad, M., Qu, Y., Zafar, A. U., Ding, X., & Rehman, S. U. (2020). Translating stakeholders' pressure into environmental practices–The mediating role of knowledge management. *Journal of Cleaner Production*, *275*, 124163. DOI: 10.1016/j.jclepro.2020.124163

Smith, T. W. (1999). Aristotle on the Conditions for and Limits of the Common Good. *The American Political Science Review*, *93*(3), 625–637. DOI: 10.2307/2585578

Teece, D. J., Pisano, G., & Shuen, A. (1997). 'Dynamic Capabilities and Strategic Management. *Strategic Management Journal*, *18*(7), 509–533. DOI: 10.1002/(SICI)1097-0266(199708)18:7<509::AID-SMJ882>3.0.CO;2-Z

Thirumalesh Madanaguli, A., Dhir, A., Kaur, P., Mishra, S., & Srivastava, S. (2023). A systematic literature review on corporate social responsibility (CSR) and hotels: Past achievements and future promises in the hospitality sector. *Scandinavian Journal of Hospitality and Tourism*, *23*(2-3), 141–175. DOI: 10.1080/15022250.2023.2221214

Uma, R., & Uma, K. (2021). Corporate social responsibility in india-an overview. *ACADEMICIA: An International Multidisciplinary Research Journal*, *11*(5), 906–912.

Van Marrewijk, M. (2003). Concept and Definitions of CSR and Corporate Sustainability: Between Agency and Corporate Sustainability: Between Agency nd Communion. *Journal of Business Ethics*, *44*(2), 95–105. DOI: 10.1023/A:1023331212247

Verma, S. (2011). Why Indian companies indulge in CSR? *Journal of Management and Public Policy*, *2*(2), 52–69.

Wang, Z., Shahid, M. S., Binh An, N., Shahzad, M., & Abdul-Samad, Z. (2022). Does green finance facilitate firms in achieving corporate social responsibility goals? *Economic research-. Ekonomska Istrazivanja*, *35*(1), 5400–5419. DOI: 10.1080/1331677X.2022.2027259

Windsor, D. (2001). The Future of Corporate Social Responsibility. *The International Journal of Organizational Analysis*, *9*(3), 225–256. DOI: 10.1108/eb028934

Wirba, A. V. (2023). Corporate social responsibility (CSR): The role of government in promoting CSR. *Journal of the Knowledge Economy*, *15*(2), 1–27. DOI: 10.1007/s13132-023-01185-0

Chapter 13
Healthcare Chatbots Using Artificial Intelligence and Sentiment Analysis

Mily Lal

Dr. D.Y. Patil School of Science and Technology, Dr. D.Y. Patil Vidyapeeth, India

S. Neduncheliyan

Department of CSE, School of Computing, Bharath Institute of Higher Education and Research, India

Arti Kaushik

https://orcid.org/0009-0008-6173-8232

G.V.M College of Education, Sonepat, India

Avinash Goswami

https://orcid.org/0009-0008-7441-9951

The LNM Institute of Information Technology, Jaipur, India

ABSTRACT

The integration of Artificial Intelligence in healthcare, especially through chatbots, has the potential to transform patient engagement and support. The authors explore how AI-powered chatbots utilize natural language processing, machine learning, and sentiment analysis to offer personalized, accessible, and continuous health-care. By tailoring interactions to individual needs, these chatbots enhance care quality, promote proactive health management, and improve patient satisfaction and adherence to treatment plans. Despite these benefits, the chapter highlights the importance of a cautious approach, including rigorous testing and ongoing

DOI: 10.4018/979-8-3693-8009-3.ch013

monitoring, to mitigate risks. Successful AI integration will hinge on maintaining trust among patients and healthcare professionals, ensuring AI remains a valuable tool in healthcare's evolving landscape.

1. INTRODUCTION

The healthcare industry has witnessed a significant transformation with the advent of artificial intelligence technology, which has introduced pioneering solutions aimed at enhancing patient experience and outcomes (Jovanovic et al., 2021). Healthcare chatbots, powered by AI and natural language processing, are among these innovative developments, offering personalized interactions, guidance, and support to patients (Iannantuono et al., 2023). These conversational agents hold the potential to redefine healthcare delivery by improving initial triage, symptom assessment, mental health support, and chronic disease management (Sun & Zhou, 2023). However, the integration of AI into healthcare chatbots also presents various challenges, including concerns about data privacy and security, algorithmic bias, and the ethical implications of automating human interactions. Addressing these issues requires a measured and well-informed approach to ensure that AI chatbots complement rather than replace human healthcare services, with a focus on rigorous testing, continuous monitoring, and maintaining the trust of patients (Altamimi et al., 2023).

This chapter aims to provide a thorough exploration of AI and sentiment analysis in healthcare chatbots. The key objectives are to: 1) Investigate the design and architecture of AI-powered healthcare chatbots, detailing their core components and functionalities. 2) Conduct a comparative analysis of various AI models and sentiment analysis techniques, evaluating their effectiveness and suitability for different healthcare applications. 3) Examine how AI technologies, including machine learning and sentiment analysis, are utilized to improve patient engagement and support, with a focus on enhancing personalized care and overall patient satisfaction. The chapter will cover the evolution of healthcare chatbots, their technological foundations, and their impact on patient care and clinical workflows, providing a comprehensive overview of the current advancements and future directions in this field.

2. OVERVIEW OF HEALTHCARE CHATBOTS

The integration of artificial intelligence within the medical field has facilitated the development of healthcare chatbots, which have enabled a transformative shift in the delivery of healthcare services. These conversational agents leverage AI capabilities to simulate human-like interactions, assisting patients and healthcare providers in a diverse range of applications, including symptom diagnosis, chronic disease management, and the provision of mental health support.

The origins of healthcare chatbots can be traced back to early AI systems that sought to mimic human decision-making. One of the pioneering examples is Eliza, a conversational agent developed in 1964 at the Massachusetts Institute of Technology's Artificial Intelligence Laboratory by Joseph Weizenbaum. Eliza was designed to simulate dialogue with a human therapist by identifying keywords within the input text and generating responses based on predefined reassembly rules. Despite its rudimentary nature, Eliza established the groundwork for more advanced chatbot systems by demonstrating the feasibility of machines engaging in meaningful interactions with humans (2023)(Fermat's Library | Eliza - A Computer Program For the Study of Natural Language Communication Between Man And Machine, 2023).

The advancement of AI technology facilitated the diversification of healthcare chatbots. The 1970s represented a pivotal period with the emergence of INTERNIST-1, an AI system developed to aid in clinical diagnosis. INTERNIST-1 employed a search algorithm to generate differential diagnoses based on patient symptomatology, underscoring the potential for AI to support healthcare professionals in their decision-making processes. Subsequently, systems such as MYCIN and DXplain further expanded the role of AI in healthcare by assisting with the diagnosis and treatment of infectious diseases and other medical conditions (Etori et al., 2023).

In the contemporary era, healthcare chatbots have become progressively more advanced, facilitated by advancements in natural language processing and machine learning. These conversational agents are now capable of executing a diverse array of functions, encompassing responding to patient inquiries, scheduling appointments, providing medication reminders, and even offering mental health support. For example, IBM's Watson, which initially gained recognition by triumphing in the game show Jeopardy!, has been adapted for healthcare applications, such as identifying RNA-binding proteins associated with amyotrophic lateral sclerosis and delivering personalized treatment recommendations (Bohr & Memarzadeh, 2020)

Healthcare conversational agents have become valuable tools for not only engaging with patients but also optimizing clinical workflows. For instance, organizations such as BotMD have developed AI-powered systems that support healthcare professionals with a variety of clinical responsibilities, including locating on-call physicians, scheduling appointments, addressing prescription-related queries, and accessing hospital protocols. These technological advancements have enhanced the efficiency of healthcare service delivery, especially in resource-constrained settings (Al-Shorbaji, 2022).

The rise of telehealth and the increasing demand for healthcare services have further fueled the adoption of healthcare chatbots. These chatbots offer a scalable solution to the growing shortage of healthcare practitioners, particularly in underserved regions. By providing on-demand healthcare services via smartphones and other devices, chatbots help reduce costs, improve access to care, and minimize the risk of exposure to contagious illnesses (Haque et al., 2023).

3. ROLE OF ARTIFICIAL INTELLIGENCE IN HEALTHCARE

AI has become a valuable tool in healthcare, driving innovation and transforming how services are delivered, managed, and optimized. AI's ability to process and learn from large amounts of data has opened new ways to improve patient outcomes, enhance healthcare system efficiency, and personalize medical care.

3.1. Enhancing Diagnostic Accuracy

One of the most significant contributions of AI in healthcare is its capacity to enhance diagnostic precision. Traditional diagnostic processes often heavily rely on the expertise and intuition of healthcare professionals, which, although valuable, can occasionally result in variable outcomes. AI addresses this challenge by analyzing extensive datasets, encompassing medical images, patient records, and genetic information, to identify patterns that might elude human perception. For example, AI models have been developed to assist in the diagnosis of conditions such as skin cancer, diabetic retinopathy, and various other diseases by analyzing medical images with a level of precision that rivals or even surpasses that of experienced clinicians. These systems possess the ability to process thousands of images in a fraction of the time required by human experts, thereby providing rapid and accurate diagnoses that can significantly enhance patient care (Quer et al., 2017).

3.2. Facilitating Personalized Medicine

The ability of AI to process and analyze complex data sets has been crucial in driving the progress of personalized medicine. Personalized medicine involves tailoring medical interventions to the unique characteristics of each patient, including their genetic profile, lifestyle, and environmental factors. AI algorithms can leverage a patient's genetic data alongside other health information to predict their potential response to various treatments, enabling healthcare providers to select the most effective therapies while minimizing the risk of adverse effects. This approach is particularly valuable in the management of complex diseases, such as cancer, where

AI can assist in identifying the most promising treatment options based on the specific genetic mutations present in a patient's tumor (Álvarez-Machancoses et al., 2020).

3.3. Optimizing Clinical Workflows

In addition to its applications in diagnostics and treatment, AI is increasingly being leveraged to optimize clinical workflows, thereby enhancing the efficiency of healthcare service delivery. AI-driven systems can automate routine administrative responsibilities, such as scheduling appointments, managing patient records, and processing insurance claims, allowing healthcare providers to devote more time and attention to patient care. For instance, AI can streamline the patient intake process by automatically populating forms and updating electronic health records, reducing the burden on administrative personnel and mitigating the risk of errors. Furthermore, AI can assist healthcare organizations in predicting patient admissions, managing staffing levels, and allocating resources more effectively, ensuring the timely provision of appropriate care(Davenport & Glaser, 2022).

3.4. Supporting Predictive Analytics and Early Intervention

The capacity of AI to analyze data in real-time has also rendered it a valuable asset for predictive analytics and early intervention in the healthcare domain. By continuously monitoring patient data, including vital signs and laboratory results, AI systems possess the capability to identify early indications of potential health concerns before they escalate into critical situations. For example, AI algorithms can forecast the likelihood of a patient experiencing postoperative complications or identify individuals at risk of developing chronic conditions such as diabetes or cardiovascular disease. This proactive approach empowers healthcare providers to intervene at an earlier stage, potentially mitigating the occurrence of severe health outcomes and reducing the overall costs of care (Stellefson et al., 2013).

3.5. Enhancing Patient Engagement and Education

AI-powered healthcare technologies have emerged as valuable tools for enhancing patient engagement and education. Conversational agents, virtual assistants, and mobile applications provide patients with convenient access to information about their medical conditions, treatment alternatives, and medication regimens. These intelligent systems can address common patient inquiries, deliver medication adherence reminders, and offer personalized health guidance tailored to individual patient data. By empowering patients with knowledge and self-management resources, AI-driven technologies foster greater patient involvement in their healthcare, which can translate to improved health outcomes and enhanced patient satisfaction (Dhopte & Bagde, 2023).

3.6. Advancing Drug Discovery and Development

The drug discovery and development pipeline is renowned for its substantial time and financial requirements, frequently spanning years and costing billions to bring a new medication to market. AI is revolutionizing this process by expediting the identification of promising drug candidates and optimizing clinical trial design. Machine learning algorithms can analyze extensive datasets encompassing chemical compounds, biological information, and clinical trial results to pinpoint novel drug targets and forecast how diverse compounds may interact with those targets. Furthermore, AI can assist in designing more efficient clinical trials by identifying the most appropriate patient cohorts and predicting potential outcomes, ultimately accelerating the drug development timeline and accelerating the availability of new treatments for patients (Askin et al., 2023).

3.7. Addressing Public Health Challenges

AI has also proven instrumental in tackling broader public health challenges. During the COVID-19 crisis, AI models were leveraged to simulate viral transmission dynamics, forecast outbreaks, and optimize the allocation of critical resources like ventilators and vaccines. Additionally, AI-driven analytical tools were employed to study vast quantities of public health data, enabling governments and healthcare organizations to make informed decisions regarding lockdown measures, testing protocols, and vaccination programs. Beyond pandemic response, AI can be further utilized to monitor and predict the spread of infectious diseases, assess the health implications of environmental factors, and evaluate the effectiveness of public health interventions (Nguyen et al., 2020).

3.8. Overcoming Challenges and Ethical Considerations

The implementation of AI in healthcare presents significant advantages, yet also raises critical challenges and ethical considerations that require careful deliberation. Concerns surrounding data privacy, algorithmic bias, and the interpretability of AI-driven decision-making processes must be diligently addressed. Ensuring the design and deployment of AI systems that uphold principles of fairness, accountability, and transparency is essential to mitigate unintended outcomes and foster trust among healthcare stakeholders, including providers and patients (Crigger et al., 2022).

4. IMPORTANCE OF SENTIMENT ANALYSIS IN HEALTHCARE

Natural language processing techniques, such as sentiment analysis, have become pivotal in healthcare by enabling the interpretation of patients' emotional states and viewpoints. This technological capability enhances various aspects of patient care, including communication and decision-making, through the analysis of feedback gathered from diverse sources like social media platforms and survey responses. By gaining deeper insights into patient sentiments, healthcare providers can implement targeted improvements, such as optimizing wait times, which ultimately leads to enhanced patient experiences and increased satisfaction. (Chiang et al., 2019)

Furthermore, sentiment analysis has become a valuable tool for monitoring patient mental health by identifying indications of anxiety or depression in their communications (McCoy et al., 2015). It also enhances patient-provider interactions by enabling the interpretation of the emotional tone of messages, facilitating more empathetic responses. Additionally, it supports improved treatment adherence by detecting negative emotions that may influence a patient's willingness to follow prescribed care plans (Devaram, 2020). Beyond individual-level applications, sentiment analysis can be utilized to analyze broader public health trends and guide the optimization of healthcare services, informing both research and policy decisions.

5. DESIGN AND ARCHITECTURE OF AI-POWERED HEALTHCARE CHATBOTS

The development of AI-powered healthcare chatbots entails a methodical process focused on designing intelligent systems that elevate patient engagement and support. At the heart of this design approach lies a clear delineation of the chatbot's purpose and scope. These conversational agents are purposefully crafted to address specific healthcare needs, such as offering patient assistance, conducting symptom assessments, or managing appointment scheduling. Comprehending the target user population and their requirements is pivotal in tailoring the chatbot's functionalities to effectively meet real-world healthcare demands.

5.1. Key Components of a Healthcare Chatbot

Healthcare chatbots consist of both the user interface and the backend infrastructure. A well-designed user interface is critical to enabling seamless user interactions, featuring an intuitive and user-friendly design that facilitates smooth communication. On the backend, robust systems manage user requests, process data, and maintain user interactions. Furthermore, integrating the chatbot with existing healthcare information systems, such as electronic health records and appointment management platforms, is essential for delivering accurate and comprehensive support to users. Figure 1. Demonstrates the architecture of the healthcare chatbot.

Figure 1. Architecture of the healthcare chatbot.

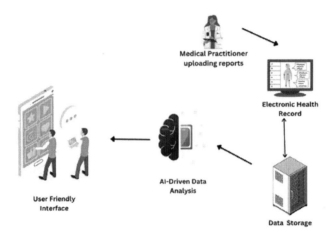

The design of effective conversational flows is another essential component of healthcare chatbot development. Crafting robust conversational scripts and implementing robust dialogue management strategies are crucial for handling a diverse range of user inputs and scenarios. By establishing well-defined conversational pathways, chatbots can guide users through their interactions and deliver pertinent responses, thereby enhancing the overall quality of the user experience.

6. AI TECHNIQUES FOR HEALTHCARE CHATBOTS

Innovations in Artificial Intelligence have facilitated the development of advanced chatbots that are revolutionizing the healthcare domain. These conversational agents harness a variety of AI techniques to offer personalized, empathetic, and efficient assistance to patients. Various AI techniques, including machine learning, natural language processing, and deep learning, collectively contribute to the chatbot's intelligence.

Natural language processing represents a fundamental AI technique leveraged in healthcare chatbots. NLP empowers these conversational agents to comprehend and interpret the natural language employed by patients, facilitating more natural and intuitive dialogues. By analyzing the semantic meanings and contextual cues within patient inputs, NLP equips the chatbots to deliver relevant and personalized responses, thereby elevating the overall user experience (Haque et al., 2023).

Machine learning algorithms, such as classification and regression models, empower chatbots to learn from data and generate informed predictions. Furthermore, the utilization of machine learning algorithms is pivotal in enabling healthcare chatbots to continuously learn and adapt over time. By analyzing extensive patient data, including medical records, chat logs, and user feedback, these machine learning models are able to identify patterns and trends that inform and refine the decision-making processes of the chatbots. This dynamic learning capability allows the chatbots to persistently enhance their functionalities and offer increasingly personalized and effective support to patients.

Furthermore, the incorporation of deep learning techniques, such as neural networks and transformer models, further enhances the language capabilities of healthcare chatbots. These advanced architectures enable the chatbots to better comprehend patient inputs and generate more contextually relevant and natural responses. Deep learning models excel at extracting complex semantic features and discerning intricate relationships from large language datasets, empowering the chatbots to engage in more nuanced and contextually aware dialogues with patients (Lal & Neduncheliyan, 2024).

Sentiment analysis represents another crucial AI technique that empowers healthcare chatbots to comprehend and respond to the emotional states of patients. This technique involves evaluating user emotions and intentions through textual data. By scrutinizing the tone, language, and other indicators within patient interactions, sentiment analysis equips the chatbots to dynamically modify their responses, fostering empathetic and emotionally supportive dialogues. This functionality not only enhances the overall user experience but also enables the chatbot to address diverse emotional states effectively (Devaram, 2020).

Studying real-world exemplars of AI-driven healthcare chatbots offers valuable insights into their design and deployment. Case studies of successful systems, such as those developed by Babylon Health and Ada Health, elucidate how these conversational agents are architected and the impact they have on patient experiences. Analyzing these practical use cases can help identify industry best practices and common challenges, providing actionable lessons to guide the development of effective healthcare chatbots (Montenegro et al., 2019).

6.1. Comparative Analysis of the Various AI Models

Understanding the advantages, limitations, and appropriate use cases of different AI models and sentiment analysis methods is crucial for developers and healthcare professionals seeking to implement chatbot solutions that meet specific patient needs and healthcare goals. comparative analysis of various AI models and sentiment analysis techniques used in healthcare chatbots, presented in Table 1.

Table 1. Comparative analysis of various AI models and sentiment analysis techniques

Aspect	Machine Learning Models	Deep Learning Models	NLP Techniques	Sentiment Analysis Techniques
Examples	Decision Trees, SVM, KNN	CNN, RNN, LSTM, Transformers	Tokenization, Named Entity Recognition (NER), Dependency Parsing	Rule-Based Methods, Machine Learning Approaches (e.g., SVM, Naive Bayes), Deep Learning Models (e.g., LSTM, Transformers)
Strengths	- Simple and interpretable (Decision Trees) - Effective for classification and regression tasks - Suitable for small to medium datasets	- High accuracy in complex tasks - Excellent for pattern recognition (CNN) - Effective in processing sequential data (RNN, LSTM) - Context-aware language understanding (Transformers)	- Facilitates understanding of user inputs - Essential for breaking down complex medical terminology - Advanced models (Transformers) offer nuanced text generation and understanding	- Rule-Based: Simple implementation - ML-Based: Learns from labeled data, improving accuracy - Deep Learning: Captures context, nuance, and complex emotions effectively
Weaknesses	- Limited in handling high-dimensional data (except SVM) - Requires feature engineering - May struggle with complex patterns	- Requires large datasets for training - Higher computational costs - Less interpretable compared to simpler models	- Basic techniques (e.g., tokenization) may not capture context - Advanced techniques require significant computational resources	- Rule-Based: Limited adaptability - ML-Based: Needs extensive training data - Deep Learning: High computational cost, requires large labeled datasets

continued on following page

Table 1. Continued

Aspect	Machine Learning Models	Deep Learning Models	NLP Techniques	Sentiment Analysis Techniques
Best Suited For	- Basic classification and regression tasks - Applications with limited data - Use cases requiring interpretability	- Complex, high-dimensional tasks - Applications needing high accuracy - Natural language processing and image analysis	- Understanding and generating medical language - Enhancing chatbot's conversational abilities - Advanced applications with nuanced language needs	- Basic sentiment detection (Rule-Based) - More complex emotion detection (ML-Based) - Advanced emotional understanding in real-time interactions (Deep Learning)
Healthcare Applications	- Symptom classification - Basic patient data analysis - Treatment recommendation systems	- Medical image analysis (CNN) -Conversational AI for complex dialogues (RNN, Transformers) - Predictive analytics and personalized medicine	- Processing and interpreting medical queries - Generating accurate responses - Handling complex dialogues with nuanced medical language	- Detecting patient emotions during chatbot interactions - Providing empathetic responses - Enhancing patient satisfaction and engagement through emotion-aware interactions

An examination of AI models and sentiment analysis techniques illuminates their relative strengths and limitations, informing developers in the selection of the most suitable tools for their specific chatbot applications. While rule-based sentiment analysis may be adequate for rudimentary applications, deep learning models demonstrate greater aptitude for more complex tasks necessitating nuanced understanding and dynamic responsiveness. Similarly, the choice between machine learning and deep learning models hinges on the task's complexity, data availability, and the required level of interpretability.

7. ENHANCING PATIENT ENGAGEMENT AND SUPPORT THROUGH AI

The integration of Artificial Intelligence has emerged as a transformative force within the healthcare domain, particularly in enhancing patient engagement and support. AI-powered healthcare chatbots are at the vanguard of this transformation, delivering innovative solutions that transcend basic interaction to actively engage patients in their healthcare experiences. The core objective is to investigate how AI technologies can be leveraged to bolster patient engagement and establish robust support systems that catalyze improved health outcomes and patient satisfaction.

The capacity to personalize interactions is central to enhancing patient engagement through AI. Traditional healthcare approaches frequently struggle to deliver individualized care due to time constraints and the standardized nature of many healthcare processes. In contrast, AI-driven chatbots can analyze extensive patient data, including medical history, preferences, and behavioral patterns, to tailor interactions to the unique needs of each individual. For instance, a chatbot may provide personalized guidance on managing a chronic condition, suggest lifestyle modifications based on a patient's health profile, or furnish information on upcoming medical appointments. This level of personalization not only makes patients feel valued and understood, but also encourages them to take a proactive role in managing their health, which is pivotal for improving adherence to treatment plans and achieving optimal health outcomes (Crutchfield & Kistler, 2017).

Furthermore, AI-powered healthcare chatbots enhance patient support by offering continuous and accessible assistance, which can be particularly advantageous for individuals managing chronic conditions or complex treatment regimens. These conversational agents can provide medication reminders, follow up on patient symptoms, and deliver educational content tailored to the specific needs of each patient. For example, a chatbot may regularly update a diabetic patient on strategies for managing blood sugar levels, remind them to monitor their glucose, and suggest tips for maintaining a healthy diet. This type of sustained support helps patients adhere to their prescribed treatment plans, thereby mitigating the risks of complications and hospital readmissions.

Additionally, AI-powered healthcare chatbots enhance patient support by offering continuous and readily available assistance, which can be particularly advantageous for individuals managing chronic conditions or complex treatment regimens. These conversational agents can provide medication reminders, follow up on patient symptoms, and deliver educational content tailored to the specific needs of each patient. For example, a chatbot may regularly update a diabetic patient on strategies for managing blood sugar levels, remind them to monitor their glucose, and suggest tips for maintaining a healthy diet. This type of sustained support helps patients adhere to their prescribed treatment plans, thereby mitigating the risks of complications and hospital readmissions (Unger, 2012).

The ability to perform sentiment analysis is a pivotal capability of AI-powered healthcare chatbots, as it enables these systems to interpret and address the emotional responses of patients. By scrutinizing the language and tone used by patients, the chatbots can detect a range of emotional states, such as anxiety, frustration, or sadness. For instance, if a patient expresses concern about a new symptom or treatment side effect, the chatbot can recognize this emotional state and respond accordingly, providing empathetic reassurance, additional relevant information, or a recommendation to consult with a healthcare provider. This capacity to adapt the chatbot's response to the patient's emotional state is crucial for cultivating trust and ensuring that patients feel genuinely supported, especially in situations where they may be experiencing heightened anxiety or uncertainty.

Patient satisfaction is a crucial measure of the efficacy of healthcare services, and AI-powered chatbots hold significant potential to enhance this metric. By furnishing timely, accurate, and empathetic responses, these conversational agents can markedly improve the patient experience. Healthcare providers can assess the success of these interactions through diverse assessment tools, including feedback surveys, user engagement metrics, and sentiment analysis outcomes. Elevated satisfaction levels indicate that patients perceive value in the chatbot's support, which may foster improved adherence to medical recommendations and a strengthened patient-provider relationship (Palanica et al., 2019).

Additionally, the application of AI in patient engagement and support transcends mere immediate interactions. AI-driven analytics can monitor and analyze patient interactions longitudinally, furnishing healthcare providers with invaluable insights into patient behavior, adherence trends, and overall satisfaction levels. These insights can then be leveraged to refine and enhance chatbot services, ensuring they persistently align with the evolving needs of patients.

8. CONCLUSION

The integration of AI-powered chatbots into healthcare holds significant potential to revolutionize patient engagement and support. These conversational agents, leveraging advanced AI techniques such as natural language processing, machine learning, and sentiment analysis, can tailor interactions to the unique needs of individual patients. This personalized approach enhances the quality of care and encourages a more proactive role in health management. By furnishing real-time responses, customized educational content, and empathetic support, healthcare chatbots can markedly improve patient satisfaction, adherence to treatment plans, and overall health outcomes.

Deploying AI in healthcare also presents several challenges that require careful management. Issues pertaining to data privacy, algorithmic bias, and the ethical ramifications of automating healthcare interactions must be addressed to ensure these technologies complement, rather than supplant, human healthcare providers. A prudent and informed approach, incorporating rigorous testing, continuous monitoring, and ongoing refinement, is essential to harness the full potential of AI chatbots while mitigating risks and preserving the trust of patients and healthcare professionals alike.

REFERENCES

Al-Shorbaji, N. (2022, February 9). Improving Healthcare Access through Digital Health: The Use of Information and Communication Technologies. IntechOpen. https://doi.org/DOI: 10.5772/intechopen.99607

Altamimi, I., Altamimi, A., Alhumimidi, A S., Altamimi, A., & Temsah, M. (2023, June 25). Artificial Intelligence (AI) Chatbots in Medicine: A Supplement, Not a Substitute. Cureus, Inc.. DOI: 10.7759/cureus.40922

Álvarez-Machancoses, Ó., DeAndrés-Galiana, E. J., Cernea, A., Viña, J F S. D. L., & Fernández-Martínez, J. L. (2020, March 1). On the Role of Artificial Intelligence in Genomics to Enhance Precision Medicine. *Pharmacogenomics and Personalized Medicine*, *13*, 105–119. DOI: 10.2147/PGPM.S205082 PMID: 32256101

Askin, S., Burkhalter, D., Calado, G., & Dakrouni, S E. (2023, February 28). Artificial Intelligence Applied to clinical trials: opportunities and challenges. Springer Science+Business Media, 13(2), 203-213. https://doi.org/DOI: 10.1007/s12553-023-00738-2

Bohr, A., & Memarzadeh, K. (2020, January 1). The rise of artificial intelligence in healthcare applications. Elsevier BV, 25-60. https://doi.org/DOI: 10.1016/B978-0-12-818438-7.00002-2

Chiang, A. L., Rabinowitz, L., Kumar, A., & Chan, W. W. (2019, September 6). Association Between Institutional Social Media Involvement and Gastroenterology Divisional Rankings: Cohort Study. *Journal of Medical Internet Research*, *21*(9), e13345–e13345. DOI: 10.2196/13345 PMID: 31493321

Crigger, E., Reinbold, K., Hanson, C., Kao, A., Blake, K., & Irons, M. (2022, January 12). Trustworthy Augmented Intelligence in Health Care. Springer Science+Business Media, 46(2). https://doi.org/DOI: 10.1007/s10916-021-01790-z

Crutchfield, T. M., & Kistler, C. E. (2017, January 1). Getting patients in the door: Medical appointment reminder preferences. *Patient Preference and Adherence*, *11*, 141–150. DOI: 10.2147/PPA.S117396 PMID: 28182131

Davenport, T H., & Glaser, J. (2022, October 31). Factors governing the adoption of artificial intelligence in healthcare providers. Springer Science+Business Media, 1(1). https://doi.org/DOI: 10.1007/s44250-022-00004-8

Devaram, S. (2020, January 1). Empathic Chatbot: Emotional Intelligence for Empathic Chatbot: Emotional Intelligence for Mental Health Well-being. Cornell University. https://doi.org//arxiv.2012.09130DOI: 10.48550

Dhopte, A., & Bagde, H. (2023, June 30). Smart Smile: Revolutionizing Dentistry With Artificial Intelligence. Cureus, Inc.. https://doi.org/DOI: 10.7759/cureus.41227

Etori, N. A., Temesgen, E., & Gini, M. (2023, January 1). What We Know So Far: Artificial Intelligence in African Healthcare. Cornell University. https://doi.org/ DOI: 10.48550/arXiv.2305

Fermat's Library | Eliza - A Computer Program For the Study of Natural Language Communication Between Man And Machine. (2023, April 14). https://fermatslibrary .com/s/a-computer-program-for-the-study-of-natural-language-communication -between-man-and-machine

Haque, A., Chowdhury, M N., & Soliman, H. (2023, June 1). Transforming Chronic Disease Management with Chatbots: Key Use Cases for Personalized and Cost-effective Care. https://doi.org/DOI: 10.1109/IS3C57901.2023.00104

Iannantuono, G. M., Bracken-Clarke, D., Floudas, C. S., Roselli, M., Gulley, J. L., & Karzai, F. (2023, September 4). Applications of large language models in cancer care: Current evidence and future perspectives. *Frontiers of Medicine*, *13*, 1268915. Advance online publication. DOI: 10.3389/fonc.2023.1268915 PMID: 37731643

Jovanovic, M., Báez, M., & Casati, F. (2021, May 1). Chatbots as Conversation-al Healthcare Services. *IEEE Internet Computing*, *25*(3), 44–51. DOI: 10.1109/ MIC.2020.3037151

Lal, M., & Neduncheliyan, S. (2024, March 11). Conversational artificial intelligence development in healthcare. Springer Science+Business Media. https://doi.org/DOI: 10.1007/s11042-024-18841-5

McCoy, T. H., Castro, V. M., Cagan, A., Roberson, A. M., Kohane, I. S., & Perl-is, R. H. (2015, August 24). Sentiment Measured in Hospital Discharge Notes Is Associated with Readmission and Mortality Risk: An Electronic Health Record Study. *PLoS One*, *10*(8), e0136341–e0136341. DOI: 10.1371/journal.pone.0136341 PMID: 26302085

Montenegro, J L Z., Costa, C A D., & Righi, R D R. (2019, September 1). Survey of conversational agents in health. Elsevier BV, 129, 56-67. https://doi.org/DOI: 10.1016/j.eswa.2019.03.054

Nguyen, T. T., Nguyen, Q. V. H., Nguyen, D. T., Yang, S., Eklund, P., Huynh-The, T., Nguyên, T. T., Pham, Q., Razzak, I., & Hsu, E. B. (2020, January 1). Artificial Intelligence in the Battle against Coronavirus (COVID-19): A Survey and Future Research Directions. Cornell University. https://doi.org//arxiv.2008.07343DOI: 10.48550

Palanica, A., Flaschner, P., Thommandram, A., Li, M. H., & Fossat, Y. (2019, April 5). Physicians' Perceptions of Chatbots in Health Care: Cross-Sectional Web-Based Survey. *Journal of Medical Internet Research*, *21*(4), e12887–e12887. DOI: 10.2196/12887 PMID: 30950796

Quer, G., Muse, E D., Nikzad, N., Topol, E J., & Steinhubl, S R. (2017, July 1). Augmenting diagnostic vision with AI. Elsevier BV, 390(10091), 221-221. https://doi.org/DOI: 10.1016/S0140-6736(17)31764-6

Stellefson, M., Dipnarine, K., & Stopka, C. (2013, February 21). The Chronic Care Model and Diabetes Management in US Primary Care Settings: A Systematic Review. *Preventing Chronic Disease*, *10*, 120180. Advance online publication. DOI: 10.5888/pcd10.120180 PMID: 23428085

Sun, G., & Zhou, Y. (2023, December 19). AI in healthcare: Navigating opportunities and challenges in digital communication. *Frontiers of Medicine*, *5*, 1291132. Advance online publication. DOI: 10.3389/fdgth.2023.1291132 PMID: 38173911

Unger, J. (2012, March 1). Uncovering undetected hypoglycemic events. Dove Medical Press, 57-57. https://doi.org/DOI: 10.2147/DMSO.S29367

Chapter 14
Social Capital and Sales Promotions:
Entrepreneurial Insights Into Network–Based Marketing

Roop Kamal

ⓘ https://orcid.org/0009-0001-8334-0101

Chandigarh University, India & RIMT University, India

ABSTRACT

This paper explores the critical role of social capital in enhancing the effectiveness of sales promotional strategies within entrepreneurial ventures. Social capital, defined as the value derived from social networks, provides entrepreneurs with access to essential resources, market information, and trust, all of which are pivotal in crafting successful promotional campaigns. The study reveals a strong positive relation between an entrepreneur's social capital and the success of their sales promotions. Key findings highlight that entrepreneurs with robust social networks can more effectively mobilize resources, gain market insights, and leverage word-of-mouth and viral marketing, ultimately leading to greater customer acquisition and market penetration. This study provides practical insights for entrepreneurs aiming to leverage their social capital for more impactful marketing efforts, offering a pathway to sustained business growth and competitive advantage.

INTRODUCTION

Being a rather broad notion, social capital became a subject of focus within different areas and disciplines such as sociology, economics, political science, and business. It means the web of social relations extended among people, communities

DOI: 10.4018/979-8-3693-8009-3.ch014

and organisations and the value arising from such a web (Yudha, 2018). Social is not like physical capital; physical capital is like money or a machine or machinery that can be seen and touched and felt, or even human capital which is human skills, human knowledge, human experience and human relations, but social capital is with the social relations through which people are able to cooperate and trust each other and be willing to help each other (Schmitt et al., 2024).

The antecedent notion of Social Capital was developed from the assertion that networks have stock value. People and organizations can use their relationships wherever for the provisions of the means, necessary information, or assistance that in other circumstance might be constrained or perhaps out of reach. This is made possible by trust, norms and obligation that are created in these networks hence the importance of social capital in realizing self and other people's agendas (Savari & Khaleghi, 2023).

1. **Bonding Social Capital:** This type of social capital is confined for instance within the realms of families, intimate circles of friends or a very small and restrictive society. It is characterized by such social capital where there are close contacts, high level of trust and mutual obligations between members. Bonding social capital is particularly crucial for support in terms of the desire to belong and maintenance of social cohesion (Muliadi et al., 2024).

2. **Bridging Social Capital:** Bridging social capital is associated with linking people or groups that are dissimilar; it may involve people that belong to different social, economic or culture class. These connections are generally less close than in bonding social capital, although essential for collecting new information, opportunities and resources. While bridging social capital creates social as well as structural integration to incorporate the wider society into one's world or into the world of a certain group (Kim & Aldrich, 2005).

3. **Linking Social Capital**: The concept of linking social capital is the relation between people or groups and organizations, entities or power structures. It is a matter of affiliation with persons or bodies in authority, for instance, the political leadership or business entities. External social capital is crucial in the sense that it opens the gates towards access to resources and opportunities that are beyond an individual or group's discretion (Chen & Lee, 2024).

Delacroix & Suire social capital theoretical framework give social capital a significant role in business success within the context of entrepreneurship. In this study, it was ascertained that attending firms depend upon their networks, to acquire information on market opportunities, resources, partners and customer trust (Purwati et al., 2021). Social capital helps the entrepreneurs to mitigate the issue of uncertainty and limited resources hence giving the entrepreneur a competitive edge in the market (Dar & Mishra, 2020).

However, social capital like any form of capital can be developed and increased in its stock over time. There is also something the entrepreneurs can do which directly involves the social capital as an asset: the latter involves networking, work with the stakeholders within and outside the organizational environment, and community or industry memberships. In this way, they can set up proper conditions which will encourage innovation, mutual cooperation and development (Sanyal et al., 2024).

Social capital is a very important and complex concept that reflects the importance of relationships and networks to the attainment of personal as well as group objectives. The application reaches across the board, and in the sphere of business, it is a valuable tool that can help entrepreneur work through various challenges of the business environment.

SALES PROMOTIONS

Sales promotion is a strategic tool in the marketing mix since it has a significant effect on the consumers' behavior, sales and brands awareness. As a temporary tool in the marketing communications mix, the aim of sales promotion is to get the consumer to take an immediate action in terms of purchase of product or availing of a service or making an immediate connection with a brand (Elias, 2024).

Promotion sales involve a range of strategies that are employed in marketing for the purpose of increasing consumer pull and achieving goal oriented marketing strategies. These techniques can be vouchers, discounts, rebates, carry out promotions, gifts, loyalty program, and time-bound lots of others. The main aim of such activities is to encourage a consumer to buy a particular product or to take a particular decision before a particular date (Schmitt et al., 2024).

Types of Sales Promotions

Figure 1. Different types of sales promotions

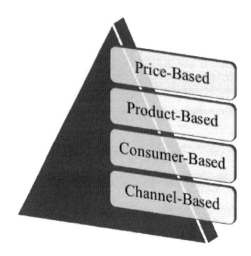

1. **Price-Based Promotions:** These are: discounted prices, reduced prices, two for price of one, and rebates. Value-appeal promotions are yet another common type of promotions that appeal to the self-interest for consumers by offering them price breaks thus getting them to purchase product (Baycan & Öner, 2023).
2. **Product-Based Promotions:** This category can include freebie deliveries, the possibility to taste certain products, and the use of offers that combine several products. For promotional tools, the focus is made on the point that they aim at the presentation of a new product or at additional consumption of other goods and services.
3. **Consumer-Based Promotions:** These promotions are aimed at appealing to the consumer taking full advantage of Contest/Sweepstakes, loyalty and other reward programs. Consumer-based promotions build and sustain a brand image and directly link consumer mandated bonuses to purchasing behavior.
4. **Channel-Based Promotions:** They are used on distributors such as the retailers and the wholesalers as a way of encouraging them take more of a specific product in their inventory, advertise it or even sell it to customers (Sanyal et al., 2024). This category consists of trade allowances, point of purchase displays as well as cooperative advertising.

Objectives of Sales Promotions

Indeed, sales promotions have multiple and sundry purposes which are dependent on the business environment and the stage in the product life-cycle (Mishchuk et al., 2023). Some of the key objectives include:

Figure 2. Objectives of sales promotions

1. Boosting Short-Term Sales
2. Encouraging Product Trial
3. Increasing Brand Awareness
4. Enhancing Customer Loyalty

1. **Boosting Short-Term Sales**: One of the main goals of a sales promotion is thus to increase sales volume through promotional offers that are time-bound. This is especially vital in the case of invulnerable products, when introducing new products in the market or during periods of low activity.
2. **Encouraging Product Trial:** It indicates that sales promotions can be also used in market introduction new products in communicating that the risk of buying the product is low (Chen & Lee, 2024). A free sample or low-priced deal makes the consumer experiment with a new product rather than investing 'real money' for the same.
3. **Increasing Brand Awareness:** Special promotions are normally meant to increase brand familiarity and recognition; more so when goods and services are many in the market. Through promoting the chosen brand, consumers may pay their attention to it, and the more likely they will make the right choice and select this brand (Elias, 2024).
4. **Enhancing Customer Loyalty:** Loyalty programs and rewards are other methods that sales promotions use to foster close relationships with the customers. This means that loyalty can be encouraged through a range of activities, which make customers come back for more in the place of going to other business establishments (Annamalah et al., 2023).

STRATEGIC ROLE OF SALES PROMOTION IN MARKETING:

Sales promotions are not single tools of marketing communication but are a subset of the marketing technique. Sales promotion, when integrated with advertising and public relation and other marketing communication activities can complement the brand's overall marketing communication. It is easily quantifiable and allows for a control of consumers' actions and is especially suitable for getting particular groups of consumers' attention or for meeting particular business goals (Yudha, 2018).

Thus, sales promotions must be managed well so that they are as effective as they should be. Promotional marketing when overused particularly on price promotions may cause 'promotion wear out' effect on the consumer or even lower perception of the brand. This means that the marketers should be very careful in their use of promotions in as much as they need to support their long term brand building strategies for growth (Sanyal et al., 2024).

Sales promotions are therefore one of the most effective and time tested strategies among all the marketing communication tools that are available to a marketer; sales promotions are always immediate and straightforward in that they provide marketers with an easy and effective way of directly reaching out to consumers and being certain of the changes that has ensued as a result of such contact (Yani et al., 2020). Also, it's worth noting that through these strategies, sales promotions used when appropriate aims at creating value for both the business and the customer where the value may be in enhancing on sales, introducing new products to the market or rewarding loyal customers (Frese & Gielnik, 2023).

ENTREPRENEURSHIP

It may also arguably be defined as the process of searching for new business and the accomplishment of engendering value in new forms (Bhandari & Bhuyan, 2023). This is a major force for economic and employment, and for the transformation of society as whole since it generates new ideas for products, services, and business models, which economic agents offer to the market in response to the needs and demand from the consumer.

In its essence, entrepreneurship is not just the creation of an enterprise but the ability to identify available opportunity and then create, build and manage new ventures by assuming certain risks (Kanini & Muathe, 2019). The term 'Entrepreneur' is generally associated with an individual's capacity to imagine, plan and design for the future, as well as the capacity to innovate and make decisions in conditions of risk. They occupy a prominent place in creating industries and niche markets that cause shifts in the conventional approaches to doing commerce (Ince et al., 2023).

Key Elements of Entrepreneurship

Figure 3. Key elements of entrepreneurship

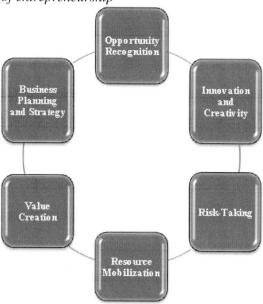

1. **Opportunity Recognition**: The journey knocked off by searching for an opportunity to exploit. Bear in mind that business people are very good at realizing that there exist 'holes' or 'niche markets' where people's needs are not fully meet or processes that exist and that needs improvement (Parthymos & Daskalopoulou, 2024). Ideation requires a creative approach, looking at markets, and an analysis of customer needs and purchasing patterns.

2. **Innovation and Creativity**: Innovation is used synonymously with entrepreneur as a process of creating and bringing into operation new goods, services, interpretations, techniques and forms of organizing the economic activity. There exists a continuum of innovation, which can be process innovation, refining existing ideas or radical innovation, which is the creation of new need solutions. Ideation involves the ability to come up with unique or original ideas which can be taken to form the basis of a business (Muliadi et al., 2024).

3. **Risk-Taking**: It is a well-known fact that any kind of business or business plan entails risk factor. Much private capital, time, and resources are put into business by entrepreneurs for which they cannot be sure of success. Risk appraisal and management and compliance with risk mitigation measures are vital skills.

Businessmen have to have thick skin and be ready to change and grow from their mistakes (Sabet & Khaksar, 2024).

4. **Resource Mobilization**: For one to be an entrepreneur, one has to have adequate skills in identifying and assembling such vital resources as capital, manpower and technology. Self starters have to be able to fund raise, assemble personnel and form partnerships to foster their new initiatives.

5. **Value Creation**: The main aim of venturing is to create new worth in the business undertakings, be it for the consumer or society. This value can be monetary as profits, or can be in terms of effecting changes within society and the physical environment. Business people require their businesses to be sensitive to the prevailing market needs and come up with products that are marketable (Yani et al., 2020).

6. **Business Planning and Strategy**: One of the keys to any entrepreneurial endeavor is an accurate business model Document and sound strategy. This process entails defining goals and objectives, description of the business strategy, analysis of the competition and a strategy map for the business (Dar & Mishra, 2020). Strategic planning enables the budding business person to understand the hurdles and the processes in establishing a business.

Types of Entrepreneurship

Entrepreneurship can take many forms, depending on the goals, scale, and impact of the venture. Some common types include:

1. **Small Business Entrepreneurship**: This type includes undertaking and operating small businesses that will target the local market. Such enterprises may comprise of stores, services providers and petty producers, including the trading business (Aulet, 2024).

2. **Scalable Startup Entrepreneurship**: These ventures especially are designed to grow quickly and to become large scale companies as quickly as possible, sometimes with the assistance of venture capital (Ince et al., 2023). A good example is technology startups where the objective is growth and, where possible, disruption of existing market players.

3. **Social Entrepreneurship**: Social entrepreneurs' main concern is in providing solutions to the social, environmental, or community challenges in unique business strategies. Instead of concentrating on the maximum generation of their profits, their main objective is to generate a positive social return (Kanini & Muathe, 2019).

4. **Corporate Entrepreneurship (Intrapreneurship)**: In the case of big firms, corporate entrepreneurship or intrapreneurship is one where invention of new value creating idea is encouraged and supported within a large firm. It lets the companies compete and be on the right edge especially due to constant market fluctuations.

5. **Lifestyle Entrepreneurship**: Lifestyle business owners create companies that they want to reflect who they are as people and where they want live. It is all about enabling people to achieve a favorable work-life balance or engaging in business for the purpose of passion as well as making profits.

Importance of Entrepreneurship

There is no doubt that the entrepreneurial initiatives play an essential role in economic growth and development. Many regard it as a pro-business model since it helps job creation, fosters competitiveness and resource efficiency. The entrepreneurs introduce new products and services into the market; most of them disrupt industries and diversify economies. However, it advances social capital as people can be employed in the process of creating value and form their own companies that will help society (Frese & Gielnik, 2023).

Besides, Like and McMahon also pointed it out, entrepreneurship has a social implication where it contributes to solving societal problems. For example, social entrepreneurs apply business to solve problems like poverty, education, health, and the environment, which bring solutions that affect the welfare of the society and its economy.

Over the years, much progress has been made in analyzing entrepreneurship as an organizational activity but still at the same time it remains a complex and changing process which involves creation of goods as well as services, risk taking and other forms of resources and planning. They are the major driving forces in innovations, economical and social development at large. That is why, especially in the modern context, the place of the entrepreneur, and thus the study of the processes taking place in the field of entrepreneurial activity, is of great importance and potential (Sabet & Khaksar, 2024).

Challenges Faced by the Entrepreneurs

The undertaking of the business venture is a fulfilling experience of possibility for innovation, autonomy over one's income and the joy of building a new enterprise. However, it also comes with challenges which might make or mar the attempts made by entrepreneurs and innovators. It is important for any entrepreneur to posses this knowledge because he will have to deal with all those issues when establishing and developing his business. Here are some of the key challenges faced by entrepreneurs: Here are some of the key challenges faced by entrepreneurs:

I. Financial Constraints
   1. Securing Funding: Among the major problems that businessmen face today is the problem of how to get the right amount of capital to invest in business. The conventional form of financing, for instance, through banks, may at times be getting a challenge due to the no revenues of new companies. A social capital is usually obtained from personal funds, loans from friends or relatives, business angels and venture capital firms. Nevertheless, it is not easy to persuade investors to fund a new concept more so when the market is already crowded (Purwati et al., 2021).
   2. Cash Flow Management: A must know factor in the management of any business is cash management. It means that start-up entrepreneurs should always have adequate amounts of cash in order to meet recurring and un-avoidable requirements such as costs of operation, payment of employees' wages, and emergencies. Inadequate management of cash flow results to the firm's cash flow being cyclical and hence it is unable to support the various activities required in operation of the business.

II. Market Competition
   1. Standing Out in a Crowded Market: Competition is usually an issue that entrepreneurs have to deal with and this is especially so where industries have low entry barriers. To gain market share it is crucial to differentiate ones product or service from that of his or her competitors (Rust et al., 2023). This should involve the need to have a good understanding on the market, the customers and good marketing techniques.
   2. Market Saturation: Sometimes markets are already developed in certain industries, and because of this, it is not easy for entrants to penetrate and grab a large market (Annamalah et al., (2023). Small business owners have to look for areas where they are adding value during product differentiation or use innovations frequently or aim at specialised markets in order to address this issue.

III. Uncertainty and Risk
   1. Market Uncertainty: They do business under the conditions of risk, as the market prospects, customer needs, and legal regimes may shift quickly. Such uncertainty tends to hamper the ability to forecast the businesses prospects and indeed their future. It is crucial for the entrepreneurs to note that they must always be ready to shift their gears as they implement their plans (Kehinde et al., 2024).

2. Risk Management: In its very nature, entrepreneurship entails risk in terms of money, operations, or brand image. There are causal factors to these risks, and the entrepreneur has to evaluate and manage them, but at the same time, s/he should be ready to assume certain risks for the growth of the business (Ince et al., 2023).

IV. Formation and Management of Team

1. Attracting Talent: Hiring and keeping talent is often a problem faced since talent is a key resource of value to almost all organizations, but a challenge more often found in startup organizations. The talent may want to work for larger and more stable organizations this poses a problem for the entrepreneur when recruiting. The candidate wants to receive competitive wages, to work at the company with friendly colleagues and face new career advancements' challenges (Schmitt et al., 2024).

2. Leadership and Management: Legitimate authority is highly important in the sense that it provide direction on what needs to be done by the team members. Management is an important function of entrepreneurship and its necessary for entrepreneurs to be able to delegate, communicate, and solve conflict among the workforce (Savari & Khaleghi, 2023).

V. Regulatory and Legal Challenges

1. Compliance with Regulations: Starting a business requires the entrepreneur to wade through numerous rules and laws and this can be rather challenging. /Among the regulatory requirements and controls, reflected at the enterprise level, it is necessary to identify and comply with industry standards and requirements, as well as taxes and tariffs; labor legislation and terms of protection of intellectual property rights are also critical since violation of these provisions entails legal liability (Kim & Aldrich, 2005). The existing legislation should be monitored by business people and legal advice should be sought at any one time.

2. Intellectual Property Protection: The preservation of inventions, marks and other assets has become increasingly important because the more an organisation innovates the more it has to protect from potential misuse. However, the procedure of applying for the patents, trademarks, as well as copyrights can take long, costly and be full of legal procedures.

VI. Marketing and Customer Acquisition

1. Creating Brand Awareness: Another factor is that for new business making people aware of their existence is very crucial. Marketing communication is another strategy that needs to be put to work effectively by the entrepreneurs to ensure that they get the right audience for their products and services. There are quite often when teeth whitening is covered by a combination of

multiple approach to the promotion, including digital marketing and public relations, and obvious traditional advertising (Ince et al., 2023).

    2. Customer Retention: People usually want to get client base and this is true but getting clients is not enough if you arm not able to retain them. Businesspeople have to provide the clients with memorable service, achieve a high level of product or service quality, and keep in touch with the clients.

VII. Time Management

    1. Balancing Multiple Roles: It is crucial for entrepreneurs to act as managers of their business as they are usually involved in mainly all aspects like the development of the product, business promotions, funds, and the business execution. This may cause the management of time, where the entrepreneur is unable to schedule time, organize work and observe work-life balance. One cannot afford to be overwhelmed by work and should make sure he or she delegates activities in order to prevent being overwhelmed and possibly burning out (Hidalgo et al., 2024).

VIII. Scaling the Business

    1. Growth Management: Growth of a business on the other hand has its own complexities. Business owners have to be concerned about operations increasing their workload, quality issues and ability of business to sustain growth (Dar & Mishra, 2020). The major danger of fast growth in organizations is that it results in Poor organizational structure, Rod services to the customers, and Strain on resources respectively.

    2. Maintaining Culture and Vision: Shareholders, employees and board of directors in growing business may have emerged with different visions and culture of the company. Owners and managers must also guarantee that the organizational culture, which encompasses their values, is sustained and well transmitted to the newly joining employees because as the organization grows, so does the risk of losing those aspects (Chen & Lee, 2024).

IX. Adapting to Technological Changes

    1. Keeping Up with Innovation: The advancement in technology is fast and an entrepreneur need to embrace the change to be in apposition to compete effectively. This encompasses developing new tools, technologies, platforms and processes that can deliver more value, generate new customer experiences or avenues of business. If an organisation fails to innovate it may become outdated or gradually gets outnumbered by other competitors (Analia et al., 2020).

X. Personal Challenges

    1. Stress and Mental Health: Realizing that being an entrepreneur is stressful, with issues such as no guaranteed income, long hours of work, and being in charge of the business, all affect the health of an enterprise (Yani et al.,

2020). Stress, anxiousness, and burnout are more typical problems, and an entrepreneur needs to mind their health, find help, and build ways to help them work through life's adversities.

Entrepreneurship is a challenging but rewarding journey that requires resilience, adaptability, and a proactive approach to problem-solving. By understanding and preparing for these challenges, entrepreneurs can better navigate the complexities of starting and growing a business, ultimately achieving their goals and contributing to economic and social progress (Rust et al., 2023).

SALES PROMOTIONS IN ENTREPRENEURSHIP

They are an indispensable weapon in the arsenal of the businessman, they allow them to increase the volume of sales and attract new clients, as well as increase the awareness of their product (Kehinde et al., 2024). In particular, a competent implementation of sales promotions can make a breakthrough when it comes to developing new entrepreneurial businesses which work with limited capital, and compete keenly in the market.

Utility of Sales Promotions to Entrepreneurs

1. Driving Immediate Sales:

Sale promotions and are used to appeal to the emotions of customers so that they will buy the product. These create demand especially to business people who are in new product or market since price cuts, low prices, and special offers create a short term wave that opens up a market share for the business (Hidalgo et al., 2024). This is very important when it comes to establishing the cash flow of a business because the period of time that follows a start up is usually characterized by a lot of expenses.

2. **Building Brand Awareness**:

One of the biggest problems in the modern environment that affects new companies is the problem of visibility. Miscellaneous sales promotions can create awareness of the brands in question through attention grabbing and attention creation. For instance, a spectacular and widely advertised promotional activity such as a tele-communication company's company promotional campaign may make the brand popular thus creating brand awareness and even brand loyalty (Yani et al., 2020).

3. Encouraging Product Trials:

Thus new product which often is outside the consumer's comfort zone is a territory that an entrepreneur explores. The likes of free samples, trial offers or money back guarantee, while being part of the sales promotions, help in reversing the decision-making process and hence, get people to try out a new product. When consumers use the product they are in a better position to continually purchasing the product hence becoming loyal customers.

4. Clearing Inventory:

Stock is one of the most sensitive factors of any business, more so for start ups with limited space and capital. Special offers such as clearance sales or a combination of related goods and services can also work for the entrepreneur in the sense that they clear the usual stock in a short period thus reducing the amount of capital tied to products and space for new products.

5. Creating Competitive Advantage:

In competitive industries, sales promotions make it easier for the entrepreneur to draw a line between his company's products and services with those of the competing firms. In particular, customers who are provided with certain incentives that would not be available for them in any other way will be attracted to the particular entrepreneur and thus be able to provide the entrepreneur with a competitive advantage over the competitors.

6. Enhancing Customer Loyalty:

Therefore, it is necessary to focus on creating a base of a stable customer following. Other sales promotion tools that may be used effectively to foster customer relations are those that entail the patronage reward where the buyer is rewarded after the purchase and these include the loyalty programmes, bonus on future purchases and any other privileges among others (Bhandari & Bhuyan, 2023) . In its turn, customer loyalty can become a continual source of income and generate word-of-mouth recommendations for other clients for an entrepreneur.

Types of Sales Promotions in Entrepreneurship

1. **Discounts and Price Promotions**: Promotional offers, which include discounts, can be considered as one of the most widespread and efficient promotional tactics, and the most frequently used are discounts in per cent, 'buy one, get one free,' or combined prices. Cost-cutting promotions are aimed at and can reach the bargain hunters and such promotional methods can produce large volumes of sales in the shortest time.

2. **Rebates:** Coupons are used to make the target consumers have a direct reason of buying a particular product at a cheaper price either through electronic commerce or physical shops (Hidalgo et al., 2024). The same holds true for rebates, which make the buyer get back a portion of the amount paid for the product after purchase; this tool also motivates the buyers while at the same time does not undermine the perceived value of the product.

3. **Sweepstakes**: Perceptively, contest and sweepstakes are other forms of consumer involvement where they are given a chance to win something for instance a product or a voucher through buying the product or even providing information through product survey or even endorsing a product or even sharing it in social networks. These promotions not only encourage sells but also builds brand loyalty and creates other forms of content from the users (Kehinde et al., 2024).

4. **Loyalty Programs:** The reward programs encourage customer subsequent purchases by creating a redeeming point system or by the use of membership levels. Of special interest for the entrepreneurs, these programs may serve as a strong stimulus for their customers to continue their patronage.

5. **Product Trials**: This is because free samples or trials help the consumers test the product before they opt to buy them. This strategy is most beneficial to new product or services or products that may not become easily popular in the market because customers are likely to be wary of such products (Frese & Gielnik, 2023).

6. **Limited-Time Offers:** Offers that are valid for a certain period of time like the flash sale or limited-time offer instill a sense of urgency on people to act. Business owners should use these limited-time offers during off-peak seasons, or during the time of release of a product into the market.

7. **Referral Programs**: A referral technique encourages consumers to introduce other consumers due to the fact that the introduction generates appealing terms for the newly brought customer. Through word of mouth, new customers can be attained by the business without making huge expenditures in marketing.

Some of the Difficulties of Sales Promotions for the Entrepreneurs

Figure 4. Entrepreneurial challenges of sales promotions

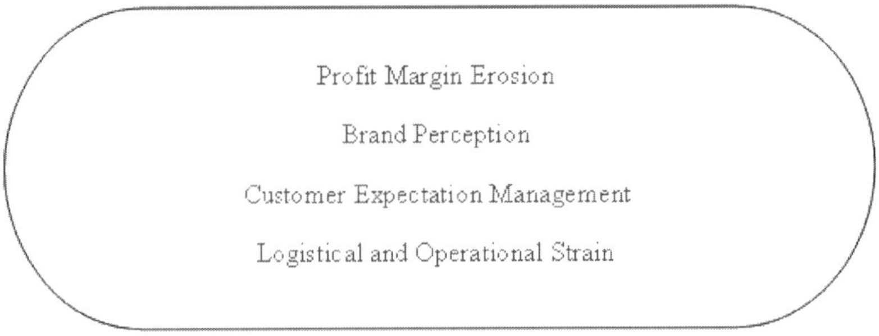

> Profit Margin Erosion
>
> Brand Perception
>
> Customer Expectation Management
>
> Logistical and Operational Strain

1. **Profit Margin Erosion:** The use of 'big or hefty offers' is not very advisable due to the impact that cutting the price often has in that any company new to the market will have to deal with low profit margins (Aulet, 2024). Kalvet and Masson have well pointed to the fact that there can be clearly seen benefits of increased sales, which, at the same time, has to be compensated within the impact on profitability.
2. **Brand Perception:** Promotion on the other hand when overused and more so price-based promotions reduces the perception of the brand thereby making the brand be seen as cheap or of a lower quality. Self employed people need to make sure that the advertisements they employ are in harmony with their brand image and objectives (Baycan & Öner, 2023).
3. **Customer Expectation** Management: If the customers get used to the promotion strategy, they might put off their purchases in order to wait for the next promotion which could cause effect on the normal cycles of sale. Promotion should therefore be applied tactfully for it does not set a bad precedence of letting customers expect exaggerated magnanimity of an organization (Parthymos & Daskalopoulou, 2024).
4. **Logistical and Operational Strain:** A good example of this is sales promotions where an entrepreneur can attract a huge traffic of buyers that they are unable to handle through supply chain and inventory. It needs to be understood that business owners should be ready to multiply the amount of work done to accommodate the demand without having to reduce service standards.

It is commonly recognized that sales promotions are one of the most important and versatile techniques in the framework of the entrepreneurial activity since they allow increasing sales, creating brands, and constructing customer relationships (Elias, 2024). If well timed and well implemented, promotions can afford the entrepreneur the push to grow his business and establish itself in the market. But the same people, who attempt to establish a start-up, need to view the possible difficulties and make sure their advertising strategy fits well into the overall goals of its further development for it to become successful in the long run (Muliadi et al., 2024).

ROLE OF SOCIAL CAPITAL AND SALES PROMOTIONS IN ENTREPRENEURSHIP

They both have important functions in determining the further destiny of the company and its ability to solve problems, use opportunities and develop stably.

Social Capital in Entrepreneurship

Social capital can be defined as on the one hand the social relations in the social structure through which the entrepreneur capitalizes on to access resources and knowledge within the network (Sabet & Khaksar, 2024). These are the networks as such, which remorseless goodwill and trust may foster cooperation, lower the transaction costs and be useful in the process of an entrepreneur (Rust et al., 2023).

To the ordinary business persons, social capital is an essential resource. It may bring an access to the funding, collaboration, guidance, and markets the person would not be able to reach otherwise. Access to information and relationships is important for the successful implementation of innovative starts-ups; the sources of knowledge and relationships are social networks (Savari & Khaleghi, 2023).

Social Capital as a Resource

It becomes clear that through social capital, entrepreneurs can be able to source some of the most vital organizational resources among them being funding, human talent, and market intelligence. For instance, direct contacts with investors or venture capitalists help to source for the funds needed while interaction with the industry practitioners may help in sourcing for funds and sound advice in the process (Purwati et al., 2021).

Social capital also helps in the creation of credibility and trust of customers, suppliers and other members of the public. It can create an improved perception of an entrepreneur hence easy to attract more customers, negotiate with suppliers for better deal or agreements and also be in a position to form associations with other parties (Mishchuk et al., 2023).

Innovation and Collaboration

One of the key reasons why social capital enhances innovation is that it provides the necessary platform for sharing of ideas. The networks a businessman has access to are crucial since they raise the chances of meeting new ideas, developing ways to fix them, and noticing trends (Mishchuk et al., 2023). Business interactions in social networks result to partnerships, co-production or partnerships in business/ product formation resulting in business expansion.

Overcoming Challenges

It is not a secret that enterprises face a lot of difficulties which are connected with the competition in the market, financial opportunities, and others. Employing social capital has the virtue of making you realise that help is available in terms of recommendations, resources and assistance within the social circle during most troubled times. As it has been seen above, it is the networks that enable the entrepreneurs to address the issues, share risks, and manage uncertainties.

Interplay Between Social Capital and Sales Promotions

1. Leveraging Networks for Effective Promotions:

Typically, sales promotions employed by business people can be enhanced using this social capital. For example, they can use the connection with influencers and acquaintances and clients who had positive experience with a promo. Social capital also helps to establish other cooperations with other companies or media, which in turn increases the efficiency of promotional activities (Hidalgo et al., 2024).

2. Trust and Credibility:

Sales promotion activities benefit from social capital in that they improve the authenticity of the promotions. Word of mouth promotion of business is more effective in delivery of the message to those who are in the target market since the message is passed through influential individuals in the society. This trust can enhance the possibility of the consumers seizing the offers that are being offered with a discount price (Kim & Aldrich, 2005).

3. Collaboration and Co-Promotion:

In some networks there are opportunities to work together to offer various specials, for example, co-op marketing where two or more businesses offer a two-for-one deal. These associations can widen the coverage of the promotion and unveil the business to new consumer platforms (Frese & Gielnik, 2023). They are easier to organize and facilitate because of social capital.

4. Feedback and Iteration:

Entreprneurial sales promotions are enhanced by social capital in that it forms a feedback loop by which the promotions can be fine tuned. Engaging with their networks, the entrepreneurs are better placed to know how promotions are received, what is positive or negative and what needs enhancing. It permits the constant refining and improving of the advertisement techniques.

CONCLUSION

Sales promotions and social capital are on separate conducts that are significant in entrepreneurial ventures. Social capital gives the networks, asset, and support required for managing most of the demands of entrepreneurship hence making use of sales promotions as a strategic means of sale stimulants, brand creation, and marketing superiority. When social capital has been combined with the use of sales promotions, one can meaningfully improve his chances of success, obtain a sustainable business development, and leave a long-term positive imprint in his industry.

REFERENCES

Analia, D., Syaukat, Y., Fauzi, A., & Rustiadi, E. (2020). The impact of social capital on the performance of small micro enterprises. *Jurnal Ekonomi Malaysia*, *54*(1), 81–96.

Annamalah, S., Paraman, P., Ahmed, S., Dass, R., Sentosa, I., Pertheban, T. R., Shamsudin, F., Kadir, B., Aravindan, K. L., Raman, M., Hoo, W. C., & Singh, P. (2023). The role of open innovation and a normalizing mechanism of social capital in the tourism industry. *Journal of Open Innovation*, *9*(2), 100056. DOI: 10.1016/j.joitmc.2023.100056

Aulet, B. (2024). *Disciplined Entrepreneurship: 24 Steps to a Successful Startup, Expanded & Updated*. John Wiley & Sons.

Baycan, T., & Öner, Ö. (2023). The dark side of social capital: A contextual perspective. *The Annals of Regional Science*, *70*(3), 779–798. DOI: 10.1007/s00168-022-01112-2

Bhandari, A., & Bhuyan, M. N. H. (2023). Social capital and capital allocation efficiency. *Journal of Business Finance & Accounting*, *50*(7-8), 1439–1466. DOI: 10.1111/jbfa.12662

Chen, Z. F., & Lee, J. Y. (2024). Relationship cultivation and social capital: Female transnational entrepreneurs' relationship-based communication on social media. In *Start-up and Entrepreneurial Communication* (pp. 59–82). Routledge. DOI: 10.4324/9781003481171-5

Dar, I. A., & Mishra, M. (2020). Dimensional impact of social capital on financial performance of SMEs. *The Journal of Entrepreneurship*, *29*(1), 38–52. DOI: 10.1177/0971355719893499

Elias, R. (2024). The influence of family social capital toward the entrepreneurial intention among prospective graduates in Tanzanian universities. *Journal of Applied Research in Higher Education*.

Frese, M., & Gielnik, M. M. (2023). The psychology of entrepreneurship: Action and process. *Annual Review of Organizational Psychology and Organizational Behavior*, *10*(1), 137–164. DOI: 10.1146/annurev-orgpsych-120920-055646

Hidalgo, G., Monticelli, J. M., & Vargas Bortolaso, I. (2024). Social capital as a driver of social entrepreneurship. *Journal of Social Entrepreneurship*, *15*(1), 182–205. DOI: 10.1080/19420676.2021.1951819

Ince, H., Imamoglu, S. Z., & Karakose, M. A. (2023). Entrepreneurial orientation, social capital, and firm performance: The mediating role of innovation performance. *International Journal of Entrepreneurship and Innovation*, *24*(1), 32–43. DOI: 10.1177/14657503211055297

Kanini, K. S., & Muathe, S. M. (2019). Nexus between social capital and firm performance: A critical literature review and research agenda. *International Journal of Business and Management*, *14*(8), 70–82. DOI: 10.5539/ijbm.v14n8p70

Kehinde, S., Oluseun, B., Moses, C., Borishade, T., Kehinde, O., Simon-ilogho, B., & Kehinde, K. (2024). Effective Sales Promotion: A Necessary Catalyst for Elongating the Product Life Cycle in A Developing Economy. *International Journal of Advances in Social Sciences and Humanities*, *3*(1), 1–7. DOI: 10.56225/ijassh.v3i1.231

Kim, P. H., & Aldrich, H. E. (2005). Social capital and entrepreneurship. Foundations and Trends® in Entrepreneurship, 1(2), 55-104.

Mishchuk, H., Bilan, Y., Androniceanu, A., & Krol, V. (2023). Social capital: Evaluating its roles in competitiveness and ensuring human development. *Journal of Competitiveness*, *15*(2).

Muliadi, M., Muhammadiah, M. U., Amin, K. F., Kaharuddin, K., Junaidi, J., Pratiwi, B. I., & Fitriani, F. (2024). The information sharing among students on social media: The role of social capital and trust. *VINE Journal of Information and Knowledge Management Systems*, *54*(4), 823–840. DOI: 10.1108/VJIKMS-12-2021-0285

Parthymos, A., & Daskalopoulou, I. (2024). Entrepreneurship and social capital: Some evidence on micro-spatial interactions. *Journal of Small Business and Entrepreneurship*, *36*(1), 108–129. DOI: 10.1080/08276331.2020.1868839

Purwati, A., Budiyanto, B., Suhermin, S., & Hamzah, M. (2021). The effect of innovation capability on business performance: The role of social capital and entrepreneurial leadership on SMEs in Indonesia. *Accounting*, *7*(2), 323–330. DOI: 10.5267/j.ac.2020.11.021

Rust, N. A., Ptak, E. N., Graversgaard, M., Iversen, S., Reed, M. S., de Vries, J. R., Ingram, J., Mills, J., Neumann, R. K., Kjeldsen, C., Muro, M., & Dalgaard, T. (2023). Social capital factors affecting uptake of sustainable soil management practices: A literature review. *Emerald Open Research*, *1*(10). Advance online publication. DOI: 10.1108/EOR-10-2023-0002

Sabet, N. S., & Khaksar, S. (2024). The performance of local government, social capital and participation of villagers in sustainable rural development. *The Social Science Journal*, *61*(1), 1–29. DOI: 10.1080/03623319.2020.1782649

Sanyal, P., Singh, R., & Singh, R. (2024). Making of a social buyer: The role of knowledge capital authenticity and inter-firm communication in B2B sales situations. *Journal of Marketing Theory and Practice*, ●●●, 1–20. DOI: 10.1080/10696679.2023.2291713

Savari, M., & Khaleghi, B. (2023). The role of social capital in forest conservation: An approach to deal with deforestation. *The Science of the Total Environment*, *896*, 165216. DOI: 10.1016/j.scitotenv.2023.165216 PMID: 37392871

Schmitt, L., Epler, R., Casenave, E., & Pallud, J. (2024). An Inquiry into Effective Salesperson Social Media Use in Multinational Versus Local Firms. *Journal of International Marketing*, *32*(1), 72–91. DOI: 10.1177/1069031X231207050

Yani, A., Eliyana, A., Hamidah, I., & Buchdadi, A. D. (2020). The impact of social capital, entrepreneurial competence on business performance: An empirical study of SMEs. *Systematic Reviews in Pharmacy*, *11*(9), 779–787.

Yudha, P. (2018). Exploring the impact of social capital on entrepreneurial orientation and business performance (Study on members of MSMEs communities in Malang). *Profit: Jurnal Adminsitrasi Bisnis*, *12*(1), 20–31. DOI: 10.21776/ub.profit.2018.012.01.3

Zhang, Y., Cai, J., Shan, X., Li, S., & Shou, Y. Sales promotion and supply chain finance for shopping days: Strategies of e-commerce platform and seller. Managerial and Decision Economics.

Chapter 15
Human–Centric AI in Islamic Finance:
Bridging Technology and Tradition for Enhanced Entrepreneurial Interactions

Early Ridho Kismawadi

https://orcid.org/0000-0002-9420-5212

IAIN Langsa, Indonesia

ABSTRACT

This study examines the application of human-cantered artificial intelligence (AI) in Islamic finance, with the aim of bridging modern technology and tradition to improve entrepreneurial interaction. In the era of digital transformation, AI is a major force that is transforming various industries including the financial sector. Islamic finance, which prioritizes ethical practices and social justice, has a unique approach in integrating AI according to sharia principles. This research aims to fill the literature gap regarding the application of AI in Islamic finance which is still limited. Through in-depth analysis and a comprehensive approach, the study offers practical and theoretical guidance for Islamic financial institutions to improve operational efficiency, service personalization, and productive and ethical entrepreneurial interactions.

DOI: 10.4018/979-8-3693-8009-3.ch015

INTRODUCTION

As we enter what has been described as the era of digital transformation, one of the most powerful forces helping to drive change in various industries is artificial intelligence (AI) and we see it play a role in finance (Ridho Kismawadi et al., 2023). Islamic finance represents ethical practices and social justice within the profile finance sector that intersects in a unique way with how AI is integrated (Kunhibava, 2011; Nienhaus, 2011; Shamsudheen et al., 2024). The investigation is based on human-cantered AI that strengthens human capabilities to containing aspects of Islamic financial ethics. This research has a unique virtue because it takes into account the need to translate those values into AI knowledge that upholds Sharia principles, which essentially improves customer engagement and operational efficiency for Islamic banks.

One thing you can point out to get more acceptance is the uniqueness of how they blend contemporary technology with traditional practices. It aims to fill a substantial void in knowledge about how Islamic financial institutions can stay up-to-date with technology without affecting their ethical values with such applications. Such integration is especially important at a time when digital technology has become ubiquitous and constantly changing, posing problems and opportunities for traditional financial structures.

In today's digital era, artificial intelligence (AI) technology has transformed various aspects of life, including the financial sector (Tigges et al., 2024). The use of AI in finance not only provides operational efficiency but also offers solutions that are more personalized and responsive to user needs. However, in the field of Islamic finance, the application of AI is still relatively new and has not been widely explored. Despite the great potential, there is still a gap in the literature on how AI can be integrated with Sharia principles and how this can support better entrepreneurial interactions.

The interaction of entrepreneurship in Islamic finance plays a crucial role in driving an economy based on social justice and ethical responsibility (Gümüsay, 2020; Ibrahim & Kahf, 2020). Muslim entrepreneurs need financial solutions that are not only effective and efficient but also in line with their religious values. In this context, AI has the potential to offer more customized services, provide in-depth analysis, and predict market trends that can support business decisions. However, the question that arises is how AI can be implemented without compromising Sharia principles and how this technology can support ethical and productive entrepreneurial interactions.

This research aims to answer this gap by exploring how AI can be integrated in Islamic finance to strengthen entrepreneurial interactions. Through in-depth analysis and a comprehensive approach, this research seeks to offer practical and theoretical guidance for Islamic financial institutions. The focus is on how AI can be used to improve efficiency, personalization, and fairness in financial services, as well as how this technology can support entrepreneurs in making better and ethical decisions.

METHOD

The study adopts a mixed-methods approach to investigate the integration of AI in Islamic finance, with an emphasis on aligning advanced technologies with Sharia principles. This research method begins with a comprehensive literature review to understand the current state of AI applications in finance, especially in the context of Islamic finance. Several case studies from Islamic financial institutions that have successfully implemented AI solutions are analysed in depth to provide insight into the implementation process, the specific AI technology used, and the results achieved. The case study also illustrates best practices and practical examples of AI integration that align with Islamic ethical values.

INTEGRATION OF AI AND SHARIA FINANCE

The era of digital transformation has brought significant changes in various industries, including the financial sector. Digital technologies, especially artificial intelligence (AI), play a key role in driving innovation and efficiency. AI is considered one of the most revolutionary technologies, allowing for more sophisticated data analysis, faster decision-making, and automation of complex processes. The integration of AI in Islamic finance offers a unique opportunity to combine advanced technology with strong ethical principles.

Islamic finance is based on sharia principles that emphasize justice, transparency, and social responsibility. This includes the prohibition of riba (interest), gharar (uncertainty), and encouraging fair financial practices (Jelili Amuda & Alabdulrahman, 2023; Konak & Demir, 2023; Sakinç, 2021). Islamic finance focuses on ethics and social responsibility, prioritizing the welfare of the community. This principle creates a unique framework for the integration of AI, where the technology must be in accordance with Islamic ethical and moral values.

AI enables deeper and more accurate data analysis, helping financial institutions to predict market trends and identify risks more effectively. In Islamic finance, this can be applied to predict the performance of Islamic financial products and manage profit-sharing risk. Business processes that were previously manual and time-consuming can now be automated with AI, improving operational efficiency. For example, AI can be used to automate the transaction verification process according to sharia principles. AI allows for better personalization of financial services, providing product recommendations tailored to individual needs. In the context of Islamic finance, AI can help identify products that match the customer's risk profile and sharia compliance.

The integration of AI in Islamic finance also presents ethical challenges, especially in ensuring that the use of AI does not violate sharia principles (Carvajal Zaera, 2024; Gaubienė, 2024; Hasan et al., 2024; Huang et al., 2024). For example, AI models should be designed to avoid excessive speculation and ensure transparency in decision-making. However, this challenge also presents an opportunity to create new products and services that are more efficient and in accordance with sharia values. Concrete examples include the use of AI in the analysis of Islamic investment performance and the development of more sophisticated risk management tools.

The future of AI integration with Islamic finance looks promising, with trends pointing to the development of new technologies and their potential to transform the industry. The study provides practical recommendations for further research and implementation of AI in Islamic financial institutions, including the development of policies, ethical frameworks, and best practices for successful AI integration. Thus, this research is expected to make a significant contribution to understanding and developing the integration of AI in Islamic finance, while maintaining the ethical and moral values that are the basis of the Islamic financial system.

The integration of AI in Islamic finance is expected to not only encourage innovation but also promote more inclusive and equitable finance. Human-centric AI will enable Islamic financial institutions to offer services that are more personalized and responsive to customer needs, while still adhering to sharia principles. For example, through sophisticated data analysis, AI can help identify individual financial needs and offer appropriate solutions, such as microfinance or sharia-based investments tailored to the customer's risk profile and financial goals.

AI has great potential in improving Islamic financial literacy. AI-based applications can be used to educate the public on the principles of Islamic finance, providing interactive guidance on sharia-compliant investment and financial management. This will not only empower individuals to make better financial decisions but also improve the general understanding of Islamic finance.

Data security and privacy are also important aspects of the integration of AI with Islamic finance (Hermann, 2022; Wangmo et al., 2019). AI can be used to improve security systems, detect and prevent financial fraud, and ensure that customer data remains safe and not misused. The use of blockchain technology alongside AI can ensure transparency and trust in financial transactions, which is in line with the principles of fairness and transparency in Islamic finance.

The development of innovative Islamic financial products with AI is also an important focus. AI can help in creating new products that are more responsive to the needs of the market and customers. For example, AI can be used to develop more flexible and efficient sharia-based financing solutions, as well as offer investment options that are in line with sharia principles and attractive profit potential.

CONTINUOUS LEARNING AND DEVELOPMENT

Human-cantered AI has great potential to support continuous learning and development for entrepreneurs and financial professionals in the Islamic finance ecosystem (Nguyen et al., 2023; Wilson & van der Velden, 2022). Through the analysis of individual data, AI can provide personalized training programs, helping each individual improve their skills and knowledge according to their specific needs. For example, an investment manager in an Islamic bank can receive a course recommendation on green sukuk tailored to their understanding and experience. By using AI technology, financial institutions can provide more relevant and effective training, allowing such managers to understand the basic concepts of green sukuk, investment strategies, and case studies from global markets. This not only improves individual competence but also assists financial institutions in offering investment products that are more innovative and in accordance with sharia principles.

AI-driven learning platforms offer educational modules and resources that can be accessed anytime and anywhere, including video tutorials, articles, case studies, and interactive simulations. For example, a young entrepreneur who wants to start a sharia-based fintech business can learn about the basics of Islamic finance and regulation through relevant modules. With this easy and flexible access, entrepreneurs can learn at their own pace and schedule, which is crucial in the fast-paced business world. AI can also provide suggestions on additional courses or resources that may be useful, based on the progress and specific needs of those entrepreneurs.

AI also provides real-time feedback and improvement recommendations based on the user's performance in courses or exercises, helping them correct mistakes and understand the material better. For example, a financial analyst who takes a risk analysis course can receive immediate feedback on their mistakes and get suggestions for improvement. AI can explain the error in detail, provide correct examples, and recommend additional resources to deepen that analyst's understanding of a particular topic. This quick and specific feedback allows individuals to learn from their mistakes more efficiently and improve their skills quickly.

AI helps individuals stay up-to-date with the latest trends and developments in Islamic finance by compiling reports and summaries from various sources, such as regulatory changes or product innovations. An Islamic finance consultant, for example, can receive notifications about changes in sukuk regulations and an analysis of their impact on their clients' investment strategies. With access to the most up-to-date and relevant information, consultants can make better decisions and provide more informed advice to their clients. AI can also analyse market data and trends to predict possible changes, helping financial professionals to plan and adapt quickly to dynamic market conditions.

AI also supports continuous learning to adapt to new technologies affecting Islamic finance, such as blockchain. AI can offer courses on the fundamentals of blockchain and its use in Islamic finance, ensuring that financial professionals can make effective use of this technology. For example, when blockchain technology begins to be applied in Islamic financial transactions, AI can provide training materials that explain the basic concepts of blockchain, how this technology can be used to improve transparency and efficiency in Islamic financial transactions, as well as the potential benefits and challenges that may be faced. With a better understanding of these new technologies, financial professionals can help their institutions to adopt and utilize such technologies in the most effective and sharia compliant way.

Human-centric AI can support continuous learning and development by automating administrative and training processes, so staff can focus on strategic and value-added tasks. For example, AI can be used to automate creditworthiness assessments, risk monitoring, and compliance reporting, allowing staff to focus on product development and customer service. By improving operational efficiency, AI not only reduces costs and processing times but also provides a competitive advantage for Islamic financial institutions.

Human-cantered AI can also increase transparency in the process of collecting and distributing zakat and waqf. By using AI technology, zakat and waqf management institutions can monitor and report the use of funds more accurately and efficiently. AI can help identify eligible beneficiaries, manage fund allocations, and monitor the impact of the distribution of those funds. Greater transparency in the management of zakat and waqf can increase public trust and participation in Islamic financial charities. For example, with the help of AI, zakat institutions can ensure that funds collected from the community are distributed fairly and effectively to those who are truly in need, as well as provide transparent and publicly accessible reports.

Data security and privacy are also major concerns in Islamic finance, especially with the increasing use of digital technologies. Human-centric AI can help strengthen data security and protect customer privacy by using advanced encryption techniques and security algorithms. AI can monitor systems to proactively detect and respond to cybersecurity threats. Better data protection not only protects customers' personal information but also ensures compliance with strict data privacy regulations. For example, an Islamic bank that uses AI to manage its customers' data can ensure that the data is safe from cyber threats and is only used for legitimate and ethical purposes, in accordance with sharia principles.

Human-centric AI can also play an important role in encouraging ethical and sustainable investment in Islamic finance. By using sophisticated data analysis, AI can help investors identify investment opportunities that meet strict ethical and environmental criteria. AI can evaluate the social and environmental impacts of various investments, ensuring that funds are invested in projects that support sustainable development and community well-being. This ethical and sustainable investment is in line with the core values of Islamic finance, which emphasizes social responsibility and community well-being. For example, AI can help investors to choose projects that not only provide financial benefits but also have a positive impact on the environment and society, such as renewable energy projects or sustainable infrastructure.

The enhanced user experience is one of the significant benefits of the application of human-centric AI in Islamic finance. By leveraging AI, financial institutions can provide a more intuitive and responsive interface, providing a better experience for users. For example, AI-based mobile banking applications can offer features that make it easier for users to access financial services, track transactions, and get customer support quickly and efficiently. This improved user experience can increase customer satisfaction and loyalty, driving sustainable business growth.

By integrating human-centric AI in Islamic finance, we can create a more collaborative, innovative, and efficient environment for entrepreneurs. AI not only supports compliance with sharia principles but also promotes financial inclusion, transparency, and sustainable growth. This integration not only benefits financial institutions and entrepreneurs but also supports broader economic and social development, in line with the core values of Islamic finance. By supporting continuous learning and development through personalized training, comprehensive educational modules and resources, real-time feedback, updates on the latest trends, and adaptation to new technologies, human-cantered AI can ensure that entrepreneurs and finance professionals in the Islamic finance ecosystem remain competent, up-to-date, and ready for future challenges. The implementation of AI not only improves individual abilities but also strengthens the entire Islamic finance ecosystem, encouraging innovation, inclusion, and sustainable growth.

DEVELOPMENT OF INNOVATIVE SHARIA FINANCIAL PRODUCTS WITH ARTIFICIAL INTELLIGENCE

The integration of artificial intelligence (AI) in the Islamic finance sector opens up great opportunities for innovation and product development that is more responsive to customer needs. This research highlights how AI can be used to develop innovative Islamic financial products that are in accordance with sharia principles, and examines the challenges and long-term benefits of using this technology.

One of the main results of this research is the ability of AI to develop innovative Islamic financial products. With in-depth and accurate data analysis, Islamic financial institutions can better understand consumer preferences and behaviour. AI enables more detailed and rapid analysis of big data, allowing financial institutions to tailor their products to customer needs. For example, by leveraging data from customer transactions and interactions, AI can identify patterns and trends that can be used to develop more flexible Islamic financing products or more profitable sharia-based investment options.

AI allows financial institutions to conduct real-time analysis of market trends and consumer preferences. This ability is crucial in creating products that are responsive to changes in market demand. For example, AI can identify the need for new, more flexible sharia financing products or more profitable sharia-based investments. Thus, Islamic financial institutions can develop products that are more in line with current and future market needs, increasing their competitiveness in the market.

AI also enables better personalization of financial services, which is in line with sharia principles. Using machine learning algorithms, AI can recommend financial products that match the risk profile and individual financial needs of customers. This not only increases customer satisfaction and loyalty, but also ensures that the products offered are in accordance with sharia principles. For example, AI can help in structuring investment portfolios that align with clients' risk preferences and sharia principles, ensuring that clients are sourcing products that align with their values.

The integration of AI in the operational process of Islamic financial institutions can improve operational efficiency. Business processes that were previously manual and time-consuming can now be automated with AI, increasing speed and accuracy. An example is the automation of the transaction verification process in accordance with sharia principles, which not only speeds up processing times but also ensures compliance with sharia law. By doing so, financial institutions can reduce operational costs and increase efficiency, which in turn can be passed on to customers in the form of lower costs and better services.

Another important aspect discussed in this study is how AI can be designed to adhere to Islamic ethical and moral principles. One example is ensuring that AI algorithms are non-discriminatory and fair in decision-making. It is important to ensure that the data used by AI is collected and managed in an ethical manner and in accordance with sharia law. For example, ensuring that customer data is protected and not misused is one of the important aspects of sharia compliance in the use of AI.

The study also found that AI can help in creating Islamic financial products that are more innovative and responsive to market needs. Examples of new products that can be developed include more flexible and efficient sharia-based financing solutions, as well as investment options that are in line with sharia principles and have attractive profit potential. For example, AI can be used to develop financing products that are more tailored to individual needs, such as financing for education or health that are in accordance with sharia principles.

The integration of AI in Islamic finance also has significant long-term benefits. To encouraging innovation, AI can also increase financial inclusion and Islamic financial literacy. AI-based applications can be used to educate people about the principles of Islamic finance and help them make better financial decisions. For example, AI-based mobile apps can educate customers about the available Islamic financial products and how they can leverage them to achieve their financial goals.

Thus, the results of this study show that AI has great potential to develop innovative Islamic financial products that are more responsive to customer needs, while still complying with sharia principles. Effective AI integration can improve operational efficiency, service personalization, and customer satisfaction, as well as pave the way for further innovation in the Islamic finance industry. However, it is important to ensure that the use of AI is carried out in an ethical manner and in accordance with sharia principles, to ensure that the benefits obtained can be felt by all parties involved.

The integration of artificial intelligence (AI) in the Islamic finance sector opens up great opportunities for innovation and product development that is more responsive to customer needs. This research highlights how AI can be used to develop innovative Islamic financial products that are in accordance with sharia principles, and examines the challenges and long-term benefits of using this technology.

This research finds that AI can be used effectively to develop innovative Islamic financial products that suit customer needs. With AI, data analysis becomes more in-depth and accurate, allowing Islamic financial institutions to better understand consumer preferences and behaviour. For example, by leveraging data from customer transactions and interactions, AI can identify patterns and trends that can be used to develop more flexible Islamic financing products or more profitable sharia-based investment options. A specific example is the development of sharia-based home financing products. AI can analyse historical data from customers to identify payment patterns, location preferences, and the type of property they are interested in. Based on this analysis, financial institutions can offer financing schemes that are more tailored to individual needs, such as more competitive margin offerings and better payment flexibility, so that customers can choose the option that best suits their financial situation.

AI enables financial institutions to analyse market trends and consumer preferences in real-time. This ability is crucial in creating products that are responsive to changes in market demand. For example, AI can analyse data from various sources, such as social media, financial transactions, and customer surveys, to identify the need for new, more flexible sharia financing products or more profitable sharia-based investments. For example, AI can be used to analyse investment trends among millennials who prefer low-risk investments but with stable returns. Based on this analysis, financial institutions can develop sharia investment products that combine sharia principles with modern investment needs, such as sharia mutual funds with diversified and well-managed portfolios.

AI enables better personalization of financial services, which is in line with sharia principles. Using machine learning algorithms, AI can recommend financial products that match the risk profile and individual financial needs of customers. This not only increases customer satisfaction and loyalty, but also ensures that the products offered are in accordance with sharia principles. A concrete example is the use of AI-based chatbots that can provide real-time sharia investment recommendations to customers. This chatbot can analyse customer risk profiles, investment preferences, and financial goals to provide recommendations for the most suitable sharia investment products. For example, if a customer has a conservative risk profile, the chatbot may recommend investing in low-risk sukuk or sharia mutual funds.

Business processes that were previously manual and time-consuming can now be automated with AI, improving operational efficiency. An example is the automation of the transaction verification process in accordance with sharia principles, which not only speeds up processing times but also ensures compliance with sharia law. By doing so, financial institutions can reduce operational costs and increase efficiency, which in turn can be passed on to customers in the form of lower costs and better services. For example, AI can be used to verify Murabaha (buying and selling) transactions by examining transaction documents and ensuring that all elements are in accordance with sharia principles. This AI system can reduce the time it takes to verify transactions from a few days to a few minutes, improving efficiency and customer satisfaction.

The integration of AI in Islamic finance opens up great opportunities for product and service innovation. However, there are challenges in ensuring that the use of AI does not violate sharia principles, such as the prohibition on riba (interest) and gharar (uncertainty). Therefore, it is important for Islamic financial institutions to work closely with Islamic scholars and experts to ensure that the AI models used are in accordance with sharia principles. For example, in the development of sharia-based financing products, AI must ensure that all aspects of transactions, such as profit margins and payment schemes, are in accordance with sharia law. This AI model must also be transparent in decision-making to avoid the uncertainty prohibited in sharia principles.

The integration of AI in Islamic finance not only encourages innovation but can also increase financial inclusion and Islamic financial literacy. AI-based applications can be used to educate people about the principles of Islamic finance and help them make better financial decisions. For example, AI-based mobile apps can educate customers about the available Islamic financial products and how they can leverage them to achieve their financial goals. For example, AI-based financial education applications can provide information on various Islamic financing and investment products, as well as help customers choose the product that best suits their financial needs and goals. The app can also provide financial simulations to help customers understand the impact of their financial decisions in the long run.

HUMAN-CENTRIC AI IN DIGITAL TRANSFORMATION AND ENTREPRENEURSHIP

Human-centric AI is an approach in the development and application of artificial intelligence that puts humans at the centre of attention, with a focus on collaboration between humans and machines, as well as an emphasis on ethics, transparency, and trust (Sun & Xie, 2024; Wu & Li, 2024). In the context of digital transformation and entrepreneurship, human-centric AI offers a range of benefits that can help businesses to improve efficiency, innovation, and sustainability. This article will discuss how human-centric AI can contribute to digital transformation and entrepreneurship by integrating AI into business processes, promoting ethics and trust, fostering human and AI collaboration, creating personalized customer experiences, and building an inclusive digital ecosystem.

One of the key ways in which human-centric AI can contribute to digital transformation and entrepreneurship is through the integration of AI into business processes. By automating time-consuming and repetitive tasks, AI allows workers and entrepreneurs to focus on the strategic and creative aspects of their business. For example, an AI-guided chatbot can handle basic customer service, while a human team can handle more complex issues and build deeper relationships with customers. This not only increases efficiency but also increases overall productivity.

AI can aid in business decision-making by providing more in-depth and accurate data analysis. With the ability to process and analyse large amounts of data quickly, AI can provide better insights into market trends, customer preferences, and business performance. This allows entrepreneurs to make more informed and data-driven decisions, which in turn can help in developing more effective business strategies. Ethics and trust are two key elements that are the focus of human-centric AI. By emphasizing transparency in algorithms and decisions made by AI, companies can build trust with customers and other stakeholders. This is especially important in a business environment where decisions made by AI can have a significant impact on individuals and society as a whole.

In the context of entrepreneurship, ethical principles in the use of AI can help in building a good and trustworthy reputation in the market. For example, by ensuring that AI algorithms are impartial and transparent, companies can avoid the risk of discrimination and bias, which can damage customer reputation and trust. By providing a clear explanation of how AI works and what decisions it makes, companies can increase accountability and build stronger relationships with customers. Human-centric AI encourages collaboration between humans and machines, where the two work together towards a common goal. This creates an environment that supports continuous learning and development. In the context of entrepreneurship, AI can be used to support continuous skill development and learning, helping entrepreneurs and employees learn new skills, identify market opportunities, and make better business decisions. For example, AI-based learning platforms can tailor educational content to individual needs, provide personalized advice, and facilitate more effective and efficient learning. This not only improves individual abilities but also improves the overall competitiveness of the company.

One of the key benefits of human-centric AI is its ability to create personalized customer experiences. By leveraging customer data, AI can provide more relevant and timely recommendations, which can improve customer satisfaction and drive loyalty.

In the e-commerce business, for example, AI can analyse customers' purchasing behaviour and provide product recommendations that match their preferences. This not only increases sales opportunities but also makes customers feel valued and cared for. Enhanced personalization can assist companies in identifying and targeting different market segments, thereby increasing the effectiveness of marketing campaigns. Human-centric AI drives innovation by creating an environment that supports collaboration and experimentation. In entrepreneurship, this means that entrepreneurs can test and implement new ideas quickly and efficiently, improving their ability to adapt to changing markets and customer needs.

With a focus on ethics and sustainability, companies can develop products and services that not only meet current needs but also consider the long-term impact on society and the environment. For example, AI can be used to develop more efficient and environmentally friendly energy solutions, or to improve efficiency in supply chains, which in turn can reduce environmental impact and improve business sustainability.

Human-centric AI ensures that digital transformation is inclusive and accessible to everyone, including those who may be underserved by traditional technologies. This is important to ensure that all sections of society can participate in the digital economy and benefit from technological advancements.

Entrepreneurs can leverage inclusive AI to create products and services that cater to a wider range of market segments, including those with special needs or those living in remote areas. For example, AI can be used to develop tools that help individuals with disabilities to more easily access information and services, or to create technological solutions that can be used by communities in remote areas with limited infrastructure.

E-COMMERCE AND THE CHANGING RETAIL LANDSCAPE

Digital transformation has fundamentally changed the retail landscape and the way consumers shop, with e-commerce being one of the key pillars of this change (Chen et al., 2024; Hokmabadi et al., 2024; Kyshakevych et al., 2024). E-commerce, or electronic commerce, refers to commercial transactions that are conducted on-line, allowing businesses and consumers to interact and transact without having to meet physically. One of the main advantages of e-commerce is its ability to connect entrepreneurs with global markets without the need for a physical presence in each location. This eliminates the need to open physical stores in various geographic locations, which can be a significant cost burden. As a result, many entrepreneurs are now able to reach consumers in different parts of the world just by utilizing digital platforms. By operating online, businesses can save on operational costs associated with renting retail space, managing inventory in-store, and various other overhead costs that often come with physical stores.

Digital transformation has also brought significant changes in the way consumers shop and interact with products (Ben Saad & Choura, 2023; Rosca et al., 2023). Digital technology allows consumers to make purchases anytime and anywhere, providing flexibility previously not available in traditional retail models. With just a few clicks, consumers can access a wide range of products and services from different brands and retailers, compare prices, read reviews, and make purchases without having to leave the comfort of their homes. This convenience not only increases convenience for consumers, but also influences the way businesses design their marketing and sales strategies. Businesses that leverage e-commerce can collect valuable data on consumer behaviour, which can be used to tailor their product and service offerings to better suit customer preferences and needs.

Advanced technologies such as Augmented Reality (AR) and Virtual Reality (VR) have brought a new dimension to the online shopping experience. AR and VR offer new ways to interact with products virtually, creating a more immersive and engaging shopping experience (Davis & Aslam, 2024; Khaldy et al., 2023; Nagy et al., 2022). Augmented Reality, for example, allows consumers to see how a product will look in their environment or on themselves before making a purchase. Using

AR apps, customers can visualize how furniture will fit in their space, or virtually try on clothes. This feature not only provides a more immersive visual experience but also helps consumers make more informed purchasing decisions. For example, customers looking to purchase a new sofa can use an AR app to view a variety of colour and design options in the context of their living room, so they can decide on the option that best suits their aesthetics and space needs.

Virtual Reality, on the other hand, offers a highly immersive shopping experience by creating a virtual store or showroom where consumers can explore products in depth. In a VR environment, customers can "step in" into a virtual store that mimics the real-world shopping experience, walk around different parts of the store, view products from different angles, and even interact with products as if they were in a physical store. This experience not only provides a fun and innovative way to shop but also overcomes some of the limitations that exist in traditional e-commerce, such as the inability to see or experience products in person before purchasing.

The use of AR and VR technology in e-commerce also has a positive impact on customer engagement and sales conversion. By creating a more interactive and engaging shopping experience, businesses can increase customer interest and attention. Consumers who feel more engaged with the product tend to be more satisfied with their shopping experience and are more likely to make a purchase. The immersive and interactive visual experiences offered by AR and VR can help reduce product return rates, as customers have a better understanding of what they are buying before they make a transaction. Thus, this technology not only improves customer satisfaction but can also improve operational efficiency and business profitability.

Furthermore, the integration of AR and VR with e-commerce opens up new opportunities for innovation in marketing and sales. Businesses can use this technology to create engaging and creative marketing campaigns, introduce products in new and exciting ways, and build stronger relationships with customers. For example, a company can hold a virtual product launch that allows customers to "attend" the launch event remotely, interact with the product, and get information directly from the speaker or product demonstration. This kind of campaign not only expands marketing reach but also creates a memorable experience for customers, which can increase customer loyalty and retention.

Digital transformation through e-commerce and advanced technologies such as AR and VR has brought profound changes in the way businesses operate and how consumers shop. E-commerce has opened the door for entrepreneurs to reach global markets at lower costs and greater flexibility, while AR and VR technologies have created a more immersive and interactive shopping experience. By leveraging this technology, businesses can increase customer engagement, increase sales conversions, and unlock new opportunities for innovation in marketing and sales. As technology continues to advance, we can expect more changes and innovations in the retail landscape, which will continue to shape the way we shop and do business in the future.

With the continued development of technology, the future of e-commerce seems to be increasingly integrated with more sophisticated digital innovations. One important direction that is emerging is the use of artificial intelligence (AI) to improve customer experience and operational efficiency. AI can be used to analyse customer data more deeply, offer more accurate product recommendations, and automate customer service through smart chatbots. AI not only simplifies the buying process by providing relevant advice but can also help in identifying market trends and customer preferences more quickly and efficiently. This allows businesses to adjust their marketing strategies in real-time and make more data-driven decisions, which in turn can improve customer satisfaction and optimize sales.

Blockchain technology is also starting to enter the world of e-commerce, offering solutions for transaction security and supply chain transparency. Blockchain can be used to trace the origin of products, ensure their authenticity and integrity, and facilitate secure and transparent transactions. With the ability to provide immutable transaction records, blockchain technology can reduce the risk of fraud and increase customer confidence in e-commerce transactions.

In terms of operations, logistics and order fulfilment have also undergone significant changes thanks to digital technology. Advanced inventory management systems and warehouse automation, such as the use of robots and drones, allow for faster and more efficient delivery of goods. One-day delivery and even same-day delivery are increasingly becoming the standard in e-commerce, increasing customer expectations for speed and convenience. Innovation in logistics also includes the development of more environmentally friendly solutions, such as recyclable packaging and shipping methods that reduce carbon footprint.

It is also important to consider how changes in consumer behaviour affect e-commerce. With increasing concern for social and environmental issues, many consumers are now preferring brands that demonstrate social responsibility and sustainability. Successful businesses will need to respond to these trends by offering ethical and sustainable products, as well as implementing transparent and responsible business practices. Not only does this help build customer trust but can also be a significant differentiating factor in an increasingly competitive market.

In a global context, e-commerce also faces challenges related to cross-border regulations and policies. Regulations regarding data privacy, tax, and regulatory compliance vary by country, and businesses must navigate the complex legal landscape to ensure compliance and avoid potential legal issues. While these challenges may seem daunting, they also open up opportunities for the development of better industry standards and best practices in global e-commerce.

CONCLUSION

This research has developed a human-cantered AI framework in Islamic finance, which not only deepens our understanding of the potential of AI in this context but also provides practical guidance for institutions looking to improve their innovations while still adhering to Sharia principles. The integration of AI in Islamic finance offers a unique opportunity to combine advanced technology with strong ethical values, ensuring that the use of AI not only improves operational efficiency but also strengthens the commitment to social justice and ethical responsibility, especially in entrepreneurial interactions.

The main findings of this study show that AI can significantly improve the efficiency and personalization of Islamic financial services. With the ability to analyse data in depth, AI helps in predicting market trends and managing risk, while ensuring that all processes remain in accordance with Sharia principles. The implementation of AI has also shown an increase in customer satisfaction and trust, which is a vital element in Sharia-based financial transactions, especially in interactions with entrepreneurs who need fast and precise financial solutions.

This research contributes significantly to the literature on the integration of technology in Islamic finance, highlighting how AI can be used without compromising the ethical values underlying the Islamic financial system. This paves the way for further studies on the application of advanced technologies in the context of ethical and responsible finance, as well as how these technologies can support more efficient and ethical entrepreneurial interactions.

Practically, the findings of this research can be applied by Islamic financial institutions to improve their services and ensure that they remain competitive in the digital age. For entrepreneurs, AI can provide better market analysis and faster financial solutions, supporting the growth of their businesses within the framework of Sharia ethics. Theoretically, this research enriches understanding of how technology can be used to reinforce ethical values in financial practice, offering new insights for academics and practitioners in this field.

Recommendations for future research include further exploration of how AI can be applied in different aspects of Islamic finance, as well as studies on the long-term impact of the use of AI in Islamic finance. Research can also be focused on developing policies and ethical frameworks to direct the implementation of these technologies. Further research can explore how AI can support more inclusive and sustainable entrepreneurial interactions.

The research is in line with existing literature that emphasizes the importance of technological innovation in Islamic finance, while offering a new perspective on how AI can be used to reinforce ethical principles. It enriches discussions about Islamic finance in the digital age and demonstrates the potential for sustainable innovation, particularly in the context of entrepreneurial interactions.

The study shows that the integration of AI in Islamic finance not only allows financial institutions to become more efficient and responsive to customer needs, but also strengthens their commitment to ethical principles. As such, human-cantered AI can play a key role in advancing Islamic finance in the digital age, supporting more inclusive, fair, and socially responsible financial practices, and enhancing ethical and efficient entrepreneurial interactions.

REFERENCE

Ben Saad, S., & Choura, F. (2023). Towards better interaction between salespeople and consumers: The role of virtual recommendation agent. *European Journal of Marketing*, *57*(3), 858–903. DOI: 10.1108/EJM-11-2021-0892

Carvajal Zaera, E. (2024). The influence of artificial intelligence on communication in healthcare. *European Public and Social Innovation Review*, *9*. Advance online publication. DOI: 10.31637/epsir-2024-312

Chen, K., Luo, S., & Kin tong, D. Y. (2024). Cross border e-commerce development and enterprise digital technology innovation—Empirical evidence from listed companies in China. *Heliyon*, *10*(15), e34744. Advance online publication. DOI: 10.1016/j.heliyon.2024.e34744 PMID: 39144960

Davis, L., & Aslam, U. (2024). Analyzing consumer expectations and experiences of Augmented Reality (AR) apps in the fashion retail sector. *Journal of Retailing and Consumer Services*, *76*, 103577. Advance online publication. DOI: 10.1016/j.jretconser.2023.103577

Gaubienė, N. (2024). Can Artificial Intelligence Engage in the Practice of Law as the Art of Good and Justice? *Filosofija, Sociologija, 35*(2 Special Issue), 54–63. DOI: 10.6001/fil-soc.2024.35.2Priedas.Special-Issue.6

Gümüsay, A. A. (2020). The Potential for Plurality and Prevalence of the Religious Institutional Logic. *Business & Society*, *59*(5), 855–880. DOI: 10.1177/0007650317745634

Hasan, H. E., Jaber, D., Khabour, O. F., & Alzoubi, K. H. (2024). Ethical considerations and concerns in the implementation of AI in pharmacy practice: A cross-sectional study. *BMC Medical Ethics*, *25*(1), 55. Advance online publication. DOI: 10.1186/s12910-024-01062-8 PMID: 38750441

Hermann, E. (2022). Artificial intelligence and mass personalization of communication content—An ethical and literacy perspective. *New Media & Society*, *24*(5), 1258–1277. DOI: 10.1177/14614448211022702

Hokmabadi, H., Rezvani, S. M. H. S., & de Matos, C. A. (2024). Business Resilience for Small and Medium Enterprises and Startups by Digital Transformation and the Role of Marketing Capabilities—A Systematic Review. *Systems*, *12*(6), 220. Advance online publication. DOI: 10.3390/systems12060220

Huang, K., Teng, Y., Chen, Y., & Wang, Y. (2024). From Pixels to Principles: A Decade of Progress and Landscape in Trustworthy Computer Vision. *Science and Engineering Ethics*, *30*(3), 26. Advance online publication. DOI: 10.1007/s11948-024-00480-6 PMID: 38856788

Ibrahim, A.-J., & Kahf, M. (2020). Instruments for investment protection when structuring Islamic venture capital. *Journal of Islamic Accounting and Business Research*, *11*(9), 1907–1920. DOI: 10.1108/JIABR-01-2019-0025

Jelili Amuda, Y., & Alabdulrahman, S. (2023). Reinforcing policy and legal framework for Islamic insurance in Islamic finance: Towards achieving Saudi Arabia Vision 2030. *International Journal of Law and Management*, *65*(6), 600–613. DOI: 10.1108/IJLMA-03-2023-0045

Khaldy, M. A., Ishtaiwi, A., Al-Qerem, A., Aldweesh, A., Alauthman, M., Almomani, A., & Arya, V. (2023). Redefining E-Commerce Experience: An Exploration of Augmented and Virtual Reality Technologies. *International Journal on Semantic Web and Information Systems*, *19*(1), 1–24. Advance online publication. DOI: 10.4018/IJSWIS.334123

Konak, F., & Demir, Y. (2023). Bibliometric Analysis on Islamic Insurance (Takaful). *Hitit Theology Journal*, *22*(1), 11–46. DOI: 10.14395/hid.1232415

Kunhibava, S. (2011). Reasons on the Similarity of Objections with Regards to Gambling and Speculation in Islamic Finance and Conventional Finance. *Journal of Gambling Studies*, *27*(1), 1–13. DOI: 10.1007/s10899-010-9201-5 PMID: 20514512

Kyshakevych, B., Maksyshko, N., Hrytsenko, K., Voronchak, I., & Demediuk, B. (2024). ANALYZING THE EFFICIENCY OF DIGITALIZATION IN SMALL AND MEDIUMSIZED ENTERPRISES ACROSS EU COUNTRIES USING DEA MODELS. *Financial and Credit Activity: Problems of Theory and Practice*, *3*(56), 215–229. DOI: 10.55643/fcaptp.3.56.2024.4344

Nagy, A. S., Bittner, B., Tuegeh, O. D. M., & Tumiwa, J. R. (2022). Augmented reality improving consumer choice confidence during COVID-19. *Issues in Information Systems*, *23*(2), 294–309. DOI: 10.48009/2_iis_2022_126

Nguyen, A., Ngo, H. N., Hong, Y., Dang, B., & Nguyen, B.-P. T. (2023). Ethical principles for artificial intelligence in education. *Education and Information Technologies*, *28*(4), 4221–4241. DOI: 10.1007/s10639-022-11316-w PMID: 36254344

Nienhaus, V. (2011). Islamic finance ethics and Shari'ah law in the aftermath of the crisis: Concept and practice of Shari'ah compliant finance. *Ethical Perspectives*, *18*(4), 591–623. DOI: 10.2143/EP.18.4.2141849

Ridho Kismawadi, E., Irfan, M., & Shah, S. M. A. R. (2023). Revolutionizing islamic finance: Artificial intelligence's role in the future of industry. In *The Impact of AI Innovation on Financial Sectors in the Era of Industry 5.0* (pp. 184–207). DOI: 10.4018/979-8-3693-0082-4.ch011

Rosca, M., Vatra, A.-D., & Avadanei, M. (2023). The digital transformation of garment product development. *Industria Textila (Bucuresti)*, *74*(1), 98–106. DOI: 10.35530/IT.074.01.2022148

Sakinç, İ. (2021). Analysis of the Working Capital Management Efficiency of the Manufacturing Companies in the Islamic Index. *Hitit Theology Journal*, *20*(3), 107–128. DOI: 10.14395/hid.930402

Shamsudheen, S. V., Mohamad, S., Muneeza, A., & Mahomed, Z. (2024). Ethical discourse of ethical (Islamic) finance: A systematic literature review (1988–2022) and the way forward. *Journal of Islamic Accounting and Business Research*. Advance online publication. DOI: 10.1108/JIABR-11-2022-0315

Sun, X., & Xie, X. (2024). How does digital finance promote entrepreneurship? The roles of traditional financial institutions and BigTech firms. *Pacific-Basin Finance Journal*, *85*, 102316. Advance online publication. DOI: 10.1016/j.pacfin.2024.102316

Tigges, M., Mestwerdt, S., Tschirner, S., & Mauer, R. (2024). Who gets the money? A qualitative analysis of fintech lending and credit scoring through the adoption of AI and alternative data. *Technological Forecasting and Social Change*, *205*, 123491. Advance online publication. DOI: 10.1016/j.techfore.2024.123491

Wangmo, T., Lipps, M., Kressig, R. W., & Ienca, M. (2019). Ethical concerns with the use of intelligent assistive technology: Findings from a qualitative study with professional stakeholders. *BMC Medical Ethics*, *20*(1), 98. Advance online publication. DOI: 10.1186/s12910-019-0437-z PMID: 31856798

Wilson, C., & van der Velden, M. (2022). Sustainable AI: An integrated model to guide public sector decision-making. *Technology in Society*, *68*, 101926. Advance online publication. DOI: 10.1016/j.techsoc.2022.101926

Wu, Y., & Li, Z. (2024). Digital transformation, entrepreneurship, and disruptive innovation: Evidence of corporate digitalization in China from 2010 to 2021. *Humanities & Social Sciences Communications*, *11*(1), 163. Advance online publication. DOI: 10.1057/s41599-023-02378-3

Chapter 16
Service Quality in the Age of Artificial Intelligence:
Evidence From Listed Telecommunication Businesses in Sub-Saharan Africa

Timilehin Olasoji Olubiyi
https://orcid.org/0000-0003-0690-7722
Babcock University, Ilishan-Remo, Nigeria

ABSTRACT

Artificial intelligence (AI) is progressively transforming service delivery by executing diverse activities, serving as a significant driver of innovation, but also posing a danger to human employment. The significance of artificial intelligence (AI) and how it affects Nigerian telecommunications companies' service quality are both emphasized in this chapter. The descriptive survey design was used, and the population of the study comprised staff of listed telecommunication companies in Nigeria. The findings of the simple linear regression analysis revealed that artificial intelligence has a positive and significant effect on the service quality of listed telecommunication companies in Nigeria. This implies that artificial intelligence is a significant predictor of service quality. The chapter suggests that firms should provide customers with more personalised services, as this has a significant impact on their overall experience with the enterprise.

DOI: 10.4018/979-8-3693-8009-3.ch016

INTRODUCTION

Recently, the telecommunication industry environment has become more turbulent and competitive due to globalisation and more openness in the economy. Topical research shows that artificial intelligence and robotics (AIR) are penetrating in different areas of businesses and industries including telecommunication sector (Yang *et al.*, 2020). Despite the potential benefits of AI, many firms are yet to implement an AI technology. The field of artificial intelligence is currently seeing rapid growth, with an increasing number of researchers investigating the effects of this technology in the field of marketing. Market researchers aim to determine how artificial intelligence (AI) may improve the client purchasing experience, resulting in increased customer loyalty and, consequently, higher profits. As a result of the ongoing increase in customer expectations, improving service quality has become a difficult undertaking for businesses. Companies aiming to gain a competitive edge in customer service should consider more than just delivering information at the correct moment through a suitable medium (Naumov, 2019). They should explore innovative approaches to address customers' pain points across the entire purchasing process, while also providing efficient customer service. For instance, providing customized and readily available information and assistance to individual customers resulted in an improved service quality (Bhattacharya & Sinha (2022).

The introduction of AI into the marketing sphere has mostly resolved these difficulties. Put simply, the development of machines capable of "thinking," generating patterns, analysing multiple factors, and making judgements has ushered humanity into a new era. You may have encountered such technology during the process of online shopping or booking, where you come across multiple advertisements that are relevant to your previous search queries. Alternatively, consider the scenario where you receive customer care from a virtual assistant instead of a human. Artificial Intelligence has significantly expanded opportunities in various commercial sectors, particularly in the realm of marketing. The company can benefit from an improved client experience by providing highly personalised customer care and on-demand customer support, while also minimising both time-consuming and expensive processes. This study aims to investigate the utilisation of artificial intelligence (AI) in the marketing industry, with a specific focus on its influence on the service quality, particularly in the areas of customer care and post-sales support. In light of the theoretical and practical significance of artificial intelligence, and its impact on the service quality post-pandemic, such research is still with inconclusive evidence. Artificial intelligence (AI) technology spans various industries, but relatively little attention is given to the use of AI technologies by telecommunication industries. The extensive field of AI research has primarily concentrated on its implementation in several industries, with minimal emphasis on the telecommunications sector

(Huynh *et al.,* 2021; Bag *et al.,* 2021; Pillai *et al.,* 2020; Keegan *et al.,* 2022; Benz, & Chatterjee 2021), among others. Liu *et al.* (2020) confirmed the influence of artificial intelligence (AI) on the processes of innovation and management practices within companies, namely in the realm of technological innovation.

Several researchers (Adejoh & Hadiza, 2015; Nebo, Nwankwo, & Okonkwo, 2015; Wallace & Deborah, 2016; Ghasemaghaei, & Calic, 2020; Prentice *et al.,* 2020; Olubiyi, 2022a) have investigated the impact of technology on performance and service quality and have concluded that it enhances customer satisfaction in organizations. However, these studies were undertaken in industries other than the telecommunication industry in developing economies. Therefore, the researcher is compelled to take a fresh look at the issues in the context of emerging economies, specifically Nigeria and in the telecommunication industry. As a result, the goal of this paper is to fill the gap in Nigeria the most populous and the largest economy in Africa. Therefore, the objective of this study is to identify artificial intelligence and its impact on the service quality of telecommunication industry post-pandemic in Nigeria.

LITERATURE REVIEW

Artificial Intelligence

Artificial intelligence (AI) refers to a collection of theories and algorithms that empower computers to carry out tasks that usually necessitate human intelligence, such as visual perception, voice recognition, or understanding and interpreting text while considering its context. In certain instances, AI can even enhance these capabilities. Machine learning is a prominent subfield of artificial intelligence that has gained significant popularity in recent times. Financial service providers presently employ AI technologies, including predictive analytics and speech recognition, to provide banks with the benefits of digitization and enhance their competitiveness against FinTech competitors (Dam, & Dam, 2021). Artificial intelligence (AI) can enhance banks' client experiences by enabling uninterrupted and effortless interactions with customer service staff. Nevertheless, the utilisation of artificial intelligence in financial applications extends far beyond the realm of conventional retail banking. AI might potentially benefit the back and middle offices of investment banking and other financial institutions, Doborjeh, *et al.,* (2022). AI's ability to identify and prevent fraudulent activities while improving compliance procedures gives the banking industry a promising outlook (Alzoubi, *et al.,*2021). An artificial intelligence programme can expedite tasks that typically require hours or days to complete, such as combatting money laundering, by accomplishing them in a matter of seconds. Banks can also profit from the AI's capacity to quickly extract useful information from vast data sets. Artificial intelligence bots, online payment advisers, and biometric fraud detection systems enhance the quality of service for a broader range of people (Naumov, 2019). Consequently, the bank experiences an increase in revenues, a decrease in expenses, and a significant rise in profits.

Service Quality

Service quality refers to the extent to which firms meet or exceed the expectations of their clients. It can align with, meet, or surpass customer preferences (Olowoporoku & Olubiyi, 2023; Naumov, 2019). As service quality improves, customer satisfaction increases, leading to higher profits. Service quality is a measure of how well an organisation delivers its services in relation to customer expectations. Customers seek out services based on their specific needs, and they have certain standards and expectations for how a company should meet those needs. Therefore, measuring and enhancing service quality can boost an organization's profits and reputation. Regardless of the industry, service quality directly impacts a company's ability to meet customer needs, remain competitive, and achieve customer satisfaction (Olowoporoku & Olubiyi, 2023). Yang et al. (2020) argue that AI and robotics provide numerous prospects for businesses in the hotel industry, allowing them to provide improved customer service while also increasing operational efficiency. The incorporation of service robots into hospitality environments has a substantial impact on customer service experiences. According to Doborjeh, *et al.* 2022 service quality is the entire perception that consumers have of their service interactions, taking into account multiple dimensions and related features. In contrast, Parasuraman et al. (1985) define employee service quality as staff's capacity to meet or exceed customers' service expectations. This dimension evaluates employees' capacity to offer prompt, responsive, reliable, empathetic, and professional assistance that either matches or exceeds customer expectations.

THEORETICAL REVIEW

Technology Acceptance Model (TAM)

The Technology Acceptance Model (TAM) was created by Davis (1989) with the aim of elucidating consumers' perceptions and utilization of technology (Teo, 2012). The TAM utilized Fishbein and Ajzen's (1975) theory of reasoned action (TRA) to establish its causal sequence: beliefs (namely, perceived ease of use and perceived usefulness)/attitude/behavioral intention (Kim et al., 2009). The purpose of this study was to provide a detailed explanation and forecast of how technology is adopted in the workplace. This was achieved by identifying the factors that influence individuals' beliefs, attitudes, intentions, and behaviors towards technology usage (Davis, 1989; Davis *et al.*, 1989; Liao *et al.*, 2009). Davis (1989) devised more accurate metrics for forecasting and elucidating use, centered around two theoretical concepts: perceived usefulness (PU) and perceived ease of use (PEOU), which were postulated as essential factors influencing system adoption. Taylor and Todd (1995) and Wang and Liu (2005) established that TAM initially proposed external variables as a means to track the influence of external circumstances on two primary internal beliefs: perceived usefulness and perceived ease of use. In addition, Venkatesh and Bala, (2008) propose that the perception of ease of use might be used to predict the perception of utility. TAM posits that the perceived usefulness (PU) and perceived ease of use (PEOU) of IT are affected by external factors and jointly shape the attitude towards its use. The statement implies that the perception of how easy and beneficial a technology is affects one's attitude towards utilizing it, which in turn has a favorable impact on their intention to actually use it (Venkatesh, & Davis, 2000). The objective of the Technology Acceptance Model (TAM), as described by Davis *et al.* (1989), is to offer economical and theoretically grounded models that elucidate the factors influencing the adoption of various information technologies by different user groups.

EMPIRICAL REVIEW

Relationship Between Artificial Intelligence and Service Quality

As noted by So *et al.* (2021), Hentzen *et al.* (2022) and Hollebeek *et al.* (2022), artificial intelligence has a significant effect on customer satisfaction. The studies noted the need to enhance customer engagement through the use of technological developments. Previous studies (Cardoso, Costa, & Novais, 2010; Chen, Esperança, & Wang, 2022) also demonstrated that in the electronic industry, artificial intelligence has a positive and significant effect on customer engagement and satisfaction. To effectively satisfy client expectations, businesses can take proactive measures to resolve complaints, provide pertinent products or services and optimize their strategies (Chen, Esperança, & Wang, 2022). By automating repetitive operations, implementing AI-powered customer services can drastically save operating expenses. Businesses are able to carefully allocate resources and invest in improving overall customer happiness because to this cost-effectiveness (Al-Mekhlal *et al.*, 2023). Trawnih *et al.* (2022) reported a positive and significant effect of artificial intelligence on customer satisfaction. Findings indicated that companies are better positioned to create enduring and lucrative relationships with their clientele when they include AI into their customer service operations (Trawnih *et al.*, 2022). Similarly, Yeo et al. (2022) showed a positive and significant effect of artificial intelligence on service quality. The pleasure of customers is influenced by their perception of the AI system's efficiency in resolving difficulties and offering relevant support (Yeo *et al.,* 2022). In the same vein, the studies of Gupta (2021), Fernando *et al.* (2023), and Floridi et al. (2021) demonstrated that artificial intelligence has a significant effect on customer satisfaction. Artificial intelligence has been shown to significantly contribute to the evaluation of overall service quality, as well as customer happiness and loyalty (Prentice *et al.,* 2020). Geetha (2021) while examining artificial intelligence in banking and financial services found that private banks and private financial institutions are using various AI services to improve customers' benefit so that customers are satisfied with their services. Also, Al-Adwan *et al.* (2022) examined the role of environmental cues and stimuli in online settings in Jordan and in their study highlighted the importance of policy service quality dimensions in establishing customer satisfaction and trust. Also, Al-Adwan *et al.* (2020) explored the factors affecting online trust, satisfaction and loyalty in Jordan and provided relevant and beneficial insights for e-commerce businesses in Jordan. Consequently, this study sought to bridge the gap and hypothesized thus:

H$_{o1:}$ There is no significant effect of artificial intelligence on service quality of listed telecommunication companies in Nigeria.

Figure 1. Author's Conceptual Model (2024)

Conceptual Model

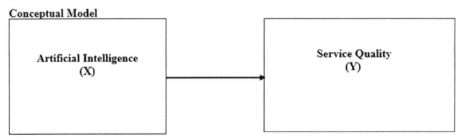

The model sheds light on artificial intelligence measured by usage of data mining, machine learning, and chatbots on service quality which is the research framework. The independent and dependent variables for this research are artificial intelligence(X) on service quality (Y) respectively.

Model Specification

The model sheds light on the effect of artificial intelligence on service quality which is the research framework. Given, the mathematical derivative function which gives the value of the slope at any value(x-------x_n) since intuition explains that as $\Delta x \to 0$, then $\Delta y \to 0$. This can be deduced mathematically since a firm's service quality is a function of artificial intelligence

$$y = f(x_1 \text{-----------} x_n).$$

Hypothesis:

$$Y = P_0 + P_1 x_1 + P_2 x_2 + P_3 x_3 + P_4 x_4 \dots\dots\dots\dots\dots\dots\dots\dots\dots \text{Regression equation}$$
(1)

METHODOLOGY

The research context of this study is the telecommunication sector and the present investigation is limited to companies listed on the Nigerian Exchange Group. For this quantitative study, the two listed telecommunication companies were considered and surveyed to investigate how the companies benefit from implementing AI. The study adopted the cross-sectional survey research design and the justification for adopting the survey is due to its usefulness in assessing the thoughts, opinions, and feelings of different groups of individuals and allowing them to give more valid and honest feedback on the area of study. This paper relied on the prior study methodology of Arokodare, and Olubiyi, (2023); Makinde, Olubiyi, and Ogundipe, (2023);Olowoporoku, and Olubiyi,(2023);Olubiyi (2023a); Adeoye,et al.,(2023); Olubiyi (2023b);Olubiyi, Adeoye, Jubril, Adeyemi, and Eyanuku, (2023); Olubiyi, and Akpa, (2023); Adeyemi, and Olubiyi, (2024); Olubiyi (2019); Olubiyi, Lawal, and Adeoye, (2022); Olubiyi, (2022a); Olubiyi, (2022b); and Omoyele, et al. (2023) with cross-sectional have adopted this method in their respective studies and found it useful. For the population of this study was from the two listed telecommunication companies in Nigeria, namely MTN Nigeria and Airtel Africa. These companies are the only two telecommunication companies listed on the Nigeria Exchange Group as at July 2024, thus justifying their inclusion. A close-ended questionnaire distributed to examine the behaviour of telecommunication companies' customers towards AI. The questionnaire contains four main sections, each taking a different aspect of the relationship between AI and service quality, focusing on the Telecommunication sector. The data was gathered and also obtained via the use of an online questionnaire facilitated by the Survey Monkey platform. Bryman and Bell (2014) argue that the use of online surveys presents a cost-efficient method for collecting considerable responses within a constrained period. According to the annual report for the year 2023 provided by MTN Nigeria, the overall count of personnel reaches 1,682. Conversely, Airtel Africa possesses a total of 3,525 personnel, distributed across different regions such as Nigeria, East Africa, and Francophonia, with Nigeria contributing 667 staff members. Given the specific focus of this particular study, solely the 667 employees in Nigeria were taken into account.

Table 1. Names of selected listed telecommunication companies and staff numbers

S/N	Firm	Number of Staff
1	MTN Nigeria	1,682
2.	Airtel Africa (Nigeria)	667
Total		2349

Source: Annual reports and Nigeria Stock Exchange Data (2023)

With the utilization of Yamane (1967) Sample Size Formula a total of size of 341 was arrived as the sample size for the study from the population of 2349 which is the total number of staff from MTN Nigeria and Airtel Africa as at the year 2023. To account for non-response and incorrect completion of questionnaire copies, the sample size of 341 was augmented by 30% of the total sample, as advised by Zikmud (2000). Thus, 30% of 341 equals 102. Consequently, the appropriate sample size is determined to be n = 341 + 102 = 443 to ensure adequate representation of all elements or groups under investigation in the sample.

Table 2. Proportionate Distribution of Sample Size

Company	Percentage representation	Sample size distribution	Sample size
MTN Nigeria	1,682/2349*100= 71.6	71.6% *443	317
Airtel Nigeria	667/2349*100= 28.4	28.4%*443	126
Total	100		443

Source: Researcher Computation, (2024)

The present research employs both descriptive and inferential statistical analyses to examine the collected data. To evaluate the study's hypotheses, the first step involves doing a descriptive analysis. This analysis entails organizing data into tables that provide the percentage distribution, mean, and standard deviation. The last step involves doing inferential analysis, using basic linear and multiple regression techniques, as well as the Pearson Correlation method of analysis, utilizing IBM SPSS Software Version 22.0 for assistance.

DATA ANALYSIS

From the 443 copies of the questionnaire distributed by the researcher and trained research assistants, a total of 421 copies of the questionnaire were filled and returned for analysis representing a response rate of ninety-nine percent (95%). Response rate is the percentage of people who responded and administered copies of the questionnaire in the survey The rest were either unreturned or had missing responses, however, the total number of questionnaires received was sufficient to represent the SMEs in Saudi Arabia, and they were analyzed. The detail of the responses is shown in Table 3

Table 3. Response Rate

	Frequency	Percentage %
Completed usable copies of the questionnaire	443	95
Unreturned/Incomplete copies of the questionnaire	22	5
Total received	421	100

Source: Researcher's computation (2024)

Restatement of Research Objective and Research Question

Objective: Find out the effect of artificial intelligence on service quality of listed telecommunication companies in Nigeria.

Research Question: What is the effect of artificial intelligence on service quality of listed telecommunication companies in Nigeria.?

Hypothesis: Artificial intelligence has no significant effect on service quality of listed telecommunication companies in Nigeria?

The objective was to find out the effect of artificial intelligence on service quality of listed telecommunication companies in Nigeria. On a six-point Likert Scale, the respondents were requested to rate their perception of various items about artificial intelligence and service quality of listed telecommunication companies in Nigeria. The findings were presented and followed with an analysis and interpretation. To test the hypothesis one, simple linear regression analysis was used with service quality as the dependent variable and artificial intelligence as the independent variable. The data for artificial intelligence was generated by adding all scores of all items for artificial intelligence, while that of service quality was generated by adding scores for all the items for the variable. Data from a total of four hundred and twenty-one respondents were analysed. The summary of the results of the simple linear regression analysis is presented in Table 4.

Table 4. Summary of simple linear regression analysis for effect of artificial intelligence on service quality of listed telecommunication companies in Nigeria

Model	Variables	B	Sig	t	R	R^2
1	(Constant)	10.737	0.000	10.053	0.432[a]	0.187
	Artificial Intelligence	0.442	0.000	7.589		

a. Dependent Variable: Service quality
Decision rule: Reject H_0 if βi = 0
Source: **Researcher's Computation (2024)**

Table 4 provides details of regression analysis results for the effect of artificial intelligence of listed telecommunication companies in Nigeria. The results reveal that artificial intelligence has a positive and significant effect on service quality of listed telecommunication companies in Nigeria ($B = 0.442$, $t = 7.589$, p < 0.05). The t-test associated with B-value was significant and artificial intelligence as the predictor was making a significant contribution to the model. The R value of 0.432 supports this result and it indicates that there is a weak positive relationship between artificial intelligence and service quality of listed telecommunication companies in Nigeria. Coefficient of determination (R^2) explains the extent to which changes in the dependent variable can be explained by the change in the independent variable or the proportion of variation in the dependent variable (service quality) that is explained by the independent variable (artificial intelligence). From the findings in the table 4 the value of $R^2 = 0.187$ indicates that about 18.7% variation that occurs in the service quality of listed telecommunication companies in Nigeria can be accounted for by the level of artificial intelligence the companies implemented while the remaining 81.3% changes that occur can be accounted for by other variables not captured in the model. From the data in Table 4, the established regression equation is:

$$Y = a_0 + \beta_1 x_1 + + e_i \text{ ----------------------- regression model 1}$$

$$SeQ = 10.737 + 0.442AI + e_i \text{---}$$
Eqn i

Where:
SeQ = Service Quality
AI = Artificial Intelligence

The above regression equation reveals that, holding artificial intelligence to a constant, the level of service quality of listed telecommunication companies in Nigeria would be 10.737 implying that without artificial intelligence, service quality of listed telecommunication companies in Nigeria will be 10.737 which is positive. The results of the simple regression analysis indicate that when artificial intelligence is improved by one unit, service quality would increase by a coefficient of 0.442 and it was significant at ($p<0.05$). This implied that for every improvement in artificial intelligence, there will be a subsequent increase in service quality of listed telecommunication companies in Nigeria. The result suggests that artificial intelligence is an important predictor of service quality of listed telecommunication companies in Nigeria. Based on these results, the null hypothesis (H_{01}) which states that artificial intelligence has no significant effect on service quality of listed telecommunication companies in Nigeria was rejected.

DISCUSSION

The hypothesis set out in this study is to determine the effect of artificial intelligence on service quality of listed telecommunication companies in Nigeria. The finding of the simple linear regression analysis revealed that artificial intelligence has a positive and significant effect on service quality of listed telecommunication companies in Nigeria ($B = 0.442$, $t = 7.589$, $p < 0.05$). This implies that artificial intelligence is a significant predictor of service quality of listed telecommunication companies in Nigeria. In consonance with the finding of this study, Hentzen *et al.* (2022) and Hollebeek *et al.* (2022) discovered that artificial intelligence has a significant effect on customer satisfaction. The studies noted the need to enhance customer engagement through the use of technological developments. Previous studies (Cardoso, Costa, & Novais, 2010; Chen, Esperança, & Wang, 2022) also demonstrated that in the electronic industry, artificial intelligence has a positive and significant effect on customer engagement and satisfaction. To effectively satisfy client expectations, businesses can take proactive measures to resolve complaints, provide pertinent products or services and optimize their strategies (Rygielski *et al.,* 2002). By automating repetitive operations, implementing AI-powered customer services can drastically save operating expenses. Businesses are able to carefully allocate resources and invest in improving overall customer happiness because to this cost-effectiveness (Al-Mekhlal *et al.,* 2023). Trawnih *et al.* (2022) reported a positive and significant effect of artificial intelligence on customer satisfaction. Findings indicated that companies are better positioned to create enduring and lucrative relationships with their clientele when they include AI into their customer service operations (Trawnih *et al.,* 2022). Similarly, Yeo *et al.* (2022) showed a positive and significant effect of artificial intelligence on service quality. The pleasure of customers is influenced by their perception of the AI system's efficiency in resolving difficulties and offering relevant support (Yeo *et al.,* 2022). In the same vein, the studies of Gupta (2021), Fernando *et al.* (2023), and Floridi *et al.* (2021) demonstrated that artificial intelligence has a significant effect on customer satisfaction. Artificial intelligence has been shown to significantly contribute to the evaluation of overall service quality, as well as customer happiness and loyalty (Prentice *et al.,* 2020). Geetha (2021) while examining artificial intelligence in banking and financial services found that private banks and private financial institutions are using various AI services to improve customers' benefit so that customers are satisfied with their services. Also, Al-Adwan *et al.* (2022) examined the role of environmental cues and stimuli in online settings in Jordan and in their study highlighted the importance of policy service quality dimensions in establishing customer satisfaction and trust. Also, Al-Adwan *et al.* (2020) explored the factors affecting online trust, satisfaction and loyalty in Jordan and provided relevant and beneficial insights for e-commerce businesses in Jordan.

CONCLUSION AND RECOMMENDATION

The major findings of the study showed that artificial intelligence had a significant positive effect on service quality ($B = 0.442$, $t = 7.589$, p < 0.05). The testing hypotheses showed that Artificial Intelligence has a statistically significant influence on service quality. This result corroborated the findings of past research about the substantial impacts of Artificial Intelligence on service quality (Chen, & Prentice, 2024; Yang,2023; Chen, Esperança, & Wang, 2022). These insights not only enhance scholarly discussions but also provide useful guidance for businesses navigating the ever-changing field of AI-driven customer interactions and service quality. This study enhanced our comprehension of the true nature of AI and its impact on businesses, customers, and society as a whole. This study is important as it offers crucial guidance for e-commerce and telecom companies to enhance their service quality and gain a competitive edge. It also provides evidence for policymakers in the field of artificial intelligence regarding the anticipated impact of AI on service quality.

The objective of this study was to examine the influence of utilising artificial intelligence (AI) to improve the overall service quality. During the research, the service quality was categorised into two distinct variables: customer service and after-sale support. A variety of methods were employed on both qualitative and quantitative data to assess the association using the most accurate measurements at hand. Upon analysing the outcomes of the executed analysis, the authors see a correspondence between the study hypotheses and their findings. The findings of the correlational and regression studies demonstrate a favourable correlation between AI and service quality. Additionally, there is a clear association between offering personalised customer service and after-sale customer care, as well as AI. Through the utilisation of descriptive analysis, in conjunction with the aforementioned analyses, the authors demonstrate that delivering personalised customer service throughout the entirety of the customer's purchasing process significantly influences the customer's overall experience. The author suggest that organisations should improve the service quality across the entire process of customers' purchasing journey, particularly during the awareness stage. Enterprises are recommended to provide more personalised services to clients, as this will enhance their entire experience by improving their comprehension of the services and goods offered. This, in turn, will help generate positive word-of-mouth about the enterprise's overall experience. Similarly, it is strongly advised to utilise AI in contact centres and other post-sales support services to reduce customer waiting time using the most popular information technology tools now accessible in the market.

This study recommend that AI can facilitate the process of reducing the need for human labour among both employees and customers. Thanks to the utilisation of machine learning, previously intricate procedures have been significantly simplified. Multiple researches provide evidence that different strategies enhance customer-telecommunication relations and provide a mutually beneficial outcome. The study further recommend that the telecommunication industry should employ artificial intelligence capabilities to ensure that service quality is smooth and effortless. AI applications have enhanced telecommunication operational efficiency, paving the way for new possibilities in the industry. Based on the research earlier research conducted in the which this current study corroborate we recommend that these companies should prioritise understanding their customers' requirements to enhance their sense of value and enable them to carry out operations conveniently, regardless of time and location.

Limitations and Future Study Direction

Despite the satisfactory results in relation to the hypothesis, the research acknowledged several limitations and also experienced significant limitations, some of these are considered to be useful precursors for future study. Primarily, the data was gathered from a limited sample within the listed telecommunication sector. The focus of the study on only telecommunication businesses in Nigeria, restricts the applicability of its results. The findings and implications of this article are specific to Nigeria and focused mostly on Telecommunication industries, which might restrict the generalizability of the results. Although limited, the findings of the current study should inspire scholars to do more comprehensive research on artificial intelligence and customer satisfaction. The paper's cross-sectional design limits the author's ability to assert causation.

Acknowledgment

Timi Olubiyi Consulting

Authors' contribution

The author proposed the research topic, model, and approach, and independently authored the paper

Funding

There was no financing received; nevertheless, personal support was available.

Declaration of Competing Interest

There are no conflicting interests, according to the author. The author further declares that he has no known competing financial interests or personal relationships that might have influenced the work disclosed in this study.

REFERENCES

Adejoh, A. M., & Hadiza, D. (2015). Impact of Effective Communication on Goal Achievement in Nigerian Polytechnics. *International Journal of Sustainability Management and Information Technologies*, *1*(2), 6–10.

Adeoye, O. O., Olubiyi, T. O., Ajiteru, W. O., & Adaranijo, L. O. (2023). Unveiling Capital Structure Determinants, Business Characteristics and High-Growth Performance of Small and Medium-Sized Enterprises: Evidence from Nigeria. *Journal of Interdisciplinary Socio-Economic and Community Study*, *3*(2), 86–98. DOI: 10.21776/jiscos.03.2.05

Adeyemi, O. S., & Olubiyi, T. O. (2024). An Analysis of Investment Trends, Innovation and Its Impact on Cabotage and Local Content Regime: Nigerian Experience from Businesses in Oil and Gas and Maritime Industries. *ABUAD Journal of Social and Management Sciences*, *5*(2), 285–301. DOI: 10.53982/ajsms.2024.0502.04-j

Adeyemi, O. S., & Olubiyi, T. O. (2023). The Impact of Brand Awareness on Customer Loyalty in Selected Food and Beverage Businesses in Lagos State Nigeria. *Jurnal Multidisiplin Madani*, *3*(3), 541–551. DOI: 10.55927/mudima.v3i3.3095

Al-Adwan, A. S., Kokash, H., Adwan, A. A., Alhorani, A., & Yaseen, H. (2020). Building customer loyalty in online shopping: The role of online trust, online satisfaction and electronic word of mouth. *International Journal of Electronic Marketing and Retailing*, *11*(3), 278–306. DOI: 10.1504/IJEMR.2020.108132

Alzoubi, H. M., Vij, M., Vij, A., & Hanaysha, J. R. (2021). What leads guests to satisfaction and loyalty in UAE five-star hotels? AHP analysis to service quality dimensions. *Enlightening Tourism:A Pathmaking Journal 11*(1), 102-135.

Arokodare, M. A., & Olubiyi, T. O. (2023). Crises as Harbingers of Opportunities: An Empirical Insight into the Moderating Effect of Knowledge Management on the Absorptive Capacity-Strategic Agility Relationship. *Malaysian Management Journal*, *27*, 83–108. DOI: 10.32890/mmj2023.27.4

Bag, J., Pretorius, C., Gupta, Y., & Dwivedi, K. (2021). Role of institutional pressures and resources in the adoption of big data analyticspowered artificial intelligence, sustainable manufacturing practices and circular economy capabilities. Technol. Forecast. Soc. Chang, 16(3), 12-20. .techfore.2020.120420DOI: <ALIGNMENT.qj></ALIGNMENT>10.1016/j

Benz, M., & Chatterjee, D. (2020). Calculated risk? A cybersecurity evaluation tool for SMEs. *Business Horizons*, *63*(4), 531–540. DOI: 10.1016/j.bushor.2020.03.010

Bhattacharya, C., & Sinha, M. (2022). The role of artificial intelligence in banking for leveraging customer experience. *Australas Accounting Business Finance Journal.*, *16*(5), 89–105. DOI: 10.14453/aabfj.v16i5.07

Cardoso, P. R., Costa, H. S., & Novais, L. A. (2010). Fashion consumer profiles in the Portuguese market: Involvement, innovativeness, self-expression and impulsiveness as segmentation criteria. *International Journal of Consumer Studies*, *34*(6), 638–647. DOI: 10.1111/j.1470-6431.2010.00891.x

Chen, D., Esperança, J. P., & Wang, S. (2022). The Impact of Artificial intelligence on Firm Performance: An application of the Resource-Based View to E-Commerce Firms. *Frontiers in Psychology*, *13*, 884830. Advance online publication. DOI: 10.3389/fpsyg.2022.884830 PMID: 35465474

Dam, S. M., & Dam, T. C. (2021). Relationships between service quality, brand image, customer satisfaction, and customer loyalty. *The Journal of Asian Finance, Economics and Business*, *8*(3), 585–593.

Davis, F. D. (1989). Perceived Usefulness, Perceived Ease of Use, and User Acceptance. *Management Information Systems Quarterly*, *13*(3), 319–340. DOI: 10.2307/249008

Davis, F. D., Bagozzi, R. P., & Warshaw, P. R. (1989). User acceptance of computer technology: A comparison of two theoretical models. *Management Science*, *35*(8), 982–1003. DOI: 10.1287/mnsc.35.8.982

Doborjeh, Z., Hemmington, N., Doborjeh, M., & Kasabov, N. (2022). Artificial intelligence: A systematic review of methods and applications in hospitality and tourism. *International Journal of Contemporary Hospitality Management*, *34*(3), 1154–1176. DOI: 10.1108/IJCHM-06-2021-0767

Fernando, E., Sutomo, R., Prabowo, Y. D., Gatc, J., & Winanti, W. (2023). Exploring customer relationship management: Trends, challenges, and innovations. *Journal of Information Systems and Informatics*, *5*(3), 984–1001. DOI: 10.51519/journalisi.v5i3.541

Floridi, L., Cowls, J., Beltrametti, M., Chatila, R., Chazerand, P., Dignum, V., Luetge, C., Madelin, R., Pagallo, U., Rossi, F., Schafer, B., Valcke, P., & Vayena, E. (2021). People—an ethical framework for a good AI society: Opportunities, risks, principles, and recommendations. *Minds and Machines*, *28*(4), 689–707. DOI: 10.1007/s11023-018-9482-5 PMID: 30930541

Geetha, A. (2021). A Study on Artificial Intelligence (Ai) in Banking and Financial Services. *International Journal of Creative Research Journal*, *9*(9), 110–114.

Ghasemaghaei, M., & Calic, G. (2020). Assessing the Impact of Big Data on Firm Innovation Performance: Big Data Is Not Always Better Data. *Journal of Business Research, 108*, 147–162. DOI: 10.1016/j.jbusres.2019.09.062

Gupta, S. (2021). Impact of artificial intelligence on financial decision 2015: A qualitative study. *Journal of Cardiovascular Disease Research, 12*(6), 2130–2137.

Hentzen, J. K., Hoffmann, A., Dolan, R., & Pala, E. (2022). Artificial intelligence in customer-facing financial services: A systematic literature review and agenda for future research. *International Journal of Bank Marketing, 40*(6), 1299–1336. DOI: 10.1108/IJBM-09-2021-0417

Hollebeek, L. D., Sharma, T. G., Pandey, R., Sanyal, P., & Clark, M. K. (2022). Fifteen years of customer engagement research: A bibliometric and network analysis. *Journal of Product and Brand Management, 31*(2), 209–309. DOI: 10.1108/JPBM-01-2021-3301

Huynh, D., Hille, E. L., & Nasir, A. (2020). Diversification in the age of the 4th industrial revolution: The role of artificial intelligence, greenbonds and crypto currencies. *Technological Forecasting and Social Change, 15*(9), 120–188. DOI: 10.1016/j.techfore.2020.120188

Kim, D. J., Ferrin, D. L., & Rao, H. R. (2009). Trust and satisfaction, the two wheels for successful e-commerce transactions: A longitudinal exploration. *Information Systems Research, 20*(2), 237–257. DOI: 10.1287/isre.1080.0188

Liao, C., Palvia, P., & Chen, J.-L. (2009). Information technology adoption behavior life cycle: Toward a Technology Continuance Theory (TCT). *International Journal of Information Management, 29*(4), 309–320. DOI: 10.1016/j.ijinfomgt.2009.03.004

Liu, N., Nikitas, A., & Parkinson, S. (2020). Exploring expert perceptions about the cyber security and privacy of connected and autonomous vehicles: A thematic analysis approach. *Transportation Research Part F: Traffic Psychology and Behaviour, 75*, 66–86. DOI: 10.1016/j.trf.2020.09.019

Makinde, O. G., Olubiyi, T. O., & Ogundipe, F. (2023). New Insights from Entrepreneurial Characteristics and Business Performance: Empirical Findings from Nigeria. *Asian Journal of Management. Entrepreneurship and Social Sciences, 3*(04), 72–100.

Naumov, N. (2019). *The Impact of Robots.* Artificial Intelligence, and Service Automation on Service Quality and Service Experience in Hospitality. In Robots, Artificial Intelligence, and Service Automation in Travel, Tourism, and Hospitality Emerald Publishing Limited.

Nebo, C., Nwankwo, P., & Okonkwo, R. (2015). The role of effective communication on organizational performance: A Study of Nnamdi Azikiwe University, Awka. *Review of public administration and management, 4*(8), 132-148.

Olowoporoku, A. A., & Olubiyi, T. O. (2023). Evaluating Service Quality and Business Outcomes Post-Pandemic: Perspective from Hotels in Emerging Market. *Sawala: Jurnal Administrasi Negara, 11*(2), 182–201. DOI: 10.30656/sawala.v11i2.6975

Olubiyi, O., Lawal, A. T., & Adeoye, O. O. (2022). Succession Planning and Family Business Continuity: Perspectives from Lagos State, Nigeria. *Organization and Human Capital Development, 1*(1), 40–52. DOI: 10.31098/orcadev.v1i1.865

Olubiyi, O. T., Lawal, A. T., & Adeoye, O. O. (2022). Succession planning and family business continuity: Perspectives from Lagos State, Nigeria. *Organization and Human Capital Development, 1*(1), 40–52. DOI: 10.31098/orcadev.v1i1.865

Olubiyi, T. O. (2019). Knowledge management practices and family business profitability: Evidence from Lagos State, Nigeria. *Global Journal of Management and Business Research, 19*(A11), 21–31. https://journalofbusiness.org/index.php/GJMBR/article/view/2872

Olubiyi, T. O. (2020). Knowledge management practices and family business profitability: Evidence from Lagos State, Nigeria. *Arabian Journal of Business and Management Review, 6*(1), 23–32.

Olubiyi, T. O. (2022a). Measuring technological capability and business performance post-COVID Era: Evidence from Small and Medium-Sized Enterprises (SMEs) in Nigeria. *Management & Marketing Journal, xx*(2), 234–248. DOI: 10.52846/MNMK.20.2.09

Olubiyi, T. O. (2022b). An investigation of sustainable innovative strategy and customer satisfaction in small and medium-sized enterprises (SMEs) in Nigeria. *Covenant Journal of Business and Social Sciences, 13*(2), 1–24.

Olubiyi, T. O. (2023a). Unveiling the role of workplace environment in achieving the Sustainable Development Goal Eight (SDG8) and employee job satisfaction post-pandemic: Perspective from Africa. *Revista Management & Marketing Craiova, XXI*(2), 212–228. DOI: 10.52846/MNMK.21.2.02

Olubiyi T. O. (2023b). Leveraging competitive strategies on business outcomes post-covid-19 pandemic: Empirical investigation from Africa. *Journal of Management and Business: Research and Practice.* DOI: 10.54933/jmbrp-2023-15-2-1

Olubiyi, T. O., Adeoye, O. O., Jubril, B., Adeyemi, O. S., & Eyanuku, J. P. (2023). Measuring Inequality in Sub-Saharan Africa Post-Pandemic: Correlation Results for Workplace Inequalities and Implication for Sustainable Development Goal ten. *International Journal of Professional Business Review*, 8(4), e01405. DOI: 10.26668/businessreview/2023.v8i4.1405

Olubiyi, T. O., & Akpa, V. A. (2023). Does innovative capability and artificial intelligence really matter for the profitability of consumer goods companies in developing economies? *SunText Review of Economics & Business*, 4(3), 191–204.

Olubiyi, T. O., Jubril, B., Sojinu, O. S., & Ngari, R. (2022). Strengthening gender equality in small business and achieving sustainable development goals (SDGs), Comparative analysis of Kenya and Nigeria. *Sawala Jurnal Administrasi Negara*, 10(2), 168–186. DOI: 10.30656/sawala.v10i2.5663

Omoyele, O. S., Olubiyi, T. O., Lanre-Babalola, F. O., Obadare, G. O., & Onikoyi, I. A. (2023). Business Model Innovation as a Catalyst for Sustainable Entrepreneurship: Empirical Findings from Small and Medium Enterprises in Nigeria. *Skyline Business Journal*, 19(2), 55–64. DOI: 10.37383/SBJ190205

Parasuraman, A., Zeithaml, V. A., & Berry, L. L. (1985). A conceptual model of service quality and its implications for future research. *Journal of Marketing*, 49(4), 41–50. DOI: 10.1177/002224298504900403

Prentice, C., Dominique Lopes, S., & Wang, X. (2020). The impact of artificial intelligence and employee service quality on customer satisfaction and loyalty. *Journal of Hospitality Marketing & Management*, 29(7), 739–756. DOI: 10.1080/19368623.2020.1722304

Trawnih, A., Al-Masaeed, S., Alsoud, M., & Alkufahy, A. (2022). Understanding artificial intelligence experience: A customer perspective. *International Journal of Data and Network Science*, 6(4), 1471–1484. DOI: 10.5267/j.ijdns.2022.5.004

Ukabi, O. B., Uba, U. J., Ewum, C. O., & Olubiyi, T. O. (2023). Measuring Entrepreneurial Skills and Sustainability in Small Business Enterprises Post-Pandemic: Empirical Study from Cross River State, Nigeria. *International Journal of Business. Management and Economics*, 4(2), 132–149. DOI: 10.47747/ijbme.v4i2.1140

Uwem, E. I., Oyedele, O. O., & Olubiyi, O. T. (2021). Workplace green behavior for sustainable competitive advantage. In *Human Resource Management Practices for Promoting Sustainability*. IGI Global.

Venkatesh, V., & Bala, H. (2008). Technology acceptance model 3 and a research agenda on interventions. *Decision Sciences*, *39*(2), 273–315. DOI: 10.1111/j.1540-5915.2008.00192.x

Venkatesh, V., & Davis, F. D. (2000). A theoretical extension of the technology acceptance model: Four longitudinal field studies. *Management Science*, *46*(2), 186–204. DOI: 10.1287/mnsc.46.2.186.11926

Wallace, C., & Chen, G. (2006). A multilevel integration of personality, climate, self-regulation, and performance. *Personnel Psychology*, *59*(3), 529–557. DOI: 10.1111/j.1744-6570.2006.00046.x

Yamane, Y. (1967). Mathematical Formulae for Sample Size Determination.

Yang, L., Henthorne, T. L., & George, B. (2020). Artificial intelligence and robotics technology in the hospitality industry: Current applications and future trends. Digital transformation in business and society: Theory and cases, 211-228.

Yang, X. (2023). The effects of AI service quality and AI function-customer ability fit on customer's overall co-creation experience. *Industrial Management & Data Systems*, *123*(6), 1717–1735. DOI: 10.1108/IMDS-08-2022-0500

Yeo, S. F., Tan, C. L., Kumar, A., Tan, K. H., & Wong, J. K. (2022). Investigating the impact of AI-powered technologies on Instagrammers' purchase decisions in digitalization era–A study of the fashion and apparel industry. *Technological Forecasting and Social Change*, *177*, 121551–121567. DOI: 10.1016/j.techfore.2022.121551

Chapter 17
Humanizing Customer Interaction:
AI-Powered Experiences in the Digital Age

Mahesh Deshpande

https://orcid.org/0009-0006-2234-6733

Genpact, USA

ABSTRACT

This chapter examines AI's transformative role in humanizing digital customer experiences, exploring how technologies like natural language processing, machine learning, and predictive analytics are reshaping customer interactions. It focuses on hyper-personalization, empathetic conversational AI, and proactive customer service, emphasizing the balance between AI capabilities and human touch. The chapter addresses ethical considerations including data privacy, transparency, and fairness in AI systems. Case studies highlight successes and challenges in implementing AI-driven customer experience strategies, while also forecasting future trends. Throughout, it advocates for a human-centric approach where AI augments rather than replaces human capabilities. This comprehensive overview serves as a valuable resource for businesses aiming to leverage AI to enhance customer experiences while maintaining ethical, human-centered practices. The chapter underscores the potential of AI to create more meaningful and efficient customer interactions when implemented thoughtfully.

DOI: 10.4018/979-8-3693-8009-3.ch017

I. INTRODUCTION

In the rapidly evolving landscape of digital transformation, businesses are increasingly turning to artificial intelligence (AI) to revolutionize customer experiences. The integration of AI technologies into customer interactions promises unprecedented levels of efficiency, personalization, and scalability. As Gartner predicts, "By 2025, 80% of customer service and support organizations will be applying generative AI technology in some form to improve agent productivity and customer experience (CX)" (Brackenbury & LoDolce, 2023). However, as we stand at the crossroads of technological advancement and human-centric design, a critical question emerges: How can we leverage AI to enhance customer experiences while maintaining the essential human elements that customers crave?

This chapter explores the multifaceted role of AI in humanizing customer interactions within the context of digital transformation. We will delve into the paradox that defines modern customer expectations: the simultaneous desire for instant, efficient service and deeply personalized, empathetic interactions. As businesses navigate this complex terrain, AI emerges as a powerful tool capable of bridging the gap between technological efficiency and the human touch.

The journey through this chapter will take us from the current landscape of customer experience to the cutting-edge AI technologies that are reshaping customer interactions. We will explore three key areas where AI is making significant strides:

1. Hyper-personalization at scale, where AI algorithms analyze vast amounts of data to create tailored experiences for individual customers.
2. Empathetic conversational AI, which leverages natural language processing and emotional AI to create more human-like interactions.
3. Predictive and proactive customer service, shifting the paradigm from reactive problem-solving to anticipatory care.

Throughout our exploration, we will maintain a critical focus on the human element. As Kai-Fu Lee, AI expert and author, emphasizes, "AI is powerful and will change our lives, but it's not nearly as powerful as humans armed with AI. We'll use our human creativity, human compassion, and human love to deploy and use AI to make our lives better" (Lee, 2018). The goal is not to replace human interactions with AI but to augment and enhance them, creating a symbiotic relationship between technology and human touch.

As we navigate this exciting yet complex field, we will not shy away from addressing the challenges and ethical considerations that arise from the implementation of AI in customer experiences. Issues such as data privacy, algorithmic bias, and the authenticity of AI-driven interactions will be thoroughly examined, providing readers with a comprehensive understanding of the landscape.

To ground our discussion in real-world applications, we will analyze case studies from various industries, highlighting both successes and failures in AI-powered customer experience initiatives. These practical examples will offer valuable insights and lessons for businesses embarking on their own AI-driven transformations.

Finally, we will cast our gaze toward the horizon, exploring emerging technologies and trends that promise to further revolutionize customer experiences. From augmented reality interactions to the potential impact of quantum computing, we will speculate on how these advancements might shape the future of customer engagement.

As we embark on this exploration, it is crucial to remember that the ultimate goal of AI in customer experience is not to create a world of automated, impersonal interactions. Rather, as Accenture's Technology Vision 2020 report states, "Artificial Intelligence (AI) should be an additive contributor to how people perform their work". (Accenture, 2020)The goal is to create a more human experience with technology. By striking the right balance between AI capabilities and human values, businesses can truly humanize their digital transformation journey and build lasting relationships in an increasingly digital world.

II. THE CURRENT LANDSCAPE OF CUSTOMER EXPERIENCE IN THE DIGITAL AGE

In today's digital-first world, customer experience (CX) has become a critical differentiator for businesses across industries. The advent of digital technologies has fundamentally altered customer expectations, creating a landscape where convenience, personalization, and immediacy are not just desired but demanded.

According to a recent PwC survey, 73% of consumers point to customer experience as an important factor in their purchasing decisions, ranking it above product quality and price (PwC, 2018). This statistic underscores the pivotal role that CX plays in modern business success.

The digital transformation of customer experience is characterized by several key trends:

1. Omnichannel Presence: Customers now expect seamless interactions across multiple channels, including social media, mobile apps, websites, and physical stores. A study by Harvard Business Review found that 73% of consumers use multiple channels during their shopping journey (Edelman & Singer, 2015).

2. Self-Service Options: There's a growing preference for self-service solutions. Gartner predicts that by 2030, a billion service tickets will be raised automatically by customer-owned bots (Gartner, 2019).
3. Personalization: Customers expect tailored experiences. According to Epsilon, 80% of consumers are more likely to make a purchase when brands offer personalized experiences (Epsilon, 2018).
4. Real-Time Responsiveness: The expectation for immediate responses has grown. A HubSpot Research report shows that 90% of customers rate an "immediate" response as important or very important when they have a customer service question (Shrivastav, 2022).
5. Data-Driven Insights: Companies are leveraging big data and analytics to gain deeper insights into customer behavior and preferences.

Figure 1. Modern Customer Experience Landscape

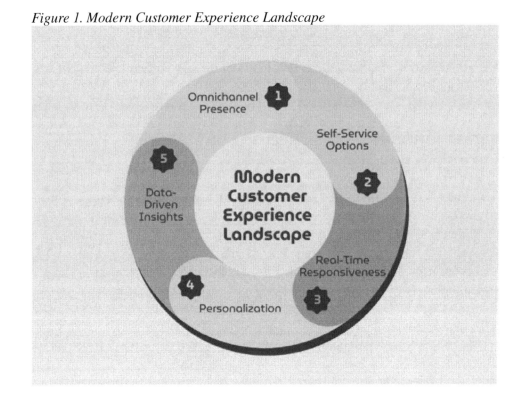

However, this digital transformation has also introduced new challenges. As businesses strive to meet these evolving expectations, they often face a paradox: how to provide highly efficient, technology-driven service while maintaining the human touch that customers still crave.

This paradox is evident in consumer attitudes. While 59% of consumers feel companies have lost touch with the human element of customer experience, 75% of the same group prefer to interact with a computer when trying to avoid long hold times or in-store lines (PwC, 2018). This dichotomy highlights the complex balancing act that businesses must perform in the digital age.

Moreover, the COVID-19 pandemic has accelerated digital transformation efforts across industries, further emphasizing the importance of digital customer experiences. McKinsey reports that the crisis has accelerated the digitization of customer interactions by several years, with the share of digital customer interactions increasing from 36% to 58% globally between December 2019 and July 2020 (McKinsey & Company, 2020).

Despite these rapid advancements, challenges persist. Many organizations struggle with data silos, legacy systems, and a lack of digital skills, hindering their ability to deliver seamless, personalized experiences. A Gartner survey found that 87% of senior business leaders say digitalization is a company priority, yet only 40% of organizations have brought digital initiatives to scale (Venkatraman, 2021).

Furthermore, as digital interactions increase, so do concerns about data privacy and security. The 2021 Entrust Global Consumer Research report reveals that 79% of consumers are concerned about data privacy, and 64% believe that organizations are not doing enough to protect their data (Kadet, 2021).

In this complex landscape, artificial intelligence emerges as a powerful tool to address these challenges and bridge the gap between digital efficiency and human-centric experiences. AI technologies offer the potential to process vast amounts of data, automate routine tasks, and provide personalized interactions at scale. However, the successful implementation of AI in customer experience requires a thoughtful approach that balances technological capabilities with human empathy and ethical considerations.

As Shep Hyken, customer service and experience expert, aptly puts it, "AI and automation will not replace human beings in the customer support world. They will, however, make humans more efficient and effective in the process" (Hyken, 2023). This perspective encapsulates the promise of AI in transforming customer experience – not as a replacement for human interaction, but as a tool to enhance and humanize digital interactions.

In the following sections, we will explore how AI is being leveraged to create more human-centric, empathetic, and personalized customer experiences in the digital age, addressing the challenges and opportunities presented by this evolving landscape.

III. AI TECHNOLOGIES TRANSFORMING CUSTOMER INTERACTIONS

Artificial Intelligence is revolutionizing customer interactions through a diverse array of technologies. These innovations are enabling businesses to provide more efficient, personalized, and empathetic customer experiences at scale. Let's explore the key AI technologies that are reshaping the customer experience landscape.

A. Natural Language Processing (NLP) and Natural Language Understanding (NLU)

NLP and NLU form the backbone of many AI-powered customer interaction systems. These technologies enable machines to understand, interpret, and generate human language, facilitating more natural and context-aware conversations.

According to a report by MarketsandMarkets, the global NLP market size is expected to grow from USD 11.6 billion in 2020 to USD 35.1 billion by 2026, at a Compound Annual Growth Rate (CAGR) of 20.3% (MarketsandMarkets, 2021). This growth underscores the increasing importance of NLP in customer interactions.

Key applications of NLP in customer experience include:

- Chatbots and virtual assistants
- Voice recognition systems
- Sentiment analysis
- Automated email responses
- Text analytics for customer feedback

Figure 2. NLP Applications in Customer Experience

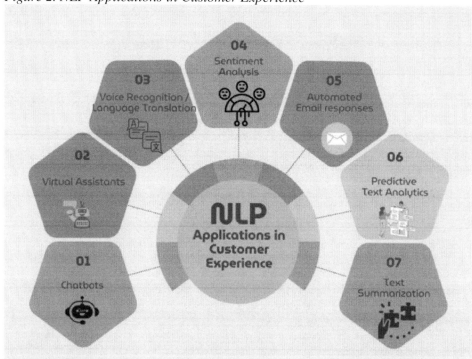

B. Machine Learning (ML) and Deep Learning

Machine Learning algorithms, particularly deep learning models, are at the heart of AI's ability to process vast amounts of customer data and derive actionable insights. These technologies enable:

- Predictive analytics for customer behavior
- Personalized product recommendations
- Churn prediction and prevention.
- Dynamic pricing optimization
- Fraud detection in customer transactions

A study by MIT Technology Review Insights found that 87% of businesses are already realizing a return on their AI investments in customer experience, with improved ML algorithms being a key factor (MIT Technology Review Insights, 2020).

C. Computer Vision

Computer vision technologies are increasingly being used to enhance customer experiences, particularly in retail and service industries. Applications include:

- Visual search capabilities
- Augmented reality (AR) for product visualization
- Facial recognition for personalized in-store experiences.
- Quality control in product manufacturing

The global computer vision market is projected to reach USD 19.1 billion by 2027, growing at a CAGR of 7.3% from 2020 to 2027 (Grand View Research, 2020), indicating its growing importance in customer interactions.

D. Emotion AI

Emotion AI, also known as affective computing, aims to detect and interpret human emotional states. This technology is particularly valuable in creating more empathetic customer interactions. Use cases include:

- Emotion detection in customer service calls
- Sentiment analysis in social media monitoring
- Personalized marketing based on emotional states
- User experience optimization

"Emotion AI represents the next frontier in customer experience," says Rana el Kaliouby, CEO of Affectiva. "It allows businesses to understand not just what customers are doing, but how they're feeling, enabling truly empathetic interactions" (el Kaliouby, 2020).

E. Robotic Process Automation (RPA)

RPA, while not strictly an AI technology, is often enhanced by AI capabilities to automate repetitive tasks in customer service. Applications include:

- Automated data entry and validation
- Processing customer requests and queries
- Order processing and tracking
- Customer onboarding processes

Gartner predicts that by 2024, organizations will lower operational costs by 30% by combining hyperautomation technologies with redesigned operational processes (Gartner, 2020).

F. Voice and Speech Recognition

Voice AI technologies are transforming how customers interact with businesses, especially through smart speakers and voice assistants. Key applications include:

- Voice-activated customer service
- Voice biometrics for authentication
- Voice-based shopping
- Interactive voice response (IVR) systems

According to Juniper Research, the number of digital voice assistants in use will triple to 8 billion by 2023 (Smith, 2019), highlighting the growing importance of voice AI in customer interactions.

G. Predictive Analytics

AI-powered predictive analytics enables businesses to anticipate customer needs and behaviors. Applications include:

- Next best action recommendations
- Predictive maintenance
- Inventory optimization.
- Customer lifetime value prediction

A report by Forrester found that insights-driven businesses are growing at an average of more than 30% annually and are on track to earn $1.8 trillion by 2021 (Forrester, 2018).

H. Autonomous Agents

Autonomous agents, or AI-powered software that can perform tasks without human intervention, are becoming increasingly sophisticated. In customer experience, they're used for:

- 24/7 customer support
- Personalized product recommendations

- Proactive customer outreach
- Complex problem-solving

"Autonomous agents represent a paradigm shift in customer service," says Kate Leggett, VP and Principal Analyst at Forrester. "They're not just responding to queries, but proactively engaging customers and solving problems before they arise" (Leggett, 2019).

I. Knowledge Graphs

Knowledge graphs are becoming crucial in organizing and utilizing vast amounts of customer data. They enable:

- Contextual understanding of customer queries
- More accurate product recommendations
- Enhanced search capabilities
- Improved customer segmentation

While these AI technologies offer immense potential for transforming customer interactions, it's crucial to remember that their true value lies in how they're implemented and integrated with human touch points. As Fei-Fei Li, Co-Director of Stanford's Human-Centered AI Institute, aptly puts it, "AI is made by humans, intended to behave like humans, and, ultimately, to impact humans' lives and human society" (Li, 2018).

The successful application of these technologies requires a deep understanding of customer needs, ethical considerations, and a commitment to maintaining the human element in customer experiences. In the following sections, we'll explore how these technologies are being applied to create more personalized, empathetic, and proactive customer experiences, while addressing the challenges and ethical considerations that arise in their implementation.

IV. HYPER-PERSONALIZATION AT SCALE

In the age of digital transformation, customers expect experiences tailored to their individual preferences, behaviors, and needs. Hyper-personalization, powered by AI, allows businesses to meet these expectations at scale, creating unique, relevant interactions for each customer across all touchpoints.

A. Defining Hyper-Personalization

Hyper-personalization goes beyond traditional segmentation, using real-time data and AI to deliver highly contextual and individualized experiences. As defined by Gartner, "Hyper-personalization is the most advanced way brands can tailor their marketing to individual customers. It involves using AI and real-time data to create highly contextual communication relevant to the customer" (Slavickova, 2022).

B. The Business Case for Hyper-Personalization

The benefits of hyper-personalization are significant:

- *Increased Customer Engagement*: According to Epsilon, 80% of consumers are more likely to make a purchase when brands offer personalized experiences (Epsilon, 2018).
- *Higher Conversion Rate:* A study by Trask found that Hyper-personalization can increase your conversion rates by up to 60% (Comarch, 2024).
- *Higher Revenue:* McKinsey reports that personalization can deliver five to eight times the ROI on marketing spend and lift sales by 10% or more (McKinsey & Company, 2021).

C. AI Technologies Enabling Hyper-Personalization

Several AI technologies work in concert to enable hyper-personalization:

- *Machine Learning:* ML algorithms analyze vast amounts of customer data to identify patterns and predict preferences.
- *Natural Language Processing:* NLP helps understand customer intent from text-based interactions.
- *Computer Vision*: In retail, computer vision can analyze images to understand customer style preferences.
- *Predictive Analytics*: This technology forecasts future customer behaviors based on historical data.

D. Key Components of AI-Driven Hyper-Personalization

1) Data Collection and Integration

The foundation of hyper-personalization is comprehensive, real-time data. This includes:

- Demographic data
- Behavioral data (browsing history, purchase history)
- Contextual data (location, time, device)
- Social media data
- Customer service interactions

AI systems integrate these diverse data sources to create a 360-degree view of each customer. However, this raises privacy concerns. As Tim Berners-Lee, the inventor of the World Wide Web, cautions, "The web has evolved into an engine of inequity and division; swayed by powerful forces who use it for their own agendas" (Berners-Lee, 2017). It's crucial for businesses to balance personalization with privacy protection.

2) AI-Powered Customer Segmentation

AI enables dynamic micro-segmentation, creating highly specific customer groups based on multiple attributes. This goes beyond traditional demographic segmentation to include behavioral and psychographic factors.

3) Real-Time Decision Making

AI algorithms can make real-time decisions about what content, offers, or experiences to present to each customer. This enables:

- Dynamic website personalization
- Personalized product recommendations
- Tailored email content
- Customized push notifications

Amazon's recommendation engine, which drives 35% of its revenue, is a prime example of this capability (McKinsey & Company, 2013).

4) Omnichannel Personalization

AI enables consistent personalization across all channels - website, mobile app, email, social media, in-store, and customer service. According to Omnisend, marketers using three or more channels in their campaigns earn a 250% higher purchase rate than those using single-channel campaigns (Omnisend, 2019).

E. Implementing Hyper-Personalization: Best Practices

1) Start with a Strong Data Foundation

Ensure you have a robust data infrastructure that can collect, integrate, and analyze data from multiple sources in real-time.

2) Focus on Customer Consent and Transparency

Be transparent about data collection and use. Implement strong data protection measures and give customers control over their data.

3) Use AI to Augment, Not Replace, Human Decision Making

As Garry Kasparov, chess grandmaster and AI commentator, notes, "The idea is not to replace human decision-making, but to enhance it" (Kasparov, 2017).

4) Continuously Test and Refine

Use A/B testing and machine learning to continuously improve personalization algorithms.

5) Balance Personalization and Privacy

Respect customer privacy preferences and provide options for customers to opt-out of personalization.

F. Challenges in Implementing Hyper-Personalization

While the benefits are clear, implementing hyper-personalization comes with challenges:

- *Data Quality and Integration*: Ensuring data accuracy and integrating data from various sources can be complex.
- *Privacy Concerns:* With regulations like GDPR and CCPA, businesses must be careful about how they collect and use customer data.
- *Algorithmic Bias*: AI systems can perpetuate or amplify existing biases if not carefully designed and monitored.
- *Customer Acceptance:* Some customers may find hyper-personalization intrusive. A study by Gartner predicts that by 2025, 80% of marketers who have

invested in personalization will abandon their efforts due to lack of ROI, the perils of customer data management or both (Gartner, 2019).

G. Future Trends in Hyper-Personalization

Looking ahead, several trends are shaping the future of hyper-personalization:

1. *Emotional AI*: Incorporating emotional state into personalization decisions.
2. *Voice-Based Personalization:* As voice interfaces become more common, personalizing voice interactions will be crucial.
3. *AR/VR Personalization*: Customizing augmented and virtual reality experiences.
4. *Federated Learning:* Enabling personalization while keeping data on user devices, addressing privacy concerns.

H. Case Study: Netflix's Hyper-Personalization Strategy

Netflix serves as an excellent example of hyper-personalization at scale. The streaming giant uses AI to personalize every aspect of the user experience, from the artwork shown for each title to personalized recommendations.

Netflix's recommendation system, which saves the company an estimated $1 billion per year in value from customer retention (McAlone, 2016), analyzes viewing history, search history, ratings, and even the time of day a user typically watches to provide highly relevant content suggestions.

As Todd Yellin, Netflix's VP of Product, explains, "There are 33 million versions of Netflix" (Adalian, 2018), referring to how the platform is personalized for each user.

In conclusion, hyper-personalization represents a powerful application of AI in customer experience, enabling businesses to create highly relevant, individualized interactions at scale. While challenges exist, particularly around data privacy and algorithmic bias, the potential benefits in terms of customer engagement, loyalty, and business outcomes are substantial. As AI technologies continue to evolve, we can expect even more sophisticated and nuanced approaches to personalization in the future.

I. Experimental Results: AI-Powered Personalization of A/B Testing Results

1) Introduction

This experiment quantifies the impact of AI-powered personalization on key e-commerce metrics, comparing a standard website experience with an AI-enhanced personalized experience.

2) Methodology

- Test duration: 30 days
- Sample size: 100,000 website visitors of a BOBA store
- Control group: Standard website experience
- Test group: AI-powered personalized content and product recommendations

3) Key Metrics

Table 1. A/B Testing Metrics for AI-Powered Personalization Experience

Metric	Control Group	Test Group	Improvement
Conversion Rate	2.5%	3.8%	+52%
Average Order Value	$65	$78	+20%
Time Spent on Site	3.2 minutes	4.7 minutes	+46.9%
Customer Satisfaction	7.2/10	8.5/10	+18.1%

All results are statistically significant ($p < 0.01$).

4) Experiment Conclusion

AI-powered personalization significantly improved all key metrics, demonstrating its effectiveness in enhancing customer experience and driving business outcomes.

5) Experiment Implications

- Potential for substantial revenue growth through increased conversion rates and order values.

- Enhanced user engagement and satisfaction may lead to improved customer retention.
- Results justify increased investment in AI personalization technologies.
- Businesses must balance personalization benefits with privacy considerations.
- Continuous refinement of AI algorithms could yield further improvements.

V. EMPATHETIC CONVERSATIONAL AI

As businesses strive to humanize digital interactions, empathetic conversational AI has emerged as a powerful tool to create more meaningful and emotionally resonant customer experiences. This section explores how AI is being used to develop conversational interfaces that can understand, respond to, and even anticipate human emotions.

A. The Importance of Empathy in Customer Interactions

Empathy, the ability to understand and share the feelings of another, is a crucial component of positive human interactions. In the context of customer experience, empathy can significantly impact customer satisfaction and loyalty. According to a survey report by Freshworks, "Today's customers want the speed and efficiency of automated engagement but also crave the warming art of human empathy and emotional connection". (Swinscoe, 2023).

B. Defining Empathetic Conversational AI

Empathetic Conversational AI refers to AI systems designed to recognize and respond to human emotions in a way that mimics human empathy. These systems go beyond simple task completion to create a more human-like interaction.

C. Key Technologies Enabling Empathetic Conversational AI

Several AI technologies work together to enable empathetic conversations:

1. Natural Language Processing (NLP) and Natural Language Understanding (NLU): For understanding the content and context of user input.
2. Sentiment Analysis: To detect the emotional tone of text or speech.
3. Speech Recognition and Text-to-Speech: For voice-based interactions.
4. Emotion AI: To recognize and interpret human emotional states.
5. Machine Learning: To continuously improve responses based on interactions.

D. Components of Empathetic Conversational AI

1) Emotion Recognition

Emotion recognition is the foundation of empathetic AI. This can be achieved through:

- Text Analysis: NLP techniques to detect emotions in written text.
- Speech Analysis: Analyzing vocal patterns and intonations.
- Facial Expression Analysis: Using computer vision to interpret facial expressions in video interactions.

According to a report by MarketsandMarkets, the emotion detection and recognition market is expected to grow from $23.5 billion in 2022 to $42.9 billion by 2027 (MarketsandMarkets, 2021).

2) Contextual Understanding

Empathetic AI needs to understand the context of a conversation, including:

- **User History:** Past interactions and preferences.
- **Situational Context:** Time, location, and current circumstances.
- **Cultural Context:** Cultural norms and expectations.

3) Emotional Response Generation

Once emotions are recognized, the AI must generate an appropriate empathetic response. This involves:

- **Emotion-Appropriate Language**: Adjusting tone and vocabulary based on the user's emotional state.
- **Personalized Responses**: Tailoring responses based on the user's history and preferences.
- **Non-Verbal Cues:** In the case of virtual agents, incorporating appropriate facial expressions and gestures.

4) Continuous Learning and Improvement

Empathetic AI systems should continuously learn from interactions to improve their empathetic responses over time.

E. Applications of Empathetic Conversational AI

1) Customer Service Chatbots and Virtual Assistants

Empathetic chatbots can provide more satisfying customer service experiences. For instance, Replika, an AI companion chatbot, uses empathetic conversational AI to provide emotional support to users (Replika, 2021).

2) Mental Health Support

AI-powered mental health apps like Woebot use empathetic conversation to provide cognitive behavioral therapy (Woebot Health, 2021).

3) Sales and Marketing

Empathetic AI can help tailor sales pitches and marketing messages based on the customer's emotional state.

4) Education

AI tutors with empathetic capabilities can provide more engaging and supportive learning experiences.

F. Challenges in Implementing Empathetic Conversational AI

While promising, empathetic AI faces several challenges:

1. **Accuracy of Emotion Recognition**: Emotions are complex and context-dependent, making accurate recognition challenging.
2. **Cultural Differences**: Emotional expressions and interpretations can vary across cultures.
3. **Ethical Concerns:** There are concerns about the manipulation of human emotions by AI systems.
4. **Uncanny Valley Effect:** As AI becomes more human-like, it may paradoxically become unsettling for some users.

As Rana el Kaliouby, CEO of Affectiva, notes, "The challenge is not just to make AI more intelligent, but to make it more human" (el Kaliouby, 2017).

G. Best Practices for Implementing Empathetic Conversational AI

1) Focus on Genuine Empathy

Avoid scripted responses that may come across as insincere. Aim for genuine understanding and appropriate responses.

2) Respect User Privacy

Be transparent about emotion recognition capabilities and allow users to opt-out if desired.

3) Combine AI with Human Touch

Use AI to augment rather than replace human empathy. For complex or sensitive situations, ensure there's an option to transfer to a human agent.

4) Continuous Training and Improvement

Regularly update and refine your AI models based on new data and user feedback.

5) Cultural Sensitivity

Ensure your empathetic AI is trained on diverse data sets and can adapt to different cultural contexts.

H. Future Trends in Empathetic Conversational AI

Several trends are likely to shape the future of empathetic AI:

1. *Multimodal Emotion Recognition*: Combining text, voice, and visual cues for more accurate emotion detection.
2. *Personalized Empathy Models:* AI that adapts its empathy style to individual user preferences.
3. *Emotion-Aware Smart Environments:* IoT devices that respond to user emotions.
4. *Advanced Emotion Synthesis:* AI capable of expressing more nuanced and complex emotions.

I. Case Study: Xiaoice - Microsoft's Empathetic AI Chatbot

Microsoft's Xiaoice, an AI-powered social chatbot popular in China, serves as an interesting case study in empathetic AI. Xiaoice is designed to be a virtual friend, capable of engaging in emotionally intelligent conversations.

Key features of Xiaoice include:

- Emotional Computing Framework: Xiaoice can recognize and respond to human emotions.
- Personality and Memory: The chatbot has a consistent personality and can remember past conversations.
- Multimodal Interactions: Xiaoice can engage via text, voice, and even create art and poetry.

According to Microsoft, Xiaoice has over 660 million users worldwide and has engaged in over 30 billion conversations (Zhou, Gao, Li, & Shum, 2020).

As Harry Shum, former Executive Vice President of Microsoft's AI and Research Group, explains, "Xiaoice is designed as an AI companion with an emotional connection to satisfy the human need for communication, affection, and social belonging" (Shum, He, & Li, 2018).

In conclusion, empathetic conversational AI represents a significant step towards more human-centric AI interactions. By recognizing and responding to human emotions, these systems can create more engaging, satisfying, and effective customer experiences. However, as we continue to develop these technologies, it's crucial to address the ethical implications and ensure that empathetic AI enhances rather than replaces genuine human connections.

J. Experimental Results – Sentiment Analysis: AI vs. Human Customer Service Interactions

1) Methodology

- Dataset: 50,000 customer service interactions (25,000 AI-powered, 25,000 human-led)
- Analysis tool: Natural Language Processing (NLP) sentiment analysis model
- Sentiment categories: Positive, Neutral, Negative

2) Results:

Table 2. Sentiment Analysis Results of AI-Powered vs. Human-Led interactions

Interaction Type	Positive	Neutral	Negative
AI-Powered	60%	25%	15%
Human-Led	65%	20%	15%

3) Key Findings:

- Overall satisfaction levels are comparable between AI and human interactions.
- AI interactions show a higher percentage of neutral sentiments, possibly due to more straightforward, fact-based responses.
- Human interactions have a slightly higher positive sentiment, potentially due to empathy and personal connection.
- Most Common Positive Keywords: Helpful, Quick, Efficient
- Most Common Negative Keywords: Confused, Frustrated, Unresolved

4) Conclusion:

While human-led interactions still have a slight edge in positive sentiments (table 1), AI-powered interactions are performing remarkably well, with room for improvement in reducing neutral responses.

VI. PREDICTIVE AND PROACTIVE CUSTOMER SERVICE

In the era of AI-driven customer experience, businesses are shifting from reactive to predictive and proactive customer service models. This paradigm shift allows companies to anticipate customer needs, prevent issues before they occur, and provide timely, relevant assistance. This section explores how AI enables predictive and proactive customer service, its benefits, challenges, and best practices.

A. Defining Predictive and Proactive Customer Service

Predictive customer service uses data analytics and AI to forecast customer needs and potential issues. Proactive customer service takes this a step further by acting on these predictions to address customer needs before they're explicitly expressed.

As Bill Gates famously said, "The first rule of any technology used in a business is that automation applied to an efficient operation will magnify the efficiency. The second is that automation applied to an inefficient operation will magnify the inefficiency" (Gates, 1999). This underscores the importance of implementing predictive and proactive service strategically.

B. Key Technologies Enabling Predictive and Proactive Customer Service

Several AI and data technologies work in concert to enable predictive and proactive service:

1. *Machine Learning and Predictive Analytics:* For identifying patterns and making predictions.
2. *Big Data Analytics:* For processing and analyzing large volumes of customer data.
3. *Internet of Things (IoT):* For real-time data collection from connected devices.
4. *Natural Language Processing:* For understanding and analyzing customer communications.
5. *Customer Data Platforms (CDPs):* For creating unified customer profiles.

C. Components of Predictive and Proactive Customer Service

1) Data Collection and Integration

The foundation of predictive service is comprehensive, real-time data. This includes:

- Customer demographic data
- Interaction history
- Purchase history
- Product usage data
- Social media data
- Contextual data (e.g., location, time)

2) Predictive Modeling

AI algorithms analyze this data to make predictions about:

- Customer churn likelihood
- Product failure probability
- Upcoming customer needs
- Optimal times for engagement

3) Automated Triggers and Actions

Based on these predictions, AI systems can trigger automated actions such as:

- Proactive customer outreach
- Preventive maintenance notifications
- Personalized offers or recommendations
- Automatic issue resolution

4) Continuous Learning and Optimization

The system should continuously learn from outcomes to improve its predictions and actions over time.

D. Benefits of Predictive and Proactive Customer Service

The benefits of this approach are significant:

1. *Improved Customer Satisfaction:* According to a study by Aberdeen Group, companies with strong predictive analytics capabilities achieve a 92% customer satisfaction rate, compared to 63% for those without (Aberdeen Group, 2012).
2. *Reduced Support Costs:* Preventing issues is typically less costly than resolving them. Gartner predicts that by 2025, proactive customer service will reduce support costs by 40% for companies that excel in its deployment • (Gartner, 2021)
3. *Increased Customer Loyalty:* By predicting what customers want before they even ask for it, organizations can provide a proactive and personalized experience that increases satisfaction and fosters loyalty. (Henry & Garg, 2024).
4. *Higher Efficiency:* Predictive models can help optimize resource allocation in customer service operations.

E. Applications of Predictive and Proactive Customer Service

1) Predictive Maintenance

In industries like manufacturing and telecommunications, AI can predict equipment failures and schedule maintenance before breakdowns occur. For instance, Rolls-Royce uses AI and IoT sensors to predict maintenance needs for its jet engines, reducing downtime and improving safety (Royce, 2022).

2) Churn Prevention

AI models can identify customers at risk of churning and trigger retention actions. Telecom company Sprint used AI-powered predictive analytics to reduce customer churn by 10% (Morgan, 2018).

3) Proactive Problem Resolution

AI can detect potential issues in product usage and provide preemptive solutions. For example, Netflix uses predictive analytics to detect and automatically resolve potential streaming quality issues before they impact the viewer's experience (Netflix Technology Blog, 2012).

4) Anticipatory Customer Support

AI can predict when a customer might need assistance based on their behavior and proactively offer help. Zappos, for instance, uses predictive analytics to anticipate customer needs and provide proactive support (Michelli, 2021).

F. Challenges in Implementing Predictive and Proactive Customer Service

Although Predictive and Proactive Customer Service is promising, this approach faces several challenges:

1. *Data Privacy Concerns*: Collecting and analyzing vast amounts of customer data raises privacy issues.
2. *Data Quality and Integration*: Ensuring data accuracy and integrating data from various sources can be complex.
3. *Balancing Proactivity and Intrusiveness*: There's a fine line between helpful proactivity and unwanted intrusion.

4. *Change Management:* Shifting from reactive to proactive service requires significant organizational change.

G. Best Practices for Implementing Predictive and Proactive Customer Service

1) Start with Clear Objectives

Define specific goals for your predictive service initiatives, aligned with overall business objectives.

2) Ensure Data Quality and Integration

Invest in data cleansing and integration efforts to ensure your predictive models are based on accurate, comprehensive data.

3) Combine AI with Human Insight

Use AI to augment, not replace, human decision-making in customer service.

4) Respect Customer Preferences

Allow customers to opt in or out of proactive services and respect their communication preferences.

5) Continuously Test and Refine

Regularly assess the accuracy of your predictive models and the effectiveness of your proactive actions.

H. Future Trends in Predictive and Proactive Customer Service

Several trends are likely to shape the future of predictive and proactive service:

1. *Edge Computing:* Enabling faster, more localized predictive analytics.
2. *Emotion AI:* Incorporating emotional state into predictive models.
3. *Augmented Reality (AR):* Using AR for proactive guided assistance.
4. *Voice of Customer (VoC) Analytics:* Deeper analysis of customer feedback for predictive insights.

I. Case Study: Amazon's Anticipatory Shipping

Amazon's "anticipatory shipping" patent provides an interesting case study in predictive customer service. The system uses predictive analytics to forecast product demand and begin the shipping process before the customer places an order (United States Patent and Trademark Office, 2013).

Key features of Amazon's anticipatory shipping include:

- *Predictive Analytics:* Using historical data, browsing behavior, and other factors to predict purchases.
- *Dynamic Rerouting:* The ability to reroute packages in transit based on real-time order data.
- *Partial Order Fulfillment:* Shipping products to local distribution centers in anticipation of orders.

While the full implementation details aren't public, this initiative demonstrates the potential of predictive analytics in revolutionizing customer service and supply chain management.

As Suresh Kumar, CTO of Walmart, notes, "The future of customer service is not just about responding to customer needs but anticipating them. AI and predictive analytics are key to making this a reality" (Kumar, 2021).

In conclusion, predictive and proactive customer service represents a significant shift in how businesses interact with their customers. By leveraging AI and advanced analytics, companies can anticipate customer needs, prevent issues, and provide more personalized, efficient service. While challenges exist, particularly around data privacy and the balance between proactivity and intrusiveness, the potential benefits in terms of customer satisfaction, loyalty, and operational efficiency are substantial. As AI technologies continue to evolve, we can expect even more sophisticated approaches to predictive and proactive customer service in the future.

VII. MAINTAINING THE HUMAN TOUCH: BALANCING AI AND HUMAN INTERACTIONS

As AI technologies become increasingly sophisticated in customer service applications, a critical challenge emerges *how to maintain the human touch in customer interactions.*

This section explores the importance of balancing AI and human interactions, strategies for achieving this balance, and the role of human employees in an AI-augmented customer service environment.

A. The Importance of Human Touch in Customer Service

Despite advancements in AI, human interaction remains a crucial element of customer service. A study by PwC found that 75% of consumers worldwide want more human interaction in the future, not less (PwC, 2018). This underscores the continued importance of the human touch, even as AI becomes more prevalent.

B. The Complementary Roles of AI and Humans

AI and human agents should be viewed as complementary rather than competitive in customer service. As Gartner analyst Olive Huang notes, "The goal is not to replace humans with AI, but to use AI to make humans better at what they do" (Huang, 2020).

C. Strategies for Balancing AI and Human Interactions

1) Tiered Support Model

Implement a tiered support model where:

- AI handles routine, straightforward queries
- Human agents manage complex issues and emotionally sensitive situations
- AI assists human agents with information retrieval and suggestion generation

2) Seamless Handoffs

Ensure smooth transitions between AI and human agents. This includes:

- Clear escalation protocols
- Context preservation during handoffs
- Warm transfers that maintain conversation continuity

3) AI-Assisted Human Interactions

Use AI to augment human capabilities:

- Real-time sentiment analysis to guide agent responses
- AI-powered knowledge bases for quick information retrieval
- Predictive analytics to anticipate customer needs during interactions

4) Human-in-the-Loop AI Systems

Implement AI systems that incorporate human oversight and input:

- Human review of AI-generated responses for quality assurance
- Continuous learning from human agent feedback
- Human-guided decision making for complex scenarios

D. The Evolving Role of Human Agents

As AI takes on more routine tasks, the role of human agents is evolving. Key aspects of this evolution include:

1) Focus on Complex Problem-Solving

Human agents increasingly focus on handling complex, non-routine issues that require creativity, empathy, and critical thinking.

2) Emotional Intelligence and Empathy

With AI handling routine queries, human agents can concentrate on providing empathetic support in emotionally charged situations.

3) AI Collaboration Skills

Human agents need to develop skills in effectively collaborating with AI systems, interpreting AI-generated insights, and knowing when to override AI recommendations.

4) Continuous Learning and Adaptability

As customer service technology evolves, human agents must continuously update their skills and adapt to new tools and processes.

E. Benefits of Balancing AI and Human Interactions

A well-balanced approach combining AI and human interactions offers several benefits:

1. *Improved Efficiency:* AI can handle high volumes of routine queries, freeing up human agents for more complex tasks. Juniper Research predicts that chatbots will save businesses $8 billion per year by 2022 (Juniper Research, 2017).
2. *Enhanced Customer Satisfaction*: By routing queries to the most appropriate channel (AI or human), businesses can provide faster, more effective service.
3. *Better Employee Satisfaction*: By handling routine tasks, AI can reduce agent burnout and allow for more engaging work. A Verint study found that 64% of agents believe AI will help them have more meaningful conversations with customers (Verint, 2019).
4. *Scalability:* AI can handle sudden spikes in query volume, ensuring consistent service quality during peak times.

F. Challenges in Balancing AI and Human Interactions

Several challenges arise when trying to balance AI and human interactions:

1. *Seamless Integration:* Ensuring smooth transitions between AI and human agents can be technically challenging.
2. *Customer Perception:* Some customers may have negative perceptions of AI in customer service. According to a Forbes survey 86% of consumers prefer to interact with a human agent (Press, 2019).
3. *Employee Concerns:* Human agents may fear job displacement due to AI. Clear communication and reskilling initiatives are crucial.
4. *Maintaining Consistency*: Ensuring consistent service quality across AI and human interactions can be challenging.

G. Best Practices for Balancing AI and Human Interactions

1) Clear Communication with Customers

Be transparent about when customers are interacting with AI versus human agents.

2) Continuous Training for Human Agents

Regularly train agents on working alongside AI and handling complex, emotionally charged situations.

3) Personalization Across Channels

Ensure that personalization efforts are consistent whether the customer is interacting with AI or a human agent.

4) Regular Evaluation and Adjustment

Continuously assess the effectiveness of your AI-human balance and adjust as needed.

H. Future Trends in AI-Human Collaboration

Several trends are likely to shape the future of AI-human collaboration in customer service:

1. *Advanced Human-AI Interfaces*: More intuitive interfaces for human agents to interact with AI systems.
2. *Emotion AI in Human-AI Collaboration:* Using emotion AI to guide when to transition from AI to human agents.
3. *Virtual Reality (VR) Customer Service:* Using VR for more immersive human-assisted service experiences.
4. *AI Coaches for Human Agents:* AI systems that provide real-time coaching to human agents during customer interactions.

As Shweta Joshi from Sendbird aptly puts it, " For today's businesses, choosing between AI vs. human customer support agents is all about striking the right balance: using AI when customers need fast answers to routine queries and having a human agent step in when customers need empathy, critical thinking, or other skills that no robot can provide " (Joshi, 2024).

In conclusion, while AI is transforming customer service, the human touch remains crucial. The key to success lies in finding the right balance between AI efficiency and human empathy, creating a synergy that enhances both customer and employee experiences. As technology continues to evolve, this balance will need constant refinement, but the goal remains the same: to provide efficient, personalized, and empathetic customer service.

VIII. ETHICAL CONSIDERATIONS AND CHALLENGES

As AI becomes increasingly integrated into customer experiences, a host of ethical considerations and challenges emerge. This section explores the key ethical issues surrounding AI in customer service, the potential risks, and strategies for addressing these challenges.

A. Key Ethical Considerations

1) Privacy and Data Protection

The use of AI in customer service often requires processing large amounts of personal data. This raises significant privacy concerns.

- According to Forbes, 32% of consumers have switched companies or providers over data privacy concerns (Daniels, 2023).
- The implementation of regulations like GDPR in Europe and CCPA in California underscores the growing importance of data protection.

2) Transparency and Explainability

As AI systems become more complex, ensuring transparency in decision-making processes becomes challenging.

- A study by Pegasystems found that 65% of consumers are uncomfortable with businesses using AI to engage with them (Pegasystems, 2017).
- The concept of "explainable AI" is gaining traction, with the EU's proposed AI Act emphasizing the importance of transparency (European Commission, 2021).

3) Bias and Fairness

AI systems can inadvertently perpetuate or amplify existing biases, leading to unfair treatment of certain customer groups.

- A study by MIT and Stanford University found that three commercial gender-classification systems had error rates of up to 34.7% for darker-skinned females, compared to 0.8% for lighter-skinned males (Buolamwini & Gebru, 2018).

4) Autonomy and Human Oversight

As AI systems become more autonomous, questions arise about the appropriate level of human oversight and intervention.

- The EU Artificial Intelligence Act has an article dedicated for Human Oversight. Article 14 states that "High-risk AI systems must be designed in a way that allows humans to effectively oversee them. The goal of human oversight is to prevent or minimize risks to health, safety, or fundamental rights that may arise from using these systems". (EU Artificial Intelligence Act, 2022).

B. Potential Risks of Unethical AI in Customer Service

1) Erosion of Customer Trust

Unethical use of AI can significantly damage customer trust. According to a study by Capgemini, 41% of consumers would stop engaging with a company if they experienced an AI interaction that they perceived as unethical (Capgemini, 2020).

2) Legal and Regulatory Risks

Non-compliance with data protection and AI regulations can result in severe penalties. For instance, under GDPR, companies can face fines of up to €20 million or 4% of global annual turnover, whichever is higher (European Commission, 2018).

3) Reputational Damage

High-profile AI ethics failures can cause significant reputational damage. The controversy surrounding Amazon's AI hiring tool, which showed bias against women, is a notable example (Dastin, 2018).

4) Perpetuation of Societal Biases

Biased AI systems in customer service can reinforce and amplify existing societal inequalities.

C. Strategies for Addressing Ethical Challenges

1) Ethical AI Framework

Develop a comprehensive ethical AI framework that guides the development and deployment of AI systems in customer service.

- IBM's "Everyday Ethics for Artificial Intelligence" provides a practical framework for AI ethics (IBM, 2019). IBM highlights 5 practices for everyday ethics:

  1. Take accountability for the outcomes of your AI system in the real world, no matter your role.
  2. Be sensitive to a wide range of cultural norms and values, not just your own.
  3. Work with your team to identify and address biases and promote inclusive representation.
  4. Ensure humans can perceive, detect, and understand an AI decision process.
  5. Preserve and fortify users' power over their own data and its uses. (IBM, 2019)

2) Privacy by Design

Implement privacy by design principles, ensuring that privacy considerations are integrated into AI systems from the outset.

- The Information Commissioner's Office (ICO) in the UK provides detailed guidance on AI and data protection (Information Commissioner's Office UK, 2023).

3) Algorithmic Fairness

Implement techniques to detect and mitigate bias in AI algorithms. This includes diverse training data and regular audits for bias.

- Google's What-If Tool is an open-source resource for investigating machine learning models for fairness (Google, 2021).

4) Transparency and Explainability

Strive for transparency in AI decision-making processes and develop methods to explain AI decisions to customers.

- The AI Explainability 360 toolkit by IBM provides resources for enhancing the explainability of AI systems (IBM, 2021).

5) Human-in-the-Loop Systems

Maintain appropriate human oversight in AI systems, especially for high-stakes decisions.

6) Continuous Monitoring and Auditing

Regularly monitor and audit AI systems for ethical compliance and performance.

D. The Role of Regulation in Ethical AI

As AI becomes more prevalent in customer service, regulatory frameworks are evolving to address ethical concerns:

- The EU's proposed AI Act aims to create a comprehensive regulatory framework for AI (European Commission, 2021).
- In the U.S., the proposed Algorithmic Accountability Act would require companies to assess their AI systems for bias and effectiveness (U.S. Congress, 2019).

E. Future Trends in AI Ethics

Several trends are likely to shape the future of AI ethics in customer service:

1. Increased focus on AI auditing and certification
2. Development of industry-specific AI ethics guidelines
3. Growing emphasis on AI literacy for customers and employees
4. Evolution of AI ethics from guidelines to enforceable standards

As Fei-Fei Li, Co-Director of Stanford's Human-Centered AI Institute, emphasizes, "The technology we create must be infused with human values such as ethics, responsibility, and compassion" (Li, 2018).

In conclusion, addressing ethical considerations is crucial for the responsible and sustainable use of AI in customer service. By proactively tackling these challenges, businesses can build trust, mitigate risks, and create AI-powered customer experiences that are not only effective but also ethical and fair.

IX. CASE STUDIES: SUCCESSES AND LESSONS LEARNED

This section examines real-world applications of AI in customer experience, highlighting both successes and challenges. These case studies provide valuable insights into the practical implementation of AI-driven customer service strategies.

A. Case Study 1: Spotify's AI-Powered Personalization

Spotify has successfully leveraged AI to create highly personalized user experiences.

Key Features:

- Discover Weekly: An AI-curated playlist of new music based on listening history.
- Daily Mix: AI-generated playlists combining familiar and new tracks.
- Release Radar: Personalized playlist of new releases from followed artists.

Results:

- Over 33% of all listening on Spotify is now driven by personalized recommendations (Spotify, 2022).
- Spotify's churn rate is significantly lower than the industry average, at around 5.5% (Anderson, 2024).

Lesson Learned: AI-driven personalization can significantly enhance user engagement and retention.

B. Case Study 2: Lemonade Insurance's AI Claims Processing

Lemonade, an Insurtech company, uses AI to streamline claims processing.

Key Features:

- AI Jim: An AI bot that can settle simple claims in seconds.
- Fraud detection algorithms to identify potentially fraudulent claims.
- Behavioral economics principles incorporated into the claims process.

Results:

- 30% of claims are settled instantly by AI (Lemonade, 2021).
- Lemonade's claims bot, AI Jim, broke a world record for the fastest claim paid out to a customer: 3 seconds! (Schreiber, 2023).

Lesson Learned: AI can dramatically improve efficiency in complex processes like insurance claims, but human oversight remains crucial for complex cases and ethical considerations.

C. Case Study 3: Bank of America's Virtual Assistant, Erica

Bank of America's AI-powered virtual assistant, Erica, showcases the potential of conversational AI in financial services.

Key Features:

- Natural language processing for understanding customer queries.
- Proactive financial insights and alerts.
- Integration with bank's mobile app for seamless experience.

Results:

- Erica surpasses 2 billion interactions, helping 42 million clients since launch (Bank of America Newsroom, 2024)
- More than 98% of clients get answers they need from Erica within 44 seconds (Bank of America Newsroom, 2024).

Lesson Learned: Well-implemented conversational AI can handle a significant portion of customer interactions, freeing human agents for more complex tasks.

D. Key Takeaways from Case Studies

1) Personalization is Key:

AI-driven personalization consistently leads to improved customer engagement and loyalty.

2) Balance AI and Human Touch:

While AI can handle many tasks, human oversight and intervention remain crucial for complex situations and maintaining customer trust.

3) Continuous Learning and Improvement:

Successful AI implementations involve continuous refinement based on user feedback and performance data.

4) Ethical Considerations are Crucial:

As seen in the insurance case, AI implementation must carefully consider ethical implications, especially in sensitive areas.

5) Integration Across Operations:

The most successful AI implementations integrate across both customer-facing and back-end operations for a seamless experience.

As Satya Nadella, CEO of Microsoft, notes, "AI is going to be one of the trends that is going to be the next big shift in technology. It's going to be AI at the edge, AI in the cloud, AI as part of every application" (Nadella, 2016). These case studies demonstrate the transformative potential of AI in customer experience, while also highlighting the importance of thoughtful, ethical implementation.

X. THE FUTURE OF AI IN CUSTOMER EXPERIENCE

As AI technologies continue to evolve at a rapid pace, the future of customer experience promises to be increasingly intelligent, personalized, and seamless. This section explores emerging trends and technologies that are likely to shape the future of AI-driven customer experiences.

A. Emerging AI Technologies in Customer Experience

1) Advanced Natural Language Processing (NLP)

Next-generation NLP models, like GPT-3 and its successors, will enable more natural and context-aware conversations between customers and AI systems.

2) Emotion AI and Sentiment Analysis

Advancements in emotion recognition technologies will allow AI systems to better understand and respond to customers' emotional states.

3) Augmented and Virtual Reality (AR/VR)

AR and VR technologies, combined with AI, will create immersive and interactive customer experiences.

4) Internet of Things (IoT) and Edge AI

The proliferation of IoT devices and edge computing will enable more real-time, context-aware customer interactions.

B. Predicted Trends in AI-Driven Customer Experience

1) Hyper-Personalization at Scale

AI will enable unprecedented levels of personalization across all customer touchpoints.

2) Predictive and Proactive Service

AI systems will increasingly anticipate customer needs and provide proactive solutions.

3) Seamless Omnichannel Experiences

AI will enable truly integrated omnichannel experiences, with consistent personalization across all channels.

4) Ethical AI and Transparency

There will be an increased focus on ethical AI practices and transparency in AI-driven customer interactions.

C. Potential Challenges and Considerations

1) Data Privacy and Security

As AI systems become more sophisticated, ensuring the privacy and security of customer data will be increasingly crucial.

2) AI Bias and Fairness

Addressing and mitigating bias in AI systems will remain a critical challenge.

3) Human-AI Collaboration

Defining the optimal balance between AI and human involvement in customer experiences will be an ongoing consideration.

4) Regulatory Landscape

Evolving regulations around AI and data privacy will shape the implementation of AI in customer experience.

D. The Role of AI in Shaping Customer Expectations

As AI-driven experiences become more common, customer expectations will evolve:

E. Preparing for the Future of AI in Customer Experience

To prepare for this AI-driven future, organizations should:

1. Invest in AI and data infrastructure
2. Foster a culture of innovation and continuous learning
3. Prioritize ethical AI practices and transparency
4. Focus on seamless integration of AI across all customer touchpoints
5. Continuously upskill and reskill their workforce

The future of AI in customer experience is not just about technology; it's about creating more human, empathetic, and valuable interactions between businesses and customers. In conclusion, the future of AI in customer experience promises to be transformative, offering unprecedented levels of personalization, efficiency, and proactive service. However, realizing this potential will require careful consideration of ethical implications, a commitment to transparency, and a focus on augmenting rather than replacing human capabilities.

XI. CONCLUSION

As we've explored throughout this chapter, AI is revolutionizing customer experience in profound ways, offering unprecedented opportunities for personalization, efficiency, and proactive service. However, the journey towards truly human-centric AI in customer experience is complex, fraught with challenges, and rich with potential.

A. Key Takeaways

1. *AI as an Enabler of Human-Centric Experiences*: AI technologies, from natural language processing to predictive analytics, are not replacing human interaction but enhancing it. They enable businesses to provide more personalized, efficient, and empathetic customer experiences at scale.
2. *The Importance of Balance:* Successful implementation of AI in customer experience requires a delicate balance between automation and human touch.
3. *Ethical Considerations are Paramount:* As AI becomes more prevalent in customer interactions, addressing ethical concerns around privacy, transparency, and fairness becomes increasingly crucial. Companies must prioritize ethical AI practices to build and maintain customer trust.
4. *Data as the Foundation:* The effectiveness of AI in customer experience is fundamentally tied to the quality and quantity of data available. Companies must invest in robust data infrastructure and practices while respecting customer privacy.
5. *Continuous Evolution:* The field of AI is rapidly evolving, and so too must our approaches to implementing AI in customer experience. Continuous learning, experimentation, and adaptation are key to staying ahead in this dynamic landscape.

The integration of AI into customer experience represents a significant shift in how businesses interact with their customers. While the potential benefits are enormous, realizing them requires a thoughtful, ethical, and human-centric approach.

As Andrew Ng, co-founder of Coursera and former chief scientist at Baidu, aptly puts it, "AI is the new electricity. Just as electricity transformed almost everything 100 years ago, today I actually have a hard time thinking of an industry that I don't think AI will transform in the next several years" (Ng, 2017).

However, in this transformation, we must not lose sight of the human element. The most successful implementations of AI in customer experience will be those that use technology to augment and enhance human capabilities, not replace them.

In conclusion, the future of customer experience lies not in AI alone, but in the synergy between artificial intelligence and human intelligence. By leveraging AI to handle routine tasks, provide data-driven insights, and enable personalization at scale, businesses can free up human agents to focus on what they do best: providing empathy, creativity, and complex problem-solving skills to deliver truly exceptional customer experiences.

As we move forward in this AI-driven era, the businesses that thrive will be those that successfully blend the efficiency and scalability of AI with the irreplaceable human touch, creating customer experiences that are not just satisfactory, but truly delightful and memorable.

REFERENCES

Aberdeen Group. (2012). *State of Service Management: Forecast for 2012*. Aberdeen Group.

Accenture. (2020). Accenture Technology Vision 2020: From Tech-Clash to Trust, the Focus Must Be on People. *Newsroom*. newsroom.accenture.com/news/2020/accenture-technology-vision-2020-from-tech-clash-to-trust-the-focus-must-be-on-people

Adalian, J. (2018). Inside Netflix's TV-Swallowing, Market-Dominating Binge Factory. *Vulture*. www.vulture.com/2018/06/how-netflix-swallowed-tv-industry.html

Anderson, A. (2024). *Making Personalization the Center of Your Customer Retention Strategy like Spotify*. Sharpen, sharpencx.com/spotify-customer-retention/

"Article 14: Human Oversight." EU Artificial Intelligence Act, 2022, artificialintelligenceact.eu/article/14/.

Bank of America Newsroom. (2024). Bofa's Erica Surpasses 2 Billion Interactions, Helping 42 Million Clients since Launch. Bank of America. newsroom.bankofamerica.com/content/newsroom/press-releases/2024/04/bofa-s-erica-surpasses-2-billion-interactions--helping-42-millio.html

Berners-Lee, T. (2017). Three challenges for the web, according to its inventor. World Wide Web Foundation. Available: https://webfoundation.org/2017/03/web-turns-28-letter/

Brackenbury, J., & LoDolce, M. (2023). Gartner Reveals Three Technologies That Will Transform Customer Service and Support by 2028. www.gartner.com/en/newsroom/press-releases/2023-08-30-gartner-reveals-three-technologies-that-will-transform-customer-service-and-support-by-2028

Buolamwini, J., & Gebru, T. (2018). Gender Shades: Intersectional Accuracy Disparities in Commercial Gender Classification. *Proceedings of Machine Learning Research*, *81*, 1–15.

Capgemini. (2020). The ethics of AI in customer experience. Capgemini Research Institute. Available: https://www.capgemini.com/research/ai-and-the-ethical-conundrum/

Daniels, J. (2023). Council Post: Why Organizations Should Invest in Privacy and CCPA Compliance. Forbes. www.forbes.com/sites/forbesbusinesscouncil/2023/06/22/why-organizations-should-invest-in-privacy-and-ccpa-compliance

Dastin, J. (2018). Amazon scraps secret AI recruiting tool that showed bias against women. Reuters. Available: https://www.reuters.com/article/us-amazon-com-jobs-automation-insight-idUSKCN1MK08G

Edelman, D., & Singer, M. (2015, November). Competing on Customer Journeys. *Harvard Business Review*.

el Kaliouby, R. (2017, October). We Need Computers with Empathy. *MIT Technology Review*.

el Kaliouby, R. (2020). *Girl Decoded: A Scientist's Quest to Reclaim Our Humanity by Bringing Emotional Intelligence to Technology*. Currency.

European Commission. (2018). General Data Protection Regulation (GDPR). European Commission. Available: https://gdpr.eu/

European Commission. (2021). Proposal for a Regulation laying down harmonised rules on artificial intelligence. European Commission. Available: https://digital-strategy.ec.europa.eu/en/library/proposal-regulation-laying-down-harmonised-rules-artificial-intelligence

European Commission. (2021). Europe fit for the Digital Age: Commission proposes new rules and actions for excellence and trust in Artificial Intelligence. European Commission Press Release. https://ec.europa.eu/commission/presscorner/detail/en/IP_21_1682

Forrester. (2018). Insights-Driven Businesses Set The Pace For Global Growth. Forrester.

Gartner. (2019). Gartner Predicts 80% of Marketers Will Abandon Personalization Efforts by 2025. Gartner Press Release. Available: https://www.gartner.com/en/newsroom/press-releases/2019-12-02-gartner-predicts-80--of-marketers-will-abandon-person

Gartner. (2021). Gartner Predicts Proactive Customer Service to Reduce Costs by 40% for Companies. Gartner Press Release. Available: https://www.gartner.com/en/newsroom/press-releases/2021-02-25-gartner-predicts-proactive-customer-service-to-reduce-costs-by-40-percent-for-companies

Gartner. (2019). Gartner Says the Future of Self-Service Is Customer-Led Automation. www.gartner.com/en/newsroom/press-releases/2019-05-28-gartner-says-the-future-of-self-service-is-customer-l

Gartner. (2020). Gartner Top Strategic Technology Trends for 2021. Available: https://www.gartner.com/smarterwithgartner/gartner-top-strategic-technology -trends-for-2021/

Gates, B. (1999). *Business @ the Speed of Thought: Succeeding in the Digital Economy*. Grand Central Publishing.

Google. (2021). What-If Tool. Google AI. https://pair-code.github.io/what-if-tool/

Grand View Research. (2020). Computer Vision Market Size. Grand View Research. Available: https://www.grandviewresearch.com/industry-analysis/computer-vision -market

Henry & Garg. (2024). Predictive Analytics Examples & Benefits on CX. *InMoment*. inmoment.com/blog/predictive-analytics/.

Huang. (2020). Gartner Keynote: The Future of Customer Service. Gartner Customer Service & Support Leader Conference.

Hyken, S. (2023). AI Will Not Eliminate Jobs. *Forbes*. www.forbes.com/sites/ shephyken/2023/08/27/ai-will-not-eliminate-jobs/

IBM. (2019). Everyday Ethics for Artificial Intelligence. IBM. https://www.ibm .com/watson/assets/duo/pdf/everydayethics.pdf

IBM. (2021). AI Explainability 360. IBM Research. https://github.com/Trusted -AI/AIX360

Information Commissioner's Office UK. (2023). Guidance on AI and Data Protection. *ICO*. ico.org.uk/for-organisations/uk-gdpr-guidance-and-resources/artificial -intelligence/guidance-on-ai-and-data-protection/

Joshi, S. (2024). Ai vs. Human Agents: How to Strike the Right Balance in AI Customer Service. *Sendbird*. sendbird.com/blog/ai-vs-human-intelligence.

Juniper Research. (2017). Chatbots: A Game Changer for Banking & Healthcare, Saving $8 billion Annually by 2022. Juniper Research. Available: https://www .juniperresearch.com/press/chatbots-a-game-changer-for-banking-healthcare

Kadet, K. (2021). Data from Entrust Reveals Contradictions in Consumer Sentiment Toward Data Privacy and Security. www.entrust.com/newsroom/press-releases/ 2021/data-from-entrust-reveals-contradictions-in-consumer-sentiment-toward-data -privacy-and-security

Kasparov, G. (2017). *Deep Thinking: Where Machine Intelligence Ends and Human Creativity Begins*. PublicAffairs.

Kumar. (2021). The Future of Retail: How Technology is Transforming the Shopping Experience. National Retail Federation Big Show Keynote.

Lee, K., & Superpowers, A. I. (2018). *China, Silicon Valley, and the New World Order*. Houghton Mifflin Harcourt.

Leggett, K. (2019). *The Future of Customer Service: 10 Trends to Watch*. Forrester.

Lemonade. (2021). Lemonade Insurance Review. Lemonade. https://www.lemonade.com/insuropedia/reviews/lemonade-insurance-review/

Li. (2018). How to Make A.I. That's Good for People. The New York Times. www.nytimes.com/2018/03/07/opinion/artificial-intelligence-human.html

MarketsandMarkets. (2021). Natural Language Processing Market. MarketsandMarkets. Available: https://www.marketsandmarkets.com/Market-Reports/natural-language-processing-nlp-825.html

MarketsandMarkets. (2021). Emotion Detection and Recognition Market. MarketsandMarkets. Available: https://www.marketsandmarkets.com/Market-Reports/emotion-detection-recognition-market-23376176.html

McAlone. (2016). Why Netflix Thinks Its Personalized Recommendation Engine Is Worth $1 Billion per Year. *Business Insider*. www.businessinsider.com/netflix-recommendation-engine-worth-1-billion-per-year-2016-6

McKinsey & Company. (2013). How retailers can keep up with consumers. McKinsey & Company.

McKinsey & Company. (2020). How COVID-19 has pushed companies over the technology tipping point—and transformed business forever. McKinsey & Company. Available: https://www.mckinsey.com/business-functions/strategy-and-corporate-finance/our-insights/how-covid-19-has-pushed-companies-over-the-technology-tipping-point-and-transformed-business-forever

McKinsey & Company. (2021). The value of getting personalization right—or wrong—is multiplying. McKinsey & Company. Available: https://www.mckinsey.com/business-functions/growth-marketing-and-sales/our-insights/the-value-of-getting-personalization-right-or-wrong-is-multiplying

Michelli, J. (2021). How Zappos Uses Customer Support as an Opportunity to WOW Customers. *Calix*. www.calix.com/blog/2021/08/how-zappos-uses-customer-support-as-an-opportunity-to-wow-customers.html

Morgan, B. (2018). 10 Examples of Predictive Customer Experience Outcomes Powered by Ai. *Forbes*. www.forbes.com/sites/blakemorgan/2018/12/20/10-examples -of-predictive-customer-experience-outcomes-powered-by-ai/

Nadella, S. (2016). Satya Nadella on Microsoft's New Age of Intelligence. Bloomberg. https://www.bloomberg.com/features/2016-satya-nadella-interview-issue/

Netflix Technology Blog. (2012). Lessons Netflix Learned from the AWS Outage. Netflix. Available: https://netflixtechblog.com/lessons-netflix-learned-from-the -aws-outage-deefe5fd0c04

New Epsilon Research Indicates 80% of Consumers Are More Likely to Make a Purchase When Brands Offer Personalized Experiences. (2018). Epsilon. www.epsilon .com/us/about-us/pressroom/new-epsilon-research-indicates-80-of-consumers-are -more-likely-to-make-a-purchase-when-brands-offer-personalized-experiences

Ng, A. (2017). AI Is the New Electricity. Stanford Graduate School of Business. https://www.gsb.stanford.edu/insights/andrew-ng-ai-new-electricity

Omnisend. (2019). Omnichannel Marketing Automation Statistics Report. Omnisend. Available: https://www.omnisend.com/resources/reports/omnichannel-marketing -automation-statistics-2019/

Pegasystems. (2017). What Consumers Really Think About AI: A Global Study. Pegasystems. Available: https://www.pega.com/ai-survey

Press, G. (2024). Ai Stats News: 86% of Consumers Prefer Humans to Chatbots. *Forbes*. www.forbes.com/sites/gilpress/2019/10/02/ai-stats-news-86-of-consumers -prefer-to-interact-with-a-human-agent-rather-than-a-chatbot/

PwC. (2018). Experience is everything: Here's how to get it right. PwC. Available: https://www.pwc.com/us/en/advisory-services/publications/consumer-intelligence -series/pwc-consumer-intelligence-series-customer-experience.pdf

PwC. (2018). Experience is everything: Here's how to get it right. PwC. Available: https://www.pwc.com/us/en/advisory-services/publications/consumer-intelligence -series/pwc-consumer-intelligence-series-customer-experience.pdf

Replika. (2021). Your AI Friend. Replika. Available: https://replika.ai/

Royce, R. (2022). How Ai Scales up IOT Capability in Turbofan Jet Engines. www .rolls-royce.com/media/our-stories/discover/2020/intelligentengine-how-ai-scales -up-iot-capability-in-turbofan-jet-engines.aspx

Schreiber, D. (2023). Lemonade Sets a New World Record. *Lemonade Blog*. www .lemonade.com/blog/lemonade-sets-new-world-record/

Shrivastav, P. (2022). Hubspot Annual State of Service in 2022. www.hubspot .com/hubfs/assets/flywheel%20campaigns/HubSpot%20Annual%20State%20of %20Service%20Report%20-%202022.pdf

Shum, H., He, X., & Li, D. (2018). From Eliza to XiaoIce: Challenges and Opportunities with Social Chatbots. *Frontiers of Information Technology & Electronic Engineering*, *19*(1), 10–26.

Slavickova, L. (2022). Hyper-Personalization Can Increase Your Conversion Rates by up to 60%. *Trask*. www.thetrask.com/blog/hyper-personalization-can-increase -your-conversion-rates-by-up-to-60

Smith, S. (2019). Digital Voice Assistants in Use to Triple to 8 Billion by 2023, Driven by Smart Home Devices. www.juniperresearch.com/press/digital-voice -assistants-in-use-to-8-million-2023/

Spotify. (2022). Discovery Mode – Spotify for Artists. Discovery Mode – Spotify for Artists. artists.spotify.com/en/discovery-mode

Swinscoe. (2023). The New Rules of Customer Engagement. www.freshworks.com/ assets/resources/freshworks/The-New-Rules-of-Customer-Engagement_US.pdf

Technology Review Insights, M. I. T. (2020). The Global AI Agenda: Promise, Reality, and a Future of Data Sharing. *MIT Technology Review*.

The Power of Hyper-Personalization: How Ai Elevates Customer Experience. (2024). *Comarch*. www.comarch.com/trade-and-services/loyalty-marketing/blog/the-power -of-hyper-personalization/

United States Patent and Trademark Office. (2013). Method and System for Anticipatory Package Shipping. Patent No. US 8,615,473 B2.

U.S. Congress. (2019). Algorithmic Accountability Act of 2019. 116th Congress. Available: https://www.congress.gov/bill/116th-congress/house-bill/2231

Venkatraman. (2021, April). Part 3: The Five Traps That Stall Digitalization. *Strategic*, *CFO360*, 26. strategiccfo360.com/the-five-traps-that-stall-digitalization/

Verint. (2019). Engagement in the Always-on Era: How Humans and Technology Work Hand-in-Hand to Meet Rising Expectations. Verint.

Woebot Health. (2021). Woebot: Your Self-Care Expert. Woebot Health. Available: https://woebothealth.com/

Zhou, L., Gao, J., Li, D., & Shum, H. (2020). The Design and Implementation of XiaoIce, an Empathetic Social Chatbot. *Computational Linguistics*, *46*(1), 53–93.

Chapter 18
A Machine Learning–Based Ensemble Model for Estimating Multiple Disease Prediction

Rasmita Kumari Mohanty
https://orcid.org/0000-0002-5828-5649
VNR Vignana Jyothi Institute of Engineering and Technology, India

ABSTRACT

This research introduces a cutting-edge approach to disease prediction using a stacking ensemble model in machine learning. Through meticulous evaluation and testing on data sourced from the National Health Portal of India, the model demonstrated superior accuracy, particularly with the utilization of the random forest ensemble technique. The success of the model lies in its ensemble learning approach, which harnesses the strengths of multiple classification algorithms and incorporates robust hyper parameter selection and cross-validation methods to ensure stability and reliability.

1. INTRODUCTION

In recent years, advancements in machine learning (ML) have revolutionized various sectors, including healthcare. The ability of ML algorithms to analyze vast amounts of data and extract meaningful patterns has led to significant improvements in disease diagnosis, prognosis, and treatment. One of the key areas where ML has shown immense promise is in the prediction of multiple diseases simultaneously. Predicting multiple diseases concurrently presents a complex challenge due to the

DOI: 10.4018/979-8-3693-8009-3.ch018

diverse nature of diseases, their interdependencies, and the intricate relationships among various risk factors. Traditional single-model approaches often struggle to capture these complexities adequately. However, ensemble learning, a technique that combines predictions from multiple models, offers a promising solution. This paper proposes a novel ensemble model designed specifically for estimating multiple disease predictions concurrently. By leveraging the strengths of diverse ML algorithms and integrating their outputs intelligently, the proposed model aims to enhance prediction accuracy, robustness, and generalizability. In this introductory section, we provide an overview of the significance of multiple disease prediction, discuss the limitations of existing approaches, and outline the objectives and structure of our proposed ensemble model.

Stacked ensemble technique with five base models Random Forest, KNN, Decision Tree, Logistic Regression, and Naive Bayes to forecast numerous diseases based on symptoms entered by the user. To improve accuracy and robustness, these models are merged using a meta-learner, which is trained on pertinent medical data. The stacked ensemble model makes advantage of the user-provided symptoms to provide quick and precise disease diagnosis. This method improves patient care and management by giving medical personnel a dependable tool for individualized medical assessments. In light of the limitations posed by traditional mental health consultation methods, there exists an alarming need to find a more accessible, cost-effective, and time-efficient alternative leveraging technology. This alternative must cater to the diverse needs of individuals, particularly those from underprivileged backgrounds, ensuring widespread accessibility while maintaining the quality and effectiveness of mental health support.

2. LITERATURE SURVEY

Solomon, D. D. Solomon et al. (2024) proposed "Extensive review on the role of machine learning for multifactorial genetic disorders prediction," examined the use of machine learning algorithms—more especially, the Decision Tree algorithm—in the prediction of diseases and their subtypes. The MapReduce Algorithm was also used in the study to improve operational efficiency when handling big datasets. The model showed only moderate performance in disease prediction tasks in spite of these efforts. To get over this problem and increase the model's usefulness in actual healthcare situations, more research and model improvement are necessary.

A Hanai et al.; Guhathakurata et al. (2022) proposed "Explainable Machine Learning Classification to Identify Vulnerable Grou Amon Parenting Mothers: Web-Based Cross-Sectional Questionnaire Study " The researchers in their study used a ten-year dataset from a clinical facility in southeast China to forecast a number of diseases. By employing a hierarchical multi-stage LSTM model with four layers which are embedding layer, contextual encoding layer, attention pooling layer, and output layer, they aimed to leverage the temporal aspect of the data for increased predictive accuracy. The data could be handled as time series data because they were longitudinal across a ten- year period. This allowed them to benefit from the temporal patterns found in medical data and provide more accurate predictions. However, a major disadvantage is that the dataset is heavily biased because it is only taken from one specific institution. The model's conclusions may not apply as well to wider populations due to this bias, which emphasizes the need of diversifying data sources for more trustworthy and inclusive prediction models in the healthcare sector. Despite this flaw, their approach highlights how crucial it is to incorporate temporal dynamics into disease prediction models, leading to more sane medical decisions and therapies.

Sharma and Vishwanath (2024) proposed "Heart disease prediction using machine learning ensemble". In order to shed light on how successfully decision tree classifiers perform the task of identifying patterns in heartbeat data, this study focuses on their application in the WEKA platform to predict heartbeats. The work offers useful information on the connection between machine learning and cardiac health and demonstrates how these models may be utilized in practice to reliably predict heartbeats using decision tree algorithms. Notably, the extensive use of WEKA as a machine learning technique has made the results easier for researchers and practitioners to understand and use. However, since users would require specific test findings, which could be costly or difficult to obtain, the practicality of providing inputs might have a drawback, making mass acceptance and deployment challenging. Notwithstanding this limitation, the study emphasizes how critical it is to foster understanding and confidence in predictive models for the assessment of cardiac health by concentrating on unambiguous and understandable insights from decision tree classifiers.

Jayanthi et al. (2024) proposed "Disease Prediction System using Machine Learning".This paper describes a system that predicts the existence of certain diseases based on user-entered symptoms using the Random Forest algorithm. While having a high average prediction probability of 95%, the method has many drawbacks. Its user interface allows users to select a particular disease and enter each related symptom individually, making it harder to use. This approach could be time-consuming and possibly ineffective, especially when treating multiple diseases or symptoms. Because of this, additional improvements in usability and practicality are required for broader applicability and usage in real healthcare settings, even though the system's prediction skills appear good.

Jayanthi et al. (2024) proposed "Mental health status monitoring for people with autism spectrum disorder using machine learning". Jayanthi et al. (2024) This paper explains about a system that will be used to gather and study data about psychological disorders under the guidance of professionals in the field of mental health. The system uses a Machine Learning classification algorithm called k-NN with TF-IDF to determine the kind of mental disease a patient might be experiencing based on their symptoms. The platform is notable for having an intuitive web interface known as "Psycho Web," which makes it easier for mental health professionals to communicate with the system. Even with its promising potential, the system has difficulties because of the noisy dataset, which only contains 58% of the instances from the standardized ICD-10 classification and the remaining examples from a variety of sources.

A new approach to predict COVID-19 using artificial neural networks. Hamsagayathri and Vigneshwaran (2021)The study develops a system where users input information on 16 predefined symptoms, and an Artificial Neural Network (ANN) model estimates the user's probability of having COVID. The ANN has limitations even though it has a reasonable accuracy rate of 84.7%. A major worry is the possibility of over fitting, which is a common problem with ANNs, especially if the model architecture is more complicated than the dataset that is available. There is a chance that the ANN will learn to generalize patterns too precisely to the training data, which would impair its capacity to predict outcomes for cases that have not yet been seen. This is because COVID-19 symptoms are dynamic and ever-changing. As such, while the ANN shows promise as a predictive tool, ongoing efforts are necessary to address over fitting and ensure its robustness and reliability in real-world applications.

In the 2020 Science Direct publication titled "Machine Learning-Based Predictive Modelling of Postpartum Depression," Sun et al. (2020) the effectiveness of nine machine learning algorithms—Naive Bayes, Support Vector Machine (SVM), Random Forest (RF), K- Nearest Neighbours (KNN), Logistic Regression, Stochastic Gradient Boosting, Recursive Partitioning, Regression Trees, and Neural Networks—in the diagnosis and treatment of postpartum depression was examined. It was noted by the study that employing Random Forest alone has a drawback and that hybrid models or ensemble approaches may produce more accurate predictive modelling of postpartum depression. The models' capacity for prediction in this situation might be improved by more research into similar methods.

Symptoms Based Disease Prediction Using Machine Learning Techniques. Ramprakash et al. (2020); Li et al. (2024) A major advancement in the field of biomedicine, this work employs machine learning techniques to enhance Computer-Aided Diagnosis (CAD) systems for disease detection. By training the model from specific instances, the technique ensures a wide and realistic learning experience while capturing real-world variability in biological data. By identifying patterns from a variety of cases, this technique not only increases the diagnostic reliability of the system but also helps to identify diseases more accurately and successfully. Nevertheless, a significant obstacle is the need for a large volume of high-quality training data in order to speed up learning and provide precise diagnosis. Despite this obstacle, the initiative has the potential to improve patient care by utilizing more advanced CAD systems in clinical settings and advancing medical technology.

J Li proposed Arthy and Prasanth (2024); Yede et al. (2021) titled "Evaluating graph neural networks for link prediction: Current pitfalls and new benchmarking," published in IEEE in 2021, presents a pioneering methodology utilizing Graph Neural Networks (GNNs) for disease prediction, which incorporates external knowledge bases to enrich electronic medical records (EMR) data. This approach demonstrates effectiveness in predicting both common and rare diseases. However, a notable drawback highlighted in the study is the complexity associated with integrating data from diverse sources, such as EMRs and external knowledge bases, which may necessitate ongoing maintenance efforts to uphold data quality and consistency for reliable predictions.

Indrakumari et al. (2020) proposed "A smart heart disease prediction system using iot and adaptive deep convolution neural network," which was published in IEEE in 2020, presents a unique approach that simultaneously learns patient representations and assesses pair wise similarity among patients by leveraging time fusion CNN architecture. With this method, we want to provide a more accurate and nuanced knowledge of patient dynamics in healthcare settings. Nevertheless, the study's main shortcoming is the technical difficulty of putting the temporal fusion CNN architecture into practice and optimizing it; for successful deployment and optimization, specialized knowledge may be required.

3. PROPOSED METHOD

3.1 Dataset Collection and Preprocessing

The data was obtained from the Kaggle website, specifically from a dataset titled "Symptoms of Disease Prediction." We loaded the dataset directly, which encompassed information on more than 261 distinct diseases, serving as the labels for our dataset. Additionally, the dataset provided comprehensive data on over 450 symptoms associated with these diseases, offering a rich and extensive resource for our predictive analysis.

In the preprocessing phase, the dataset was judiciously split in a ratio of 9:1, a strategic division ensuring a balance between the training and testing subsets. This partitioning strategy is integral to the subsequent stages of model evaluation, allowing for a thorough assessment of the model's predictive capabilities. The training subset is employed to impart knowledge to the model, enabling it to learn and generalize patterns from the data, while the testing subset serves as an independent validation set to assess the model's performance on unseen data.

The utilization of this split dataset for both training and testing phases enhances the model's reliability and ensures its adaptability to new, unseen instances. This methodological approach aligns with best practices in machine learning, fostering a robust and dependable predictive model.

Model Architecture:

Diving deeper into our disease prediction model's architecture, we find a well-thought-out framework that smoothly combines user interaction, historical data storage, and the collaborative power of various machine learning algorithms.

User Interface:

Users start by inputting symptoms or disease-related data through a User UI, acting as the model's gateway. This information then flows through the system, landing in a Database that holds historical data and patient records, forming a critical reservoir for generating insightful predictions.

Base Learners:

At the core of the model, five distinct base models (M1 to M5) play pivotal roles, each utilizing different algorithms. M1 uses K-Nearest Neighbors, M2 employs Decision Tree- based classification, M3 opts for Naive Bayes' probabilistic approach, M4 leverages an ensemble of Decision Trees in the form of Random Forest, and M5 adopts a Binary Classification model through Logistic Regression. Importantly, these models not only receive input from the User UI but also tap into the wealth of knowledge stored within the Database, providing a comprehensive understanding of presented symptoms.

Figure 1. System Architecture

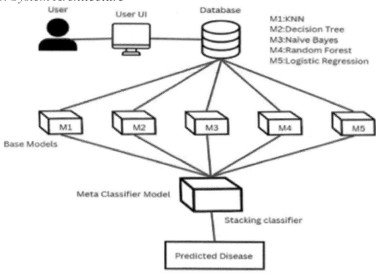

Meta Classifier

The outputs from these base models converge into a Meta Classifier Model (RF), a decision-maker leveraging the robust capabilities of Random Forest as a stacking classifier. This Meta classifier carefully assimilates diverse predictions, weighing individual strengths, and coordinates a unified final prediction for the disease. In summary, the model's architectural symphony concludes with the Predicted Disease – a result of user inputs, insights from historical data, and the collaborative predictive capabilities of multiple algorithms, all guided by the discerning meta classifier. This well-coordinated architecture speaks to the model's ability to distill a wealth of information into a singular, accurate, and clinically relevant prediction.

Training and Evaluation

The training and evaluation stages of our project were critical for developing a powerful predictive model for disease detection. Initially, in order to make sure that our machine learning algorithms would work with our dataset, we carefully prepared it, handling missing values and encoding symptom attributes. We integrated five different base learners such as decision trees, SVM, Random forest, KNN, and logistic regression using a stacking ensemble model architecture, each of these offered a distinct viewpoint on the data. These base learners were trained on the dataset to identify various relationships and patterns between symptoms and diseases during the training phase. The predictions generated by the base learners were then put into a stacking classifier, which developed the ability to efficiently combine these predictions and based on these it is able to predict the disease. We used key metrics such as accuracy, precision, recall, and F1-score to evaluate the model's performance.

3.2 Functionalities

Extracting Data

- Extracting data in machine learning refers to the process of acquiring, collecting, and preparing the data that will be used to train, validate, and test machine learning models.
- High-quality data is crucial for the success of machine learning projects because the accuracy and generalization ability of a model heavily depend on the quality and relevance of the data used.
- We extract data of the stock using web scraping from yahoo finance website for the required time period.

Preprocessing Data

- Removing inconsistencies in data like null values and outliers
- Normalize the data using min-max scalar to bring different features or variables to a similar scale or range.

Testing Model

- Testing in machine learning refers to the process of evaluating the performance and generalization ability of a trained machine learning model using a separate dataset that the model has never seen during training.
- Testing is a critical step in the machine learning pipeline as it helps assess how well the model is likely to perform on new, unseen data.
- Test model in various constraints and choose best model according to accuracies.

Predict Disease

- The symptoms entered by the user will get processed and give the predicted disease.

- The user can enter the symptoms and also see the output on the interface.

Data Preprocessing

Data preprocessing is a process of preparing the raw data and making it suitable for a machine learning model. It is the first and crucial step while creating a machine learning model. When creating a machine learning project, it is not always a case that we come across the clean and formatted data. And while doing any operation with data, it is mandatory to clean it and put in a formatted way. So for this, we use data preprocessing task. In this project, the preprocessing step we are using is creating the features and label for training the models.

Developing Model

Here we are developing 5 base models. The machine learning models used are Multinomial naïve bayes classifier, Random forest classifier, K-nearest neighbors, Logistic regression and Decision tree. Here we will train each of them separately and use these models for training meta classifiers. We are training the models by using sklearn. In sklearn we can directly import the models that we want to train. This minimizes lot of code lines very drastically.

Evaluation of Base Learners

The testing of the model is the most necessary step after training the models so checking the accuracy of the model is not only thing that we need while testing the model that we have build. Here we are using F1 score, recall and precision. Precision is a measure of the accuracy of the positive predictions made by a classifier. Recall, also known as sensitivity or true positive rate, measures the ability of a classifier to identify all relevant instances in the dataset. The F1 score is the harmonic mean of precision and recall. It provides a balance between precision and recall, considering both false positives and false negatives.

Training and Testing Meta Classifier

As we have the data ready for training meta classifier we have to train the meta classifier. We have tried different model for using as meta classifier but random forest worked best in our case. While training the meta classifier here we have used the grid search method to fine tune the random forest with correct parameters to get the accuracy. Grid search will help us to get the correct hyper parameters to train the model for our problem statement. Choosing the write parameters for training is the most important part during training. In search the model will be trained for every possible combination of hyper parameter and finalize the parameters which has the best.

4. MODEL COMPARISION USING EVALUATION METRICS

Using important assessment metrics including F1 score, precision, and recall, we carefully evaluated the performance of the stacking classifiers and individual base learners during the testing phase of our illness prediction project. We were able to obtain a thorough grasp of the models' prediction abilities by utilizing these metrics. The F1 score, which strikes a balance between recall and precision, shed light on the classifiers' overall accuracy. We were able to assess the classifiers' capacity to reduce false positives and false negatives, respectively, which is critical for accurate medical diagnosis, using precision and recall metrics. Furthermore, accuracy was a key indicator of how well the classifiers predicted illnesses from symptoms. We established the validity and efficacy of our disease prediction system by means of methodical testing and analysis with these metrics, which helped us to improve and optimize the ensemble approach for precise disease identification.

Figure 2. Bar graph displaying the F1_score

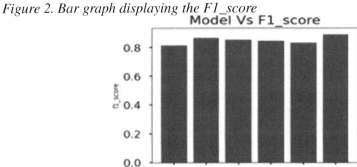

A thorough evaluation of a models accuracy is gauged by the F1 score, which finds a ground, between recall and precision. The stacking ensemble model stood out with an F1 score of 0.8897 showcasing its ability to make disease predictions by balancing recall and precision. Random Forest also performed admirably with an F1 score of 0.8660 demonstrating its skill in maintaining a balance, between recall and precision. The logistic regression and KNN models showed F1 scores of 0.8447 and 0.8537 respectively highlighting their contributions to the predictive capabilities of our system. While decision trees and Naive Bayes had F1 scores they still played crucial roles in enhancing the overall predictive performance.

Figure 3. Bar graph displaying the precision

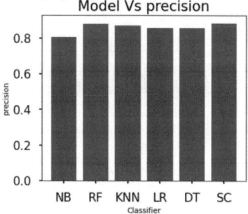

The accuracy of each model's disease prediction, depending on the input symptoms, varied. With precision ratings of 0.8799 and 0.8848, respectively, the Random Forest and stacking classifiers stood out as having the best capacity to accurately identify meaningful instances while minimizing false positives. With a precision score of 0.8706, the KNN model demonstrated strong performance as well, indicating its efficacy in precisely categorizing disorders. Even while decision trees, logistic regression, and Naive Bayes had somewhat lower precision values, they nevertheless made a substantial contribution to the system's total predictive performance. This evaluation gave us important information about the advantages and disadvantages of each model, which helped us decide which ones to incorporate into our illness prediction system.

Figure 4. Bar graph displaying the recall

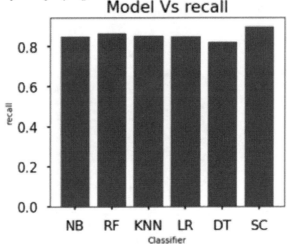

Out of all real positive cases, these recall values show how well each model can detect relevant instances. With a recall score of 0.8997, the stacking ensemble model stood out as having the highest accuracy in identifying true positive cases. With a recall value of 0.8665, Random Forest demonstrated exceptional performance as well. KNN and logistic regression, with recall values of 0.8518 each, were in close second place. Despite having somewhat lower recall values, decision trees and Naive Bayes nevertheless made a substantial contribution to the disease prediction system's overall efficacy. This evaluation gave us important information about how well the models recognized diseases from the input symptoms, which helped us decide how best to integrate and optimize them into our predictive framework.

Test Cases

Table 1. Test cases

Sl. no	Testcase	Expected Result	Actual Result	Pass/ Fail
1.	After Training, testing with test sample data	Accuracy on the test data is 91%.	Predicts the disease	PASS

This model aimed to develop an accurate and reliable machine learning model for disease prediction based on symptoms. After training & building the individual machine learning models(KNN, decision tree, naïve bayes, random forest, logistic

418

regression) individually with the data set and then used stacking mechanism in ensembled learning of machine learning, we fed all 5 individual models to one ensembled machine learning model through various combinations and finally saved the ensembled model as Random Forest because Random Forest demonstrated an efficient & superior accuracy in all the performance metrics than other machine learning models as ensembled model for the mentioned dataset. The performance of the final ensembled model is highly competitive than the other models achieving an accuracy of 91.07%, surpassing the other models in terms of predictive accuracy.

Performance Comparison

K-Nearest Neighbors (KNN): KNN is an algorithm that classify a new data point by the majority class of neighbors that are the k nearest to it. KNN is also known to encounters some difficulties in handling of high-dimensional data as well as noisy or imbalanced class distributions. In our dataset, kNN obtained a slightly better precision (84.16%), yet the best results came from the ensembled model.

Decision Tree: A tree obtains its decision-making capability from encoding the feature values such that the resulting model closely resembles a flow chart. The complexity of intrinsic diagnosed correlations between the symptoms and diseases renders them incompetent with prediction performance and considerably obsolete them to the training data. Therefore, the major reason of lacking even 2% of accuracy of the decision tree to that of ensembled model is responsible for the comparison of the decision tree model's 80.2% accuracy with that of the ensembled model.

Naive Bayes: We provided a data model where symptoms may act as independent units or can influence each other and then these units are very different from those assumed by the Bayesian theorem of independent features. Model of Naïve Bayes applied to this case was able to discover hard to catch connections with the accuracy of 84.62%, however it is inferior in terms of accuracy to the ensembled model.

Random Forest: A consensus prediction is created by averaging the results of all decision trees and every Random Forest makes use of several decision tree models. In the accuracy of 84.39%, Random Forest was working properly; as a base model to be used in the ensembled model, it can be believed.

Logistic Regression: Logistic regression makes a prediction concerning the connection between the symptoms and the binary result (sickness or non-sickness) through the logistic function. Although logistic regression is, indeed, often the preferred method for solving binary classification problems, things such as complicated interactions and non- linear features can present problems. In comparison to the logistic regression model, which yielded 87.78% as accuracy, the ensembled model showed higher accuracies. The outcome of each model was surprisingly good, but it definitely could not dig deeper into the complicated networks and interconnections within the data set. On the other said, the use of the ensembled learning model managed to mitigate the inherent flaws of each method in this manner, therefore, we could employ the strengths. Meaning by the way of stacking up these individual model outputs as I/O and to predict the given output, this machine kept on learning gradually what to execute as the task assigned to that task is completed. The work we do and we employed all kinds of modeling techniques on our problem definition until the randomly forest provided us the most functional basis model for our ensemble.

Several variables contribute to the ensemble model's superiority: Some of the factors that determine ensemble model's dominance are:

- Aspiring that the Collaboration of Bookstore and Cafe Will be the Best.
- The automatic inspection would mull over peculiarities of model defects.
- Enhanced Stability
- Improved Generalization
- Using the stacking method, the ensemble model based on Random Forest achieved 91.07% accuracy.

5. CONCLUSION

The implications of our findings extend beyond academic circles, showcasing the potential of machine learning in real-world medical applications. By enabling early disease detection and diagnosis, our model not only alleviates the burden on healthcare facilities but also enhances patient care by facilitating timely and accurate interventions. This efficiency not only reduces healthcare costs but also streamlines patient pathways, ensuring optimal resource utilization and improving overall healthcare quality. However, we acknowledge the challenges encountered during our research, including limited data availability and specificity in problem-solving areas. To address these limitations, we propose several avenues for future research, including the expansion of feature sets, data augmentation techniques, model fine-tuning, and validation through collaboration with healthcare organizations. Moving forward, our objective is to continue refining and expanding our model to contribute to improved public health outcomes. By focusing on early disease detection and better healthcare management, we aim to empower both patients and healthcare professionals with a powerful tool for proactive health monitoring and intervention. Through ongoing collaboration and innovation, we envision a future where machine learning plays a pivotal role in enhancing healthcare delivery and improving patient outcomes on a global scale.

REFERENCES

Hamsagayathri, P., & Vigneshwaran, S. (2021). Symptoms based disease prediction using machine learning techniques. 2021 Third international conference on intelligent communication technologies and virtual mobile networks (ICICV). IEEE. DOI: 10.1109/ICICV50876.2021.9388603

Hanai, A., Ishikawa, T., Sugao, S., Fujii, M., Hirai, K., Watanabe, H., ... Kawakami, E. (2024). Explainable Machine Learning Classification to Identify Vulnerable Groups Among Parenting Mothers: Web-Based Cross-Sectional Questionnaire Study. *JMIR Formative Research, 8*(1), e47372.

Indrakumari, R., Poongodi, T., & Jena, S. R. (2020). Heart Disease Prediction using Exploratory Data Analysis. *Procedia Computer Science*, *173*, 130–139. DOI: 10.1016/j.procs.2020.06.017

Jayanthi, S., Priyadharshini, V., Kirithiga, V., & Premalatha, S. (2024). Mental health status monitoring for people with autism spectrum disorder using machine learning. *International Journal of Information Technology : an Official Journal of Bharati Vidyapeeth's Institute of Computer Applications and Management, 16*(1), 43–51. DOI: 10.1007/s41870-023-01524-z

Pal, A. K., Rawal, P., Ruwala, R., & Patel, V. (2019). Generic disease prediction using symptoms with supervised machine learning. *Int. J Sci. Res. Comput. Sci. Eng. Inf. Technol, 5*(2), 1082–1086. DOI: 10.32628/CSEIT1952297

Ramprakash, P. (2020). Heart disease prediction using deep neural network. 2020 international conference on inventive computation technologies (ICICT). IEEE.

Sharma, D., & Vishwanath, V. (2024, February). Heart disease prediction using machine learning ensemble. In *AIP Conference Proceedings* (Vol. 2742, No. 1). AIP Publishing. DOI: 10.1063/5.0191651

Solomon, D. D., Sonia, , Kumar, K., Kanwar, K., Iyer, S., & Kumar, M. (2024). Extensive review on the role of machine learning for multifactorial genetic disorders prediction. *Archives of Computational Methods in Engineering, 31*(2), 623–640. DOI: 10.1007/s11831-023-09996-9

Sun, Z., Yin, H., Chen, H., Chen, T., Cui, L., & Yang, F. (2020). Disease prediction via graph neural networks. *IEEE Journal of Biomedical and Health Informatics*, *25*(3), 818–826. DOI: 10.1109/JBHI.2020.3004143 PMID: 32749976

Chapter 19
A Bibliometric Analysis of Blended Learning in the Hospitality and Tourism Industry

Monika Pandey
https://orcid.org/0009-0003-9150-9581
Amity University, Jaipur, India

Jai Sonker
https://orcid.org/0000-0001-8559-5485
Amity University, Jaipur, India

Shweta Misra
IHM Pusa, New Delhi, India

ABSTRACT

Blended learning is a type of educational learning that incorporates traditional face-to-face learning with online learning. In the hospitality and tourism industry, the blended learning model is utilized for both training and education to ensure workforce competency. The growing interest in blended learning is drawing increased attention from researchers, leading to a rise in publications on the topic within the hospitality and tourism sector. This chapter aims to add to the existing literature with an overview using the bibliometric analysis, a methodology that quantitatively analyzes academic literature and identifies significant trends, patterns, and research publications in the field of hospitality and tourism industry. This analysis was conducted using research articles published between 2014 and 2024, totaling 2,174 samples. The study emphasizes the distinctive insights provided by the bib-

DOI: 10.4018/979-8-3693-8009-3.ch019

liometric approach and identifies areas for future exploration. It concludes with a summary of the findings, acknowledges limitations, and offers recommendations for future research.

1. INTRODUCTION

The development in technology is leading to changes everywhere in the world. Digitalization has improved the learning and training processes significantly. The impact of digitalization is visible in almost every industry. The technology driven education and training have been there since a long time however covid-19 has caused a surge in blended learning, hybrid learning and flipped classroom learning. Learners across the world benefit from blended learning due to various reasons like flexibility and accessiblity.

The rapid growth of the internet has entered into the hospitality and tourism industry as well. It has a positive impact on the day to day operations leading to higher customer satisfaction. The hospitality and tourism industry is adopting the technology to remain competent and profitable. This has made a catalytic effect on the hospitality industry by incorporating technology based education and training (Viglia et al., 2018). Some challenges are also associated with the use of technology based learning which have been highlighted in many research studies. The adoption of using technology while training and teaching have also been reported low due to several reasons like difficulty to adapt to change and lack of training among the educators (Long et al,. 2019).

Learning technology has a better satisfaction among hospitality students and equips them to manage the latest technological trends (Frawley et al., 2019). Covid-19 has impacted a lot in terms of making blended learning popular. All the institutions and universities developed the blended learning models due to the restrictions covid-19 posed. Blended learning was found to be effective in terms of providing employment opportunities as the students gained a different set of skills through the blended approach in education (Guden & Safaeimanesh, 2024). The use of a blended learning model was found effective and showed improvement in students' learning not only in gaining theoretical knowledge but in practical sessions as well (Zgraggen, 2021).

The various perspectives on blended learning in hospitality and tourism have been an interesting area of research by multiple scholars however multiple areas have not been explored. Hence with the existing literature on the topic of blended learning in hospitality and tourism and with the developments of creative approaches; a bibliometric analysis was decided on the available literature.

2. LITERATURE REVIEW

Hospitality and Tourism Industry

The tourism industry is a major contributor to the global economy. Hospitality and tourism are interdependent on each other. While the tourism industry covers all the sectors associated with the tourists movement, the hospitality industry is an inseparable part of it. Hospitality is welcoming and assisting the guests and visitors. The industry revolves around people meeting other people, interaction between a host and a guest. In view of service quality, automation is required to be competent in this current age (Naumov, 2019). Tourism and hospitality industry provides employment to a huge number of people engaged in hotels, restaurants, transportation, tourist sites management and many other related areas like souvenir shops, art & craft and local shops. The industry helps to achieve a large sum of foreign exchange earnings (Skripak et al., 2016). The industry is constantly evolving with the developments in technology and academic research must progress accordingly (McKercher, 2018).

Digitalization in Hospitality and Tourism

The internet has revolutionized operations in every industry, and hospitality and tourism are no different. Digital transformation in tourism boosts a nation's appeal to tourists, as modern travelers often search for information online. This digital shift helps attract more visitors and enhances their experience at tourist destinations (Alam et al., 2022). Digital technologies are considered primary sources of efficiency and competitive edge in the hotel sector (Shin et al.,2019). Given the focus on people in the hospitality sector, the integration of technology in the industry operations is particularly significant. Technology plays a critical role in operations, management, and security. Keeping this into consideration the academicians should strategically design and train the students to meet the demands of the industry (Quinn & Buzzetto-Hollywood, 2019). Nanda et al. (2023) analyzed the tourism trend during covid-19 pandemic and digital based tourism has been recommended for sustainable tourism. Advancements in digital technology have led to major changes in tourism education. Digital technology-enhanced education fosters a more effective learning environment and provides greater opportunities for both teachers and students (Zheng, 2023).

Blended Learning in Hospitality and Tourism

The role of tourism educators is crucial and quite challenging with the integration of technology in learning with the constantly evolving learning environment (Hsu, 2018). The blended language learning model was proved to be highly effective for tourism education students when adequate support was provided (Li, 2015). Use of blended learning has also been advocated in recreation management studies (O'Boyle, 2014). Deale (2015) in his case study found that the experience of the students were positive for collaborative learning projects using technology among United States and Mexico students despite some challenges posed due to technical issues and recommended applying similar learning models in future. Shih et al. (2015) tried to create a mobile application to enrich the students' learning experience in hospitality and tourism and develop their competency to deal with real world situations in the current time. Recognizing the significance of online learning, the Malaysian Ministry of Higher Education has made it a mandatory component of higher education and as a result of this; multiple undergraduate courses were transformed into Massive Open Online Courses (Safri et al.,2020). It was proposed that an interactive and innovative teaching plan supported by technology could enhance student satisfaction in their learning experience (Kou & Liu, 2020).

Employee Training in Hospitality and Tourism

Employee training boosts staff productivity and enhances the quality of service within the organization, while also expanding professional knowledge. In today's fast-evolving world, hotel industry employees must continuously enhance their skills and competencies. Tailored training programs that address industry-specific needs can be highly beneficial, supporting their professional growth and improving service delivery (Nestoroska & Petrovska 2014). Andersen (2020) suggests that regular training can improve knowledge and skills, helping to overcome the challenge of insufficiently skilled human resources. Hotels differentiate themselves with guests by offering superior service quality. (Mola & Jusoh 2011). Effective human resource management and training are essential for the long-term growth and sustainability of the hospitality organizations. To improve guest satisfaction and strengthen employee loyalty, luxury hotels should adopt formally organized training programs (Belias et al., 2020).

3. METHODOLOGY

The research study is a review paper on the given topic hence the secondary sources of data have been used for the study. The data was collected through dimensions.ai tool which helped in retrieving the data from the global published research articles. The study has selected only the journal published research articles for the analysis.

Blended learning involves combining online education and technology with traditional in-person education and training. Therefore, to grasp the concept of blended learning in hospitality and tourism, it is important to analyze not only papers specifically about blended learning but also those focusing on the use of technology or technology-driven education and training within the hospitality and tourism industry.

The document type was a key criterion for filtering the published data on the topic. As a result, only research articles published in journals were selected when retrieving data from dimensions.ai. The data was retrieved using the keywords "Blended Learning", "Tourism" and "Hospitality & Tourism" for all the published open access research articles from the last 10 years 2014-2024. Total 2,12,345 journal research articles were published. Further filtration was done using research categories by selecting "tourism" under field of research and "quality education" under sustainable development goals. After using these filters, a total of 2174 journal research articles were retrieved and selected as samples for the study. The quantitative bibliometric analysis is used for the data analysis.

4. RESULTS

This section provides a comprehensive overview of the results from the bibliometric analysis. The presentation of the results begins with an examination of the current status of blended learning in the hospitality and tourism sectors in the first sub-section. The second sub-section covers the keywords analysis followed by co-authorship and co-citation analysis respectively in the third and fourth sub-sections.

The bibliometric analysis of the existing large volume of scientific data (research papers) helps to gain insights into the emerging trends, themes and developments in the field of research related to a given topic (Yu et al., 2020). Various network models chart the relationships among journals, researchers, organizations, and countries, providing valuable insights for the global research community. They also help identify research gaps and potential areas for further investigation within specific topics. VOSviewer was the software tool selected to facilitate the study.

Meng et al. (2020) explain in their study that VOSviewer, developed by Van Eck and Waltman at Leiden University in the Netherlands, is a software tool used to construct and visualize bibliometric networks. It can create networks for journals, researchers, keywords, and publications based on co-citation, coupling, and co-authorship relationships, and provides visual representations of these networks. Van Eck and Waltman (2010) emphasize that the tool's value lies not only in its ability to create the described networks but also in its capability to gather data from multiple scientific databases, including Web of Science, Scopus, Dimensions, and PubMed, as well as from reference management files such as RIS, EndNote, and RefWorks.

4.1 Current Status of Blended Learning in the Hospitality and Tourism Industry

4.1.1 Annual Trends in Publications

The analysis of annual trends of publications on blended learning in hospitality and tourism shows that there has been an increase every year since the last 10 years. The chart shows a significant increase in the number of publications. On an average the growth in terms of number of papers published is more than 10% each year since 2014 except the year 2022 where there are almost the same number of publications as the year 2021. As of now, with only half the year 2024 completed; there have already been 200 papers published. The exponential growth in the paper publication on blended learning is clearly visible since 2020 due to the pandemic of covid-19. The analysis indicates that the research topic of blended learning is gaining attention of researchers across the world.

Figure 1. Annual Trends of publications

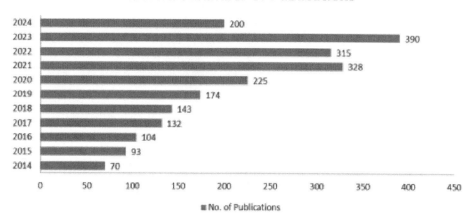

4.1.2 Distribution of Organizations

The analysis of affiliated organizations of authors reveals that the top 10 organizations that published papers on blended learning in hospitality and tourism have contributed 129 papers (Fig. 2) accounting for 5.8% of the total published papers since 2014. Hong Kong polytechnic University has the highest publications with a total of 32 papers. In the referred publications the authors have focussed on the need for developing technology based curriculum for tourism and hospitality students and people who are already employed in the industry. The use of blended learning, augmented and virtual reality tools in order to form professional competency have been discussed.

Figure 2. Top 10 institutions according to the number of publications

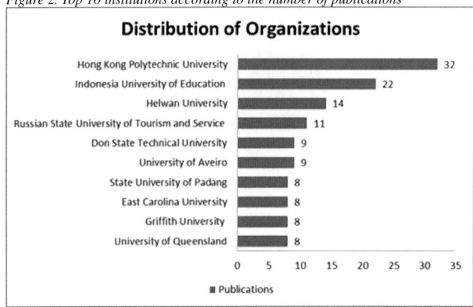

4.1.3 Distribution of Published Journals

Analysis of the study sample shows that the top 5 journals have collectively published 252 papers on the referred topic. "Journal of Hospitality Leisure Sport & Tourism Education'' is the one with the highest publications with a total of 67 articles representing 3% of the total publications globally. All the top 5 journals have published more than 30 papers each on the topic.

Figure 3. Top 5 journals according to the number of publications

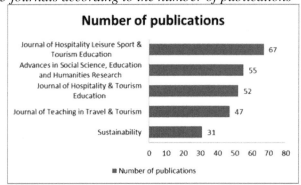

4.2 Keywords Analysis of Research Hotspots

The contexts of the selected articles were thoroughly analyzed using the relevant keywords.The keywords co-occurrence analysis helps to understand the critical points in research on the blended learning in hospitality and tourism; its uses, future potential and challenges. The keywords co-occurrence network model (Fig. 4) labels the keywords with different colored circles and the respective size of the circle and text explains the correlation with the frequency of the words in the titles and abstracts of the articles. It reveals that the highest frequently used keywords are a) "research" (108); b) "educational tourism" (66); c) "covid" (59); "impact" (59); d) "effect" (48); e) "hospitality education" (36); f) "teaching" (36); g) "pandemic" (35); h) "field" (34); i) "perception" (34).

The distance between two nodes indicates the strength among the relationship of the two nodes. The closer the nodes; higher the strength. The lines represent the correlation among the two words. The thicker the line, the more frequently the words co-occur.

Figure 4. The keywords co-occurrence network of blended learning in hospitality and tourism industry related published articles

4.3 Co-Authorship Analysis on Blended Learning in the Hospitality and Tourism Industry

Research project is a complex endeavor and to ensure valid results, it necessitates collaboration among researchers from various disciplines (Kahn, 2018). The co-authorship network models help to understand the global research landscapes including the frequency and nature of research collaborations. The co-authorship analysis provides insights into the effects on research output and quality, as well as the patterns associated with a specific topic.

4.3.1 Country Co-Authorship Analysis

The country co-authorship analysis helps to understand the countries that have the maximum contribution to the area of investigation. It's also beneficial to gauge the extent of their communication. The country co-authorship network on blended learning in hospitality and tourism industry related published articles is shown in Fig. 5. The size of the nodes indicates which countries have the most extensive work in the field of investigation. The connections between the nodes illustrate the collaborative relationships among various countries' institutions. The thickness of the links and the distance between the nodes reflect the degree of proximity of research on the given area among the countries.

The top 5 countries with the highest number of publications are Indonesia (119), China (103), United States (85), Russia (71) and the United Kingdom (57). The analysis indicates that, among the top 10 countries with the most publications in the field, 40% are from developed countries (US, UK, Australia and Spain), while 60% are from developing countries (Indonesia, China, Russia, Turkey, Ukraine and Egypt). Among the top five countries with the most publications are Indonesia, China, and Russia from the developing world, as well as the US and the UK from the developed world. The top 3 countries with the maximum citations are China (2450), United States (2315) and Australia (2057). The three countries with the greatest total link strength values are the United Kingdom (62), United States (59) and China (55).

Figure 5. The country co-authorship network model of blended learning in hospitality and tourism industry related published articles

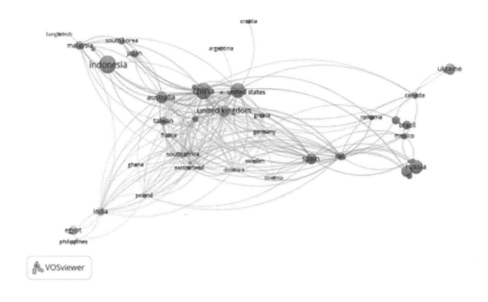

4.3.2 Highly Cited Publications

The quality of the research papers among the research community is analyzed by the number of citations it holds. A greater number of citations typically indicates a higher quality paper. To assess how research has shaped the advancement of blended learning in the hospitality and tourism industry, we analyzed the top 10 papers with the highest citation counts from our study sample. Table **1** shows the top 10 journal research articles with the highest number of publications.

Kim & Jeong (2018) outlined the recent trends in hospitality and tourism education. They focused on six key areas: leadership and human capital development, online education, diversity, active and experiential learning methods, internationalization and industry experience. The authors also proposed future research related to industry needs in hospitality and tourism education. With over 127 citations this paper is the second most cited paper on the topic of blended learning in the hospitality and tourism industry.

Dwivedi et al., in their study examined the role of AI tool chatgpt where the study presents the opportunities and challenges of chatgpt across different industries covering computer science, information systems, hospitality and tourism, education, marketing, policy, nursing, management and publishing. The paper puts forward further research scope like training of human resources, related legal and ethical issues, biases and accuracy assessments. This paper ranks first in citation count, with over 1,205 citations. Recent publication data indicates that, after the COVID-19 pandemic, there has been significant growth in the research area, resulting in a rise in related publications.

Table 1. Papers with highest citations

Title	Journal	Author (Reference sequence no.)	Year	Citation
Opinion Paper: "So what if ChatGPT wrote it?" Multidisciplinary perspectives on opportunities, challenges and implications of generative conversational AI for research, practice and policy	International Journal of Information Management	(7)	2023	1205
Research on hospitality and tourism education: Now and future	Tourism Management Perspectives	(14)	2018	127
Examining the usability of an online virtual tour-guiding platform for cultural tourism education	Journal of Hospitality Leisure Sport & Tourism Education	(4)	2018	122
"Women cannot lead": Empowering women through cultural tourism in Botswana	Journal of Sustainable Tourism	(23)	2014	105
Incorporating TikTok in higher education: Pedagogical perspectives from a corporal expression sport sciences course	Journal of Hospitality Leisure Sport & Tourism Education	(8)	2021	105
Game of algorithms: ChatGPT implications for the future of tourism education and research	Journal of Tourism Futures	(13)	2023	101
Tourism Education: What about entrepreneurial skills?	Journal of Hospitality and Tourism Management	(5)	2017	97
Tourism education on and beyond the horizon	Tourism Management Perspectives	(12)	2018	90
Online Teaching and Learning Experiences During the COVID-19 Pandemic – A Comparison of Teacher and Student Perceptions	Journal of Hospitality & Tourism Education	(16)	2021	90
The MOOC dropout phenomenon and retention strategies	Journal of Teaching in Travel & Tourism	(10)	2020	81

4.4 Co-Citation Analysis

Co-citation means when two research documents are cited together in another third research document. This indicates a possible thematic relationship between the two co-cited documents (Small, 1973). A detailed co-citation analysis was conducted to highlight the potential relationships between articles, authors and journals concerning the blended learning in hospitality and tourism topic.

4.4.1 Reference Co-Citation Analysis

When a paper references two other papers, it signifies that those two papers share a co-citation relationship. Co-citation analysis helps researchers comprehend the structure and progression of a particular research topic. By incorporating the identified papers into a network model for analysis, we can uncover the comprehensive features of a research topic. The network model, with nodes of varying sizes and colors, emphasizes the papers most closely associated with the topic of blended learning in the hospitality and tourism industry. From Fig. 6, it is evident that the largest nodes are Tribe (51), Inui (41), Sheldon et al. (40), Kim et al. (40) and Ayikoru et al. (38). The highest link strength values are of Tribe (185), Ayikoru (181), Inui (173), Sheldon et al. (173) and Belhassen et al. (167).

Figure 6. The reference co-citation network of blended learning in hospitality and tourism industry related published articles

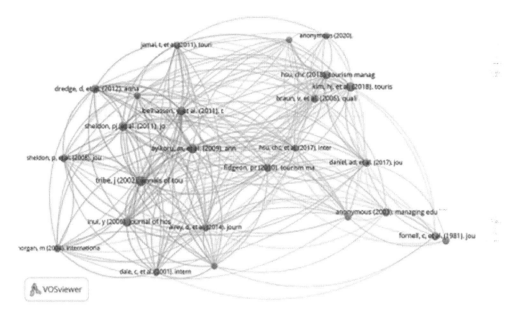

4.4.2 Journal Co-Citation Analysis

The journal co-citation network depicts the overall construct of the subject and the attributes of the journals involved. Fig. 7 shows the network model of journal co-citation analysis. The varying sizes of the nodes represent the volume of papers published on the topic. The proximity of the nodes reflects the citation frequency between journals; the closer the nodes are, the more frequently they are cited together. The journals "Tourism Management", "Annals of Tourism Research", "Journal of Hospitality Leisure Sport & Tourism Education", "Journal of Teaching in Travel & Tourism" and "Journal of Hospitality & Tourism Education" are the top 5 journals with the highest number of published papers that were co-cited. The different color clusters represent the range of journals across various fields of hospitality and tourism. The top 3 journals with highest citations are Tourism Management (1473), Annals of Tourism Research (1321) and Journal of Hospitality Leisure Sport & Tourism Education (1026).

Figure 7. The journal co-citation network of blended learning in hospitality and tourism industry related published articles

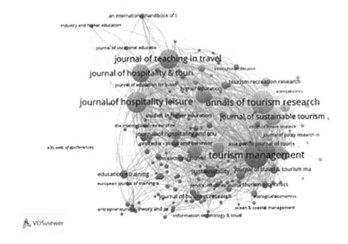

5. DISCUSSION AND CONCLUSION

This research study aimed to perform a bibliometric analysis of research articles on blended learning within the hospitality and tourism industry. The analysis clearly demonstrates that the number of publications in this field has steadily increased over the past decade, with a significant surge occurring after the COVID-19 pandemic, specifically from 2020 onwards. Indonesia, China and the United States are the top 3 countries with the highest number of papers published on blended learning in the hospitality and tourism industry. Analysis reveals that, among the top 10 countries with the highest number of publications in the field, 40% are from developed countries and 60% are from developing countries. "Journal of Hospitality Leisure Sport & Tourism Education'', "Advances in Social Science, Education and Humanities Research" and "Journal of Hospitality & Tourism Education" are the top 3 journals with the highest number of publications on the research topic. Regarding keyword co-occurrence, it has been noted that the most frequently used keywords were "research," "educational tourism," and "COVID." The average citation count for the study sample was 4.08, meaning that, on average, each paper in the sample has been cited over four times. "International Journal of Information Management", "Tourism Management Perspectives" and Journal of "Hospitality Leisure Sport & Tourism Education" are the top 3 journals with the highest citations of papers published on blended learning in hospitality and tourism related topics.

The current study provides further evidence that bibliometric analysis is a valid scientific method for conducting retrospective reviews across broad research areas. The bibliometric method is recognized as a valuable approach at all levels, from senior to junior scholars. The results of this study reveal that the topic of blended learning in the hospitality and tourism industry—encompassing technology, skill development, training, and competency—is highly relevant. The number of paper publications has grown exponentially, with over 1,033 papers on the topic published between 2021 and 2023. The study spans from 2014 to 2024, and by mid-2024, there have already been 200 papers published in this research area. The number of citations has notably risen since 2020, indicating that researchers worldwide are increasingly drawn to this area of study. Indonesia and China are the leading countries contributing the most to the topic mentioned, and this is also evident in the institutions to which the authors of the most published papers are affiliated. The top three countries with the highest number of citations are China (2,450), the United States (2,315), and Australia (2,057). The top 5 journals in terms of papers published are responsible for publishing 252 articles contributing 11.59% of the total study sample of 2174 papers.

From a theoretical standpoint, while blended learning in the hospitality and tourism industry is widely regarded as a valuable approach to learning and training, the analysis indicates that its full effectiveness has not yet been conclusively demonstrated as certain challenges persist that can occasionally hinder the positive impact of this approach.

6. LIMITATIONS AND FUTURE RESEARCH

The current study provides valuable insights through a bibliometric analysis of the existing literature on blended learning in the hospitality and tourism industry. However, it also acknowledges certain limitations. The study includes only the published research articles ignoring the conference proceedings and book chapters which could also bring some interesting insights into the topic. The study has taken the past 10 years data since 2014 in the field of investigation to understand the various areas in the given research topics. Further investigations can be performed with the data prior to 2014 by not only covering the journal articles but also conference proceedings, book chapters, theses and dissertations.

The findings also indicate that research on this topic has garnered increasing interest from developing countries with significant tourism potential, such as Indonesia, China, Turkey, and Egypt. Other developing nations could similarly leverage this research area to enhance their own tourism industries. Future researchers could conduct additional bibliometric analyses focused on training needs in the hospitality and tourism industry, as well as contemporary training models beyond those related to hospitality and tourism education. Based on the analyzed results, it is also recommended to conduct a qualitative analysis for this research area consisting of inputs collected from the industry experts.

REFERENCES

Alam, M. R., Milon, M. R. K., Rahman, M. K., & Hassan, A. (2022). Technology Application in Tourism Event, Education and Training for Making a Nation's Image. In *Technology Application in Tourism Fairs, Festivals and Events in Asia* (pp. 149–163). Springer Singapore. DOI: 10.1007/978-981-16-8070-0_9

Andersen, I. M. V. (2020). Tourism employment and education in a Danish context. *Tourism Employment in Nordic Countries: Trends, Practices, and Opportunities*, 37-56.

Belias, D., Vasiliadis, L., & Mantas, C. (2020). The human resource training and development of employees working on luxurious hotels in Greece. In *Cultural and Tourism Innovation in the Digital Era: Sixth International IACuDiT Conference, Athens 2019* (pp. 639-648). Springer International Publishing. DOI: 10.1007/978-3-030-36342-0_49

Chiao, H. M., Chen, Y. L., & Huang, W. H. (2018). Examining the usability of an online virtual tour-guiding platform for cultural tourism education. *Journal of Hospitality, Leisure, Sport and Tourism Education*, *23*, 29–38. DOI: 10.1016/j.jhlste.2018.05.002

Daniel, A. D., Costa, R. A., Pita, M., & Costa, C. (2017). Tourism Education: What about entrepreneurial skills? *Journal of Hospitality and Tourism Management*, *30*, 65–72. DOI: 10.1016/j.jhtm.2017.01.002

Deale, C. S. (2015). Hospitality and tourism education in the international classroom: A case study. *Journal of Teaching in Travel & Tourism*, *15*(4), 301–324. DOI: 10.1080/15313220.2015.1096474

Dwivedi, Y. K., Kshetri, N., Hughes, L., Slade, E. L., Jeyaraj, A., Kar, A. K., Baabdullah, A. M., Koohang, A., Raghavan, V., Ahuja, M., Albanna, H., Albashrawi, M. A., Al-Busaidi, A. S., Balakrishnan, J., Barlette, Y., Basu, S., Bose, I., Brooks, L., Buhalis, D., & Wright, R. (2023). Opinion Paper:"So what if ChatGPT wrote it?" Multidisciplinary perspectives on opportunities, challenges and implications of generative conversational AI for research, practice and policy. *International Journal of Information Management*, *71*, 102642. DOI: 10.1016/j.ijinfomgt.2023.102642

Escamilla-Fajardo, P., Alguacil, M., & López-Carril, S. (2021). Incorporating TikTok in higher education: Pedagogical perspectives from a corporal expression sport sciences course. *Journal of Hospitality, Leisure, Sport and Tourism Education*, *28*, 100302. DOI: 10.1016/j.jhlste.2021.100302

Frawley, T., Goh, E., & Law, R. (2019). Quality assurance at hotel management tertiary institutions in Australia: An insight into factors behind domestic and international student satisfaction. Journal of Hospitality and Tourism Education, 31(1), 1–9. .1480961DOI: 10.1080/10963758.2018

Goopio, J., & Cheung, C. (2021). The MOOC dropout phenomenon and retention strategies. *Journal of Teaching in Travel & Tourism, 21*(2), 177–197. DOI: 10.1080/15313220.2020.1809050

Guden, N., & Safaeimanesh, F. (2024). How has blended learning impacted tourism and hospitality graduates in employability? Feedback from graduates and tourism industry employers. *Worldwide Hospitality and Tourism Themes, 16*(1), 72–81. DOI: 10.1108/WHATT-01-2024-0007

Hsu, C. H. (2018). Tourism education on and beyond the horizon. *Tourism Management Perspectives, 25*, 181–183. DOI: 10.1016/j.tmp.2017.11.022

Ivanov, S., & Soliman, M. (2023). Game of algorithms: ChatGPT implications for the future of tourism education and research. *Journal of Tourism Futures, 9*(2), 214–221. DOI: 10.1108/JTF-02-2023-0038

Kahn, M. (2018). Co-authorship as a proxy for collaboration: A cautionary tale. *Science & Public Policy, 45*(1), 117–123. DOI: 10.1093/scipol/scx052

Kim, H. J., & Jeong, M. (2018). Research on hospitality and tourism education: Now and future. *Tourism Management Perspectives, 25*, 119–122. DOI: 10.1016/j.tmp.2017.11.025

Kou, I. T., & Liu, T. (2020). Could the adoption of Quick Response (QR) code in lectures enhance University students' satisfaction? A case study of hospitality and tourism programs in Macau. In *Cultural and Tourism Innovation in the Digital Era: Sixth International IACuDiT Conference, Athens 2019* (pp. 161-170). Springer International Publishing.

Lei, S. I., & So, A. S. I. (2021). Online teaching and learning experiences during the COVID-19 pandemic–A comparison of teacher and student perceptions. *Journal of Hospitality & Tourism Education, 33*(3), 148–162. DOI: 10.1080/10963758.2021.1907196

Li, Y. (2015). Blended learning in English for tourism: A case study. *Exploring Learning & Teaching in Higher Education*, 331-345.

Long, T., Cummins, J., & Waugh, M. (2019). Investigating the factors that influence higher education instructors' decisions to adopt a flipped classroom instructional model. *British Journal of Educational Technology*, *50*(4), 2028–2039. DOI: 10.1111/bjet.12703

McKercher, B. (2018). What is the state of hospitality and tourism research–2018? *International Journal of Contemporary Hospitality Management*, *30*(3), 1234–1244. DOI: 10.1108/IJCHM-12-2017-0809

Meng, L., Wen, K. H., Brewin, R., & Wu, Q. (2020). Knowledge atlas on the relationship between urban street space and residents' health—A bibliometric analysis based on VOSviewer and CiteSpace. *Sustainability (Basel)*, *12*(6), 2384. DOI: 10.3390/su12062384

Mola, F., & Jusoh, J. (2011). Service quality in Penang hotels: A gap score analysis. *World Applied Sciences Journal*, *12*(1), 19–24.

Moswete, N., & Lacey, G. (2015). "Women cannot lead": Empowering women through cultural tourism in Botswana. *Journal of Sustainable Tourism*, *23*(4), 600–617. DOI: 10.1080/09669582.2014.986488

Nanda, W. D., Widianingsih, I., & Miftah, A. Z. (2023). The linkage of digital transformation and tourism development policies in Indonesia from 1879–2022: Trends and implications for the future. *Sustainability (Basel)*, *15*(13), 10201. DOI: 10.3390/su151310201

Naumov, N. (2019). The impact of robots, artificial intelligence, and service automation on service quality and service experience in hospitality. In *Robots, artificial intelligence, and service automation in travel, tourism and hospitality* (pp. 123-133). Emerald Publishing Limited. DOI: 10.1108/978-1-78756-687-320191007

Nestoroska, I., & Petrovska, I. (2014). Staff training in hospitality sector as benefit for improved service quality. In *Tourism and Hospitality Industry 2014–Congress Proceedings* (pp. 437-448).

O'Boyle, I. (2014). Mobilising social media in sport management education. *Journal of Hospitality, Leisure, Sport and Tourism Education*, *15*, 58–60. DOI: 10.1016/j.jhlste.2014.05.002

Quinn, K., & Buzzetto-Hollywood, N. (2019). Faculty and student perceptions of the importance of management skills in the hospitality industry.

Safri, S. N., Mohi, Z., & Hanafiah, M. H. M. (2020). Conceptualization of MOOC E-Learning Service Quality Dimensions in Faculty of Hotel and Tourism Management, UITM, Malaysia. In *3rd Asia Pacific International Conference of Management and Business Science (AICMBS 2019)* (pp. 268-271). Atlantis Press. DOI: 10.2991/aebmr.k.200410.041

Shih, R. C., Hsin, H. T., Huang, H. C., & Cheng, Y. M. (2015). The Development of a mobile App for a Hospitality and Tourism course at technological universities. In *2015 Third International Conference on Robot, Vision and Signal Processing (RVSP)* (pp. 252-255). IEEE. DOI: 10.1109/RVSP.2015.67

Shin, H., Perdue, R. R., & Kang, J. (2019). Front desk technology innovation in hotels: A managerial perspective. *Tourism Management, 74*, 310–318. DOI: 10.1016/j.tourman.2019.04.004

Skripak, S. J. (2016). *Fundamentals of business.* Virginia Tech.

Small, H. (1973). Co-citation in the scientific literature: A new measure of the relationship between two documents. *Journal of the American Society for Information Science, 24*(4), 265–269. DOI: 10.1002/asi.4630240406

Van Eck, N., & Waltman, L. (2010). Software survey: VOSviewer, a computer program for bibliometric mapping. *Scientometrics, 84*(2), 523-538.

Viglia, G., Pelloia, M., & Buhalis, D. (2018). Information technology in hospitality education. *Innovation in Hospitality Education: Anticipating the Educational Needs of a Changing Profession*, 87-100.

Yu, Y., Li, Y., Zhang, Z., Gu, Z., Zhong, H., Zha, Q., Yang, L., Zhu, C., & Chen, E. (2020). A bibliometric analysis using VOSviewer of publications on COVID-19. *Annals of Translational Medicine, 8*(13), 816. DOI: 10.21037/atm-20-4235 PMID: 32793661

Zgraggen, M. (2021). Blended learning model in a vocational educational training hospitality setting: From teachers' perspectives. *International Journal of Training Research, 19*(3), 202–228. DOI: 10.1080/14480220.2021.1933568

Zheng, X. (2023). Application and Prospects of Digital Technology in Teaching Tourism Landscape. *Contemporary Education and Teaching Research, 4*(09), 467–471. DOI: 10.61360/BoniCETR232014900910

Compilation of References

Abbas, S., Ojo, S., Al Hejaili, A., Sampedro, G. A., Almadhor, A., Zaidi, M. M., & Kryvinska, N. (2024). Artificial intelligence framework for heart disease classification from audio signals. *Scientific Reports*, *14*(1), 3123. DOI: 10.1038/s41598-024-53778-7 PMID: 38326488

Aberdeen Group. (2012). *State of Service Management: Forecast for 2012*. Aberdeen Group.

Abu-Bakar, H., Charnley, F., Hopkinson, P., & Morasae, E. K. (2024). Towards a typological framework for circular economy roadmaps: A comprehensive analysis of global adoption strategies. *Journal of Cleaner Production*, *434*, 140066. Advance online publication. DOI: 10.1016/j.jclepro.2023.140066

Accenture. (2020). Accenture Technology Vision 2020: From Tech-Clash to Trust, the Focus Must Be on People. *Newsroom*. newsroom.accenture.com/news/2020/accenture-technology-vision-2020-from-tech-clash-to-trust-the-focus-must-be-on-people

Adalian, J. (2018). Inside Netflix's TV-Swallowing, Market-Dominating Binge Factory. *Vulture*. www.vulture.com/2018/06/how-netflix-swallowed-tv-industry.html

Adejoh, A. M., & Hadiza, D. (2015). Impact of Effective Communication on Goal Achievement in Nigerian Polytechnics. *International Journal of Sustainability Management and Information Technologies*, *1*(2), 6–10.

Adeoye, O. O., Olubiyi, T. O., Ajiteru, W. O., & Adaranijo, L. O. (2023). Unveiling Capital Structure Determinants, Business Characteristics and High-Growth Performance of Small and Medium-Sized Enterprises: Evidence from Nigeria. *Journal of Interdisciplinary Socio-Economic and Community Study*, *3*(2), 86–98. DOI: 10.21776/jiscos.03.2.05

Adeyemi, O. S., & Olubiyi, T. O. (2023). The Impact of Brand Awareness on Customer Loyalty in Selected Food and Beverage Businesses in Lagos State Nigeria. *Jurnal Multidisiplin Madani*, *3*(3), 541–551. DOI: 10.55927/mudima.v3i3.3095

Adeyemi, O. S., & Olubiyi, T. O. (2024). An Analysis of Investment Trends, Innovation and Its Impact on Cabotage and Local Content Regime: Nigerian Experience from Businesses in Oil and Gas and Maritime Industries. *ABUAD Journal of Social and Management Sciences*, *5*(2), 285–301. DOI: 10.53982/ajsms.2024.0502.04-j

Adlakha, Y. K. (2023). Human 3D brain organoids: Steering the demolecularization of brain and neurological diseases. *Cell Death Discovery*, *9*(1), 221. Advance online publication. DOI: 10.1038/s41420-023-01523-w PMID: 37400464

Afzal, A., Rasoulinezhad, E., & Malik, Z. (2022). Green finance and sustainable development in Europe. *Ekonomska Istrazivanja*, *35*(1), 5150–5163. DOI: 10.1080/1331677X.2021.2024081

Agarwal, P. (2009). *Indian Higher Education: Envisioning the Future*. SAGE Publications. DOI: 10.4135/9788132104094

Aggrawal, S., & Thomas, P. J. (2024). Investigating the Industry Perceptions and Use of AI Tools in Project Management: Implications for Educating Future Engineers. In *2024 ASEE Annual Conference & Exposition*. DOI: 10.18260/1-2--47700

Ahammed, I., Kim, B., Song, S., An, J., & Chen, Z. (2024). Acoustic-based Multitask Construction Equipment and Activity Recognition Using Customized ResNet-18.

Ahuja, L., Thakur, A., Seth, A., & Seth, K. (2023). Integrating Cloud, Blockchain and AI Technologies—Challenges and Scope. In *International Conference on Entrepreneurship, Innovation, and Leadership* (pp. 377-386). Singapore: Springer Nature Singapore.

Akhtar, N., Ahmad, W., Siddiqi, U. I., & Akhtar, M. N. (2019). Predictors and outcomes of consumer deception in hotel reviews: The roles of reviewer type and attribution of service failure. *Journal of Hospitality and Tourism Management*, *39*, 65–75. DOI: 10.1016/j.jhtm.2019.03.004

Akram, M. W., Yang, S., Hafeez, M., Kaium, M. A., Zahan, I., & Salahodjaev, R. (2023). Eco-innovation and environmental entrepreneurship: Steps towards business growth. *Environmental Science and Pollution Research International*, *30*(23), 63427–63434. DOI: 10.1007/s11356-023-26680-4 PMID: 37022542

Al Wael, H., Abdallah, W., Ghura, H., & Buallay, A. (2023). Factors influencing artificial intelligence adoption in the accounting profession: The case of public sector in Kuwait. *Competitiveness Review*, *34*(1), 3–27. DOI: 10.1108/CR-09-2022-0137

Al-Adwan, A. S., Kokash, H., Adwan, A. A., Alhorani, A., & Yaseen, H. (2020). Building customer loyalty in online shopping: The role of online trust, online satisfaction and electronic word of mouth. *International Journal of Electronic Marketing and Retailing*, *11*(3), 278–306. DOI: 10.1504/IJEMR.2020.108132

Alam, M. R., Milon, M. R. K., Rahman, M. K., & Hassan, A. (2022). Technology Application in Tourism Event, Education and Training for Making a Nation's Image. In *Technology Application in Tourism Fairs, Festivals and Events in Asia* (pp. 149–163). Springer Singapore. DOI: 10.1007/978-981-16-8070-0_9

Albitar, K., Nasrallah, N., Hussainey, K., & Wang, Y. (2024). Eco-innovation and corporate waste management: The moderating role of ESG performance. *Review of Quantitative Finance and Accounting*, *63*(2), 781–805. DOI: 10.1007/s11156-024-01281-5

Almeida, M., Lima, E., & Lee, D. (2022). Leveraging AI for Women's Entrepreneurship in Developing Economies. *Development Studies Research*, *9*(1), 41–55.

AlRyalat, S. A. S., Malkawi, L. W., & Momani, S. M. (2019). Comparing Bibliometric Analysis Using PubMed, Scopus, and Web of Science Databases. *Journal of Visualized Experiments*, *152*(152). Advance online publication. DOI: 10.3791/58494 PMID: 31710021

Al-Shorbaji, N. (2022, February 9). Improving Healthcare Access through Digital Health: The Use of Information and Communication Technologies. IntechOpen. https://doi.org/DOI: 10.5772/intechopen.99607

Alsoud, M., Trawnih, A., Yaseen, H., Majali, T., Alsoud, A. R., & Jaber, O. A. (2024). How could entertainment content marketing affect intention to use the Metaverse? Empirical findings. *International Journal of Information Management Data Insights*, *4*(2), 100258. DOI: 10.1016/j.jjimei.2024.100258

Altamimi, I., Altamimi, A., Alhumimidi, A S., Altamimi, A., & Temsah, M. (2023, June 25). Artificial Intelligence (AI) Chatbots in Medicine: A Supplement, Not a Substitute. Cureus, Inc.. DOI: 10.7759/cureus.40922

Álvarez-Machancoses, Ó., DeAndrés-Galiana, E. J., Cernea, A., Viña, J F S. D. L., & Fernández-Martínez, J. L. (2020, March 1). On the Role of Artificial Intelligence in Genomics to Enhance Precision Medicine. *Pharmacogenomics and Personalized Medicine*, *13*, 105–119. DOI: 10.2147/PGPM.S205082 PMID: 32256101

Alzoubi, H. M., Vij, M., Vij, A., & Hanaysha, J. R. (2021). What leads guests to satisfaction and loyalty in UAE five-star hotels? AHP analysis to service quality dimensions. *Enlightening Tourism:A Pathmaking Journal 11*(1), 102-135.

Amacher, G. S., Koskela, E., & Ollikainen, M. (2004). Environmental quality competition and eco-labelling. *Journal of Environmental Economics and Management, 47*(2), 284–306. DOI: 10.1016/S0095-0696(03)00078-0

Analia, D., Syaukat, Y., Fauzi, A., & Rustiadi, E. (2020). The impact of social capital on the performance of small micro enterprises. *Jurnal Ekonomi Malaysia, 54*(1), 81–96.

Andersen, I. M. V. (2020). Tourism employment and education in a Danish context. *Tourism Employment in Nordic Countries: Trends, Practices, and Opportunities,* 37-56.

Anderson, A. (2024). *Making Personalization the Center of Your Customer Retention Strategy like Spotify*. Sharpen, sharpencx.com/spotify-customer-retention/

Andrews, M. G., & Nowakowski, T. J. (2019). Human brain development through the lens of cerebral organoid models. *Brain Research, 1725,* 146470. DOI: 10.1016/j. brainres.2019.146470 PMID: 31542572

Angmo, P., Mahajan, R., & da Silva Oliveira, A. B. (2024). *Do they look human? Review on virtual influencers*. Management Review Quarterly., DOI: 10.1007/ s11301-024-00438-9

Annamalah, S., Paraman, P., Ahmed, S., Dass, R., Sentosa, I., Pertheban, T. R., Shamsudin, F., Kadir, B., Aravindan, K. L., Raman, M., Hoo, W. C., & Singh, P. (2023). The role of open innovation and a normalizing mechanism of social capital in the tourism industry. *Journal of Open Innovation, 9*(2), 100056. DOI: 10.1016/j. joitmc.2023.100056

Anomah, S., Ayeboafo, B., Owusu, A., & Aduamoah, M. (2024). Adapting to AI: exploring the implications of AI integration in shaping the accounting and auditing profession for developing economies. *EDPACS,* 1-25.

Aria, M., & Cuccurullo, C. (2017). Bibliometrix: An R-tool for comprehensive science mapping analysis. *Journal of Informetrics, 11*(4), 959–975. DOI: 10.1016/j. joi.2017.08.007

Arokodare, M. A., & Olubiyi, T. O. (2023). Crises as Harbingers of Opportunities: An Empirical Insight into the Moderating Effect of Knowledge Management on the Absorptive Capacity-Strategic Agility Relationship. *Malaysian Management Journal, 27,* 83–108. DOI: 10.32890/mmj2023.27.4

Ashley, K. D. (2020). Accounting for legal values. In *Computational Legal Studies* (pp. 190–214). Edward Elgar Publishing. DOI: 10.4337/9781788977456.00014

Askin, S., Burkhalter, D., Calado, G., & Dakrouni, S E. (2023, February 28). Artificial Intelligence Applied to clinical trials: opportunities and challenges. Springer Science+Business Media, 13(2), 203-213. https://doi.org/DOI: 10.1007/s12553-023-00738-2

Aulet, B. (2024). *Disciplined Entrepreneurship: 24 Steps to a Successful Startup, Expanded & Updated.* John Wiley & Sons.

Awad, N. F., & Ragowsky, A. (2008a). Establishing Trust in Electronic Commerce Through Online Word of Mouth: An Examination Across Genders. *Journal of Management Information Systems, 24*(4), 101–121. DOI: 10.2753/MIS0742-1222240404

Aydin, M., Degirmenci, T., Erdem, A., Sogut, Y., & Demirtas, N. (2024). From public policy towards the green energy transition: Do economic freedom, economic globalization, environmental policy stringency, and material productivity matter? *Energy, 311*, 133404. Advance online publication. DOI: 10.1016/j.energy.2024.133404

Ayub, B., Thaheem, M. J., & Din, Z. (2016). Dynamic management of cost contingency: Impact of KPIs and risk perception. *Procedia Engineering, 145*, 82–87. DOI: 10.1016/j.proeng.2016.04.021

Bag, J., Pretorius, C., Gupta, Y., & Dwivedi, K. (2021). Role of institutional pressures and resources in the adoption of big data analyticspowered artificial intelligence, sustainable manufacturing practices and circular economy capabilities. Technol. Forecast. Soc. Chang, 16(3), 12-20. .techfore.2020.120420DOI: <ALIGNMENT. qj></ALIGNMENT>10.1016/j

Baiod, W., & Hussain, M. M. (2024). The impact and adoption of emerging technologies on accounting: perceptions of Canadian companies. *International Journal of Accounting & Information Management.*

Bakhtawar, B., Thaheem, M. J., & Arshad, H. (2019). Integrating sustainability into project risk management; an application in PPP projects. In *Proceedings* (Vol. 2019). Annual Conference-Canadian Society for Civil Engineering.

Bale, A. S., Ghorpade, N., Hashim, M. F., Vaishnav, J., & Almaspoor, Z. (2022). A Comprehensive Study on Metaverse and Its Impacts on Humans. *Advances in Human-Computer Interaction, 2022*, 1–11. DOI: 10.1155/2022/3247060

Ballav, S., Ranjan, A., Sur, S., & Basu, S. (2024). *Organoid Intelligence: Bridging Artificial Intelligence for Biological Computing and Neurological Insights.* IntechOpen., DOI: 10.5772/intechopen.114304

Ball, C., & Kittler, M. (2019). Removing environmental market failure through support mechanisms: Insights from green start-ups in the British, French and German energy sectors. *Small Business Economics*, *52*(4), 831–844. https://www.jstor.org/stable/48701962. DOI: 10.1007/s11187-017-9937-8

Bank of America Newsroom. (2024). Bofa's Erica Surpasses 2 Billion Interactions, Helping 42 Million Clients since Launch. Bank of America. newsroom.bankofamerica.com/content/newsroom/press-releases/2024/04/bofa-s-erica-surpasses-2-billion-interactions--helping-42-millio.html

Barron, A. E. (1998). Integrating computer-based multimedia training into microcomputer software training. *Computers in Human Behavior*, *14*(2), 193–206.

Barron, A. E., & Iyer, R. (2019). Microlearning: Challenges, opportunities, and best practices. In Graziano, K. (Ed.), *Proceedings of Society for Information Technology &*.

Batat, W. (2024). Phygital customer experience in the Metaverse: A study of consumer sensory perception of sight, touch, sound, scent, and taste. *Journal of Retailing and Consumer Services*, *78*, 103786. DOI: 10.1016/j.jretconser.2024.103786

Baycan, T., & Öner, Ö. (2023). The dark side of social capital: A contextual perspective. *The Annals of Regional Science*, *70*(3), 779–798. DOI: 10.1007/s00168-022-01112-2

Bayraktar, M. E., Hastak, M., Gokhale, S., & Safi, B. (2011). Decision tool for selecting the optimal techniques for cost and schedule reduction in capital projects. *Journal of Construction Engineering and Management*, *137*(9), 645–655. DOI: 10.1061/(ASCE)CO.1943-7862.0000345

Belias, D., Vasiliadis, L., & Mantas, C. (2020). The human resource training and development of employees working on luxurious hotels in Greece. In *Cultural and Tourism Innovation in the Digital Era: Sixth International IACuDiT Conference, Athens 2019* (pp. 639-648). Springer International Publishing. DOI: 10.1007/978-3-030-36342-0_49

Belloch, J. A., Coronado, R., Valls, O., del Amor, R., Leon, G., Naranjo, V., Dolz, M. F., Amor-Martin, A., & Piñero, G. (2024). Urban sound classification using neural networks on embedded FPGAs. *The Journal of Supercomputing*, *80*(9), 1–11. DOI: 10.1007/s11227-024-05947-8

Ben Saad, S., & Choura, F. (2023). Towards better interaction between salespeople and consumers: The role of virtual recommendation agent. *European Journal of Marketing*, *57*(3), 858–903. DOI: 10.1108/EJM-11-2021-0892

Benabdellah, A. C., Zekhnini, K., Cherrafi, A., Garza-Reyes, J. A., & Kumar, A. (2021). Design for the environment: An ontology-based knowledge management model for green product development. *Business Strategy and the Environment*, *30*(8), 4037–4053. DOI: 10.1002/bse.2855

Benz, M., & Chatterjee, D. (2020). Calculated risk? A cybersecurity evaluation tool for SMEs. *Business Horizons*, *63*(4), 531–540. DOI: 10.1016/j.bushor.2020.03.010

Berners-Lee, T. (2017). Three challenges for the web, according to its inventor. World Wide Web Foundation. Available: https://webfoundation.org/2017/03/web-turns-28-letter/

Berry, M., & Rondinelli, D. (1998). Proactive corporate environmental management: A new industrial revolution. *The Academy of Management Perspectives*, *12*(2), 38–50. Advance online publication. DOI: 10.5465/ame.1998.650515

Bersin, J. (2022), https://joshbersin.com/wp-content/uploads/2021/12/WT-21_12 - HR-Predictions-for-2022-Report.pdf

Bhadra, S. Mostafizur Rahaman and P. Noorunnisa Khanam "Electrical and Electronic Application of Polymer–Carbon Composites", In book: Carbon-Containing Polymer Composites, pp.397-455.DOI: 10.1007/978-981-13-2688-2_12

Bhandari, A., & Bhuyan, M. N. H. (2023). Social capital and capital allocation efficiency. *Journal of Business Finance & Accounting*, *50*(7-8), 1439–1466. DOI: 10.1111/jbfa.12662

Bhatia, M. S., & Jakhar, S. K. (2021). The effect of environmental regulations, top management commitment, and organizational learning on green product innovation: Evidence from automobile industry. *Business Strategy and the Environment*, *30*(8), 3907–3918. DOI: 10.1002/bse.2848

Bhat, S., & Chakraborty, U. (2018). Online reviews and its impact on brand equity. *International Journal of Internet Marketing and Advertising*, *12*(2), 159. DOI: 10.1504/IJIMA.2018.10011683

Bhattacharya, A., & Srivastava, M. (2020). A Framework of Online Customer Experience: An Indian Perspective. *Global Business Review*, *21*(3), 800–817. DOI: 10.1177/0972150918778932

Bhattacharya, A., Srivastava, M., & Verma, S. (2019). Customer Experience in Online Shopping: A Structural Modeling Approach. *Journal of Global Marketing*, *32*(1), 3–16. DOI: 10.1080/08911762.2018.1441938

Bhattacharya, C., & Sinha, M. (2022). The role of artificial intelligence in banking for leveraging customer experience. *Australas Accounting Business Finance Journal.*, *16*(5), 89–105. DOI: 10.14453/aabfj.v16i5.07

Bianchi, R., & Noci, G. (1998). "Greening" SMEs' Competitiveness. *Small Business Economics*, *11*(3), 269–281. https://www.jstor.org/stable/40228980. DOI: 10.1023/A:1007980420087

Bi, M., Yu, X., Jin, Z., & Xu, J. (2024). IG-Based Method for Voiceprint Universal Adversarial Perturbation Generation. *Applied Sciences (Basel, Switzerland)*, *14*(3), 1322. DOI: 10.3390/app14031322

Binns, R. (2018). Fairness in Machine Learning: Lessons from Political Philosophy. *Proceedings of the 2018 Conference on Fairness, Accountability, and Transparency*, 149-159.

Boffa, J. (2023). *AI Assisted Business Analytics: Techniques for Reshaping Competitiveness.* Springer Nature. DOI: 10.1007/978-3-031-40821-2

Bohr, A., & Memarzadeh, K. (2020, January 1). The rise of artificial intelligence in healthcare applications. Elsevier BV, 25-60. https://doi.org/DOI: 10.1016/B978-0-12-818438-7.00002-2

Borgman, H., & Raza, S. (2020). Leveraging microlearning for just-in-time performance support. *Journal of Workplace Learning*, *32*(5), 313–328.

Brackenbury, J., & LoDolce, M. (2023). Gartner Reveals Three Technologies That Will Transform Customer Service and Support by 2028. www.gartner.com/en/newsroom/press-releases/2023-08-30-gartner-reveals-three-technologies-that-will-transform-customer-service-and-support-by-2028

Brands, K., & Elam, D. (2015). Identifying quality enablers for online graduate accounting courses using an appreciative inquiry case study. *International Journal of Human Resources Development and Management*, *15*(2-4), 128–141. DOI: 10.1504/IJHRDM.2015.071164

BRASS Program Planning Committee. (2010). Clean, Green, and Not So Mean: Can Business Save the World? *Reference and User Services Quarterly*, *50*(2), 135–140. https://www.jstor.org/stable/20865382. DOI: 10.5860/rusq.50n2.135

Brock, J. K. U., & Von Wangenheim, F. (2019). Demystifying AI: What digital transformation leaders can teach you about realistic artificial intelligence. *California Management Review*, *61*(4), 110–134. DOI: 10.1177/1536504219865226

Brynjolfsson, E., & McAfee, A. (2014). *The Second Machine Age: Work, Progress, and Prosperity in a Time of Brilliant Technologies*. W.W. Norton & Company.

Buhalis, D., & Law, R. (2008). Progress in information technology and tourism management: 20 years on and 10 years after the Internet—The state of eTourism research. *Tourism Management*, *29*(4), 609–623. DOI: 10.1016/j.tourman.2008.01.005

Buhalis, D., Leung, D., & Lin, M. (2023). Metaverse as a disruptive technology revolutionising tourism management and marketing. *Tourism Management*, *97*, 104724. DOI: 10.1016/j.tourman.2023.104724

Bukka, S. R., Lalam, N., Bhatta, H., & Wright, R. (2024, June). Lab scale demonstration of pipeline third-party damage classification using convolutional neural networks. In *Signal Processing, Sensor/Information Fusion, and Target Recognition XXXIII* (Vol. 13057, pp. 323–330). SPIE. DOI: 10.1117/12.3014005

Buolamwini, J., & Gebru, T. (2018). Gender Shades: Intersectional Accuracy Disparities in Commercial Gender Classification. *Proceedings of Machine Learning Research*, *81*, 1–15.

Byun, K. J., & Ahn, S. J. (2023). A Systematic Review of Virtual Influencers: Similarities and Differences between Human and Virtual Influencers in Interactive Advertising. *Journal of Interactive Advertising*, *23*(4), 293–306. DOI: 10.1080/15252019.2023.2236102

Caboni, F., Basile, V., Kumar, H., & Agarwal, D. (2024). A holistic framework for consumer usage modes of augmented reality marketing in retailing. *Journal of Retailing and Consumer Services*, *80*, 103924. DOI: 10.1016/j.jretconser.2024.103924

Cainelli, G., D'Amato, A., & Mazzanti, M. (2020). Resource efficient eco-innovations for a circular economy: Evidence from EU. *Research Policy*, *49*(1), 103827. DOI: 10.1016/j.respol.2019.103827

Calp, M. H. (2020). The role of artificial intelligence within the scope of digital transformation in enterprises. In *Advanced MIS and digital transformation for increased creativity and innovation in business* (pp. 122–146). IGI Global. DOI: 10.4018/978-1-5225-9550-2.ch006

Camilleri, M. (2021). E-Commerce Websites, Consumer Order Fulfillment and After-Sales Service Satisfaction: The Customer Is Always Right, even after the Shopping Cart Check-Out! SSRN *Electronic Journal*. DOI: 10.2139/ssrn.3853156

Capgemini. (2020). The ethics of AI in customer experience. Capgemini Research Institute. Available: https://www.capgemini.com/research/ai-and-the-ethical-conundrum/

Cappellazzo, U., Falavigna, D., & Brutti, A. (2024). Efficient Fine-tuning of Audio Spectrogram Transformers via Soft Mixture of Adapters. *arXiv preprint arXiv:2402.00828*. DOI: 10.21437/Interspeech.2024-38

Cardoso, P. R., Costa, H. S., & Novais, L. A. (2010). Fashion consumer profiles in the Portuguese market: Involvement, innovativeness, self-expression and impulsiveness as segmentation criteria. *International Journal of Consumer Studies*, *34*(6), 638–647. DOI: 10.1111/j.1470-6431.2010.00891.x

Carvajal Zaera, E. (2024). The influence of artificial intelligence on communication in healthcare. *European Public and Social Innovation Review*, *9*. Advance online publication. DOI: 10.31637/epsir-2024-312

Casaló, L. V., Flavián, C., & Guinalíu, M. (2008). The role of satisfaction and website usability in developing customer loyalty and positive word-of-mouth in the e-banking services. *International Journal of Bank Marketing*, *26*(6), 399–417. DOI: 10.1108/02652320810902433

Castro-Ospina, A. E., Solarte-Sanchez, M. A., Vega-Escobar, L. S., Isaza, C., & Martínez-Vargas, J. D. (2024). Graph-Based Audio Classification Using Pre-Trained Models and Graph Neural Networks. *Sensors (Basel)*, *24*(7), 2106. DOI: 10.3390/s24072106 PMID: 38610318

Chaidarun, N., Tepsuporn, S., Hayes, R., Beling, P., Scherer, W., & Grazioli, S. (2014, December). Computational Intelligence in Financial Engineering Trading Competition: A system for project-based learning. [IEEE.]. *Proceedings of the ... Winter Simulation Conference. Winter Simulation Conference*, *2014*, 3552–3560. DOI: 10.1109/WSC.2014.7020185

Chakraborty, D., Polisetty, A., & Rana, N. P. (2024). Consumers' continuance intention towards Metaverse-based virtual stores: A multi-study perspective. *Technological Forecasting and Social Change*, *203*, 123405. DOI: 10.1016/j.techfore.2024.123405

Chand, N., Dwivedi, U. K., & Sharma, M. K. (2007). Navin Chand*, U.K. Dwivedi, M.K. Sharma "Development and tribological behaviour of UHMWPE filled epoxy gradient composites" [Science Direct.]. *Wear*, *262*(1-2), 184–190. DOI: 10.1016/j.wear.2006.04.012

Chatterjee, S., Chaudhuri, R., & Chatterjee, S. (2023). Entrepreneurial behavior of family firms in the Indian community: Adoption of a technology platform as a moderator. *Journal of Enterprising Communities*, *17*(2), 433–453. DOI: 10.1108/JEC-08-2021-0122

Chatterjee, S., Chaudhuri, R., Vrontis, D., & Basile, G. (2022). Digital transformation and entrepreneurship process in SMEs of India: A moderating role of adoption of AI-CRM capability and strategic planning. *Journal of Strategy and Management*, *15*(3), 416–433. DOI: 10.1108/JSMA-02-2021-0049

Chaturvedi, V., & Raja Mohammed, K. B. N. (2023). Dynamic Inventory Management Using AI: A Case on Datarobot. In *International Conference on Artificial Intelligence and Knowledge Processing* (pp. 3-14). Cham: Springer Nature Switzerland.

Chaudhury, S., & Thakur, A. (2020). Policy Frameworks for Women-Centric AI: Bridging the Gender Digital Divide. *Digital Policy. Regulation & Governance*, *22*(3), 213–231.

Chaurasia, A., Dwivedi, U. K., Kumari, N., Meena, S., Rathore, D., Hashmi, S. A. R., & Jain, D. (2023). Effect of Graded Dispersion of SiC Particles on Dielectric Behavior of SiC/Epoxy Composite. *Silicon*, *15*(2), 913–923. DOI: 10.1007/s12633-022-02057-z

Chen, D., Esperança, J. P., & Wang, S. (2022). The Impact of Artificial intelligence on Firm Performance: An application of the Resource-Based View to E-Commerce Firms. *Frontiers in Psychology*, *13*, 884830. Advance online publication. DOI: 10.3389/fpsyg.2022.884830 PMID: 35465474

Cheng, M. Y., & Roy, A. F. (2011). Evolutionary fuzzy decision model for cash flow prediction using time-dependent support vector machines. *International Journal of Project Management*, *29*(1), 56–65. DOI: 10.1016/j.ijproman.2010.01.004

Chen, H. L., & Wang, H. C. (2020). Evaluating the Effectiveness of Microlearning in Enhancing Employees' Technical Competencies: An Empirical Study. *Journal of Information & Knowledge Management*, *19*(02), 2050012.

Chen, H., & Hsieh, C. (2021). Understanding the factors influencing users' acceptance of microlearning. *Computers in Human Behavior*, *114*, 106571.

Chen, J. V., Ha, Q.-A., & Vu, M. T. (2023). The Influences of Virtual Reality Shopping Characteristics on Consumers' Impulse Buying Behavior. *International Journal of Human-Computer Interaction*, *39*(17), 3473–3491. DOI: 10.1080/10447318.2022.2098566

Chen, J., Jun, S. W., Hong, S., He, W., & Moon, J. (2024). Eciton: Very low-power recurrent neural network accelerator for real-time inference at the edge. *ACM Transactions on Reconfigurable Technology and Systems*, *17*(1), 1–25. DOI: 10.1145/3629979

Chen, K., Luo, S., & Kin tong, D. Y. (2024). Cross border e-commerce development and enterprise digital technology innovation—Empirical evidence from listed companies in China. *Heliyon*, *10*(15), e34744. Advance online publication. DOI: 10.1016/j.heliyon.2024.e34744 PMID: 39144960

Chen, L., Zhou, X., & Chen, H. (2024, March). Audio Scanning Network: Bridging Time and Frequency Domains for Audio Classification. *Proceedings of the AAAI Conference on Artificial Intelligence*, *38*(10), 11355–11363. DOI: 10.1609/aaai. v38i10.29015

Chen, S., Meng, N., Li, H., & Fang, W.CHEN. (2024). Efficient Deep Neural Network Compression for Environmental Sound Classification on Microcontroller Units. *Turkish Journal of Electrical Engineering and Computer Sciences*, *32*(4), 501–515. DOI: 10.55730/1300-0632.4084

Chen, Y., & Xie, J. (2008). Online Consumer Review: Word-of-Mouth as a New Element of Marketing Communication Mix. *Management Science*, *54*(3), 477–491. DOI: 10.1287/mnsc.1070.0810

Chen, Z. F., & Lee, J. Y. (2024). Relationship cultivation and social capital: Female transnational entrepreneurs' relationship-based communication on social media. In *Start-up and Entrepreneurial Communication* (pp. 59–82). Routledge. DOI: 10.4324/9781003481171-5

Cheung, C. M. K., Lee, M. K. O., & Thadani, D. R. (2009). *The Impact of Positive Electronic Word-of-Mouth on Consumer Online Purchasing Decision.*, DOI: 10.1007/978-3-642-04754-1_51

Chiang, A. L., Rabinowitz, L., Kumar, A., & Chan, W. W. (2019, September 6). Association Between Institutional Social Media Involvement and Gastroenterology Divisional Rankings: Cohort Study. *Journal of Medical Internet Research*, *21*(9), e13345–e13345. DOI: 10.2196/13345 PMID: 31493321

Chiao, H. M., Chen, Y. L., & Huang, W. H. (2018). Examining the usability of an online virtual tour-guiding platform for cultural tourism education. *Journal of Hospitality, Leisure, Sport and Tourism Education*, *23*, 29–38. DOI: 10.1016/j. jhlste.2018.05.002

Chioatto, E., Zecca, E., & D'Amato, A. (2020). Which innovations for Circular Business Models?: A Product Life-Cycle Approach. *Fondazione Eni Enrico Mattei (FEEM)*. https://www.jstor.org/stable/resrep27688

Choi, Y., & Lee, S. (2019). The effects of microlearning on motivation and learning satisfaction in a mobile learning environment. *Sustainability*, *11*(21), 6003.

Chun, C., Park, H. J., & Seo, M. B. (2024). Static Sound Event Localization and Detection Using Bipartite Matching Loss for Emergency Monitoring. *Applied Sciences (Basel, Switzerland)*, *14*(4), 1539. DOI: 10.3390/app14041539

Clarivate. (2024). *Journal Citation Report*. https://clarivate.com/academia-government/scientific-and-academic-research/research-funding-analytics/journal-citation-reports

Cobo, S., & Angel, I. A. D.-R. (2018). From linear to circular integrated waste management systems: A review of methodological approaches. *Resources, Conservation and Recycling*, *135*, 279–295. DOI: 10.1016/j.resconrec.2017.08.003

Collins, E. M., & Kearins, K. (2010). Delivering on Sustainability's Global and Local Orientation. *Academy of Management Learning & Education*, *9*(3), 499–506. https://www.jstor.org/stable/25782033

Corbitt, B. J., Thanasankit, T., & Yi, H. (2003a). Trust and e-commerce: A study of consumer perceptions. *Electronic Commerce Research and Applications*, *2*(3), 203–215. DOI: 10.1016/S1567-4223(03)00024-3

Crawford, J., & Nilsson, F. (2023). Integrating ESG risks into control and reporting: Evidence from practice in Sweden. In *Handbook of Big Data and Analytics in Accounting and Auditing* (pp. 255–277). Springer Nature Singapore. DOI: 10.1007/978-981-19-4460-4_12

Crigger, E., Reinbold, K., Hanson, C., Kao, A., Blake, K., & Irons, M. (2022, January 12). Trustworthy Augmented Intelligence in Health Care. Springer Science+Business Media, 46(2). https://doi.org/DOI: 10.1007/s10916-021-01790-z

Cristofoli, C., & Clemmensen, T. (2023). Underlying Factors of Technology Acceptance and User Experience of Machine Learning Functions in Accounting Software: A Qualitative Content Analysis. In *International Conference on Human-Computer Interaction* (pp. 413-433). Cham: Springer Nature Switzerland. DOI: 10.1007/978-3-031-48060-7_31

Crutchfield, T. M., & Kistler, C. E. (2017, January 1). Getting patients in the door: Medical appointment reminder preferences. *Patient Preference and Adherence*, *11*, 141–150. DOI: 10.2147/PPA.S117396 PMID: 28182131

Damiano, S., Cramer, B., Guntoro, A., & van Waterschoot, T. (2024). Synthetic data generation techniques for training deep acoustic siren identification networks. *Frontiers in Signal Processing*, *4*, 1358532. DOI: 10.3389/frsip.2024.1358532

Dam, S. M., & Dam, T. C. (2021). Relationships between service quality, brand image, customer satisfaction, and customer loyalty. *The Journal of Asian Finance, Economics and Business, 8*(3), 585–593.

Daniel, A. D., Costa, R. A., Pita, M., & Costa, C. (2017). Tourism Education: What about entrepreneurial skills? *Journal of Hospitality and Tourism Management, 30*, 65–72. DOI: 10.1016/j.jhtm.2017.01.002

Daniels, J. (2023). Council Post: Why Organizations Should Invest in Privacy and CCPA Compliance. Forbes. www.forbes.com/sites/forbesbusinesscouncil/2023/06/22/why-organizations-should-invest-in-privacy-and-ccpa-compliance

Dantas, R. M., Ilyas, A., Martins, J. M., & Rita, J. X. (2022). Circular Entrepreneurship in Emerging Markets through the Lens of Sustainability. *Journal of Open Innovation, 8*(4), 211. Advance online publication. DOI: 10.3390/joitmc8040211

Dar, I. A., & Mishra, M. (2020). Dimensional impact of social capital on financial performance of SMEs. *The Journal of Entrepreneurship, 29*(1), 38–52. DOI: 10.1177/0971355719893499

Darmody, A., & Zwick, D. (2020). Manipulate to empower: Hyper-relevance and the contradictions of marketing in the age of surveillance capitalism. *Big Data & Society, 7*(1), 205395172090411. DOI: 10.1177/2053951720904112

Dastin, J. (2018). Amazon scraps secret AI recruiting tool that showed bias against women. Reuters. Available: https://www.reuters.com/article/us-amazon-com-jobs-automation-insight-idUSKCN1MK08G

Dastin, J. (2018). *Amazon Scraps Secret AI Recruiting Tool That Showed Bias Against Women*. Reuters.

Davenport, T H., & Glaser, J. (2022, October 31). Factors governing the adoption of artificial intelligence in healthcare providers. Springer Science+Business Media, 1(1). https://doi.org/DOI: 10.1007/s44250-022-00004-8

Davenport, T. H., & Ronanki, R. (2018). Artificial intelligence for the real world. *Harvard Business Review, 96*(1), 108–116.

Davis, F. D. (1989). Perceived Usefulness, Perceived Ease of Use, and User Acceptance. *Management Information Systems Quarterly, 13*(3), 319–340. DOI: 10.2307/249008

Davis, F. D., Bagozzi, R. P., & Warshaw, P. R. (1989). User acceptance of computer technology: A comparison of two theoretical models. *Management Science, 35*(8), 982–1003. DOI: 10.1287/mnsc.35.8.982

Davis, L., & Aslam, U. (2024). Analyzing consumer expectations and experiences of Augmented Reality (AR) apps in the fashion retail sector. *Journal of Retailing and Consumer Services*, *76*, 103577. Advance online publication. DOI: 10.1016/j.jretconser.2023.103577

De Jongh, D., Massey, E. K., Berishvili, E., Fonseca, L. M., Lebreton, F., Bellofatto, K., Bignard, J., Seissler, J., Buerck, L. W., Honarpisheh, M., Zhang, Y., Lei, Y., Pehl, M., Follenzi, A., Olgasi, C., Cucci, A., Borsotti, C., Assanelli, S., & Bunnik, E. M. (2022). Organoids: A systematic review of ethical issues. *Stem Cell Research & Therapy*, *13*(1), 337. Advance online publication. DOI: 10.1186/s13287-022-02950-9 PMID: 35870991

de Jong, M., Albers, C. J., & Koster, R. (2015). Effects of microlearning, game-based learning, and gamification on learners' motivation and learning performance. *Journal of Educational Psychology*, *107*(4), 1083–1099.

De Marchi, V., Cainelli, G., & Grandinetti, R. (2022). Multinational subsidiaries and green innovation. *International Business Review*, *31*(6), 102027. Advance online publication. DOI: 10.1016/j.ibusrev.2022.102027

Deale, C. S. (2015). Hospitality and tourism education in the international classroom: A case study. *Journal of Teaching in Travel & Tourism*, *15*(4), 301–324. DOI: 10.1080/15313220.2015.1096474

Dellarocas, C. N. (2003). The Digitization of Word-of-Mouth: Promise and Challenges of Online Feedback Mechanisms. SSRN *Electronic Journal*. DOI: 10.2139/ssrn.393042

Demirel, P., Li, Q. C., Rentocchini, F., & Tamvada, J. P. (2019). Born to be green: New insights into the economics and management of green entrepreneurship. *Small Business Economics*, *52*(4), 759–771. https://www.jstor.org/stable/48701958. DOI: 10.1007/s11187-017-9933-z

Devaram, S. (2020, January 1). Empathic Chatbot: Emotional Intelligence for Empathic Chatbot: Emotional Intelligence for Mental Health Well-being. Cornell University. https://doi.org//arxiv.2012.09130DOI: 10.48550

Dhopte, A., & Bagde, H. (2023, June 30). Smart Smile: Revolutionizing Dentistry With Artificial Intelligence. Cureus, Inc.. https://doi.org/DOI: 10.7759/cureus.41227

Dickinger, A. (2011). The Trustworthiness of Online Channels for Experience- and Goal-Directed Search Tasks. *Journal of Travel Research*, *50*(4), 378–391. DOI: 10.1177/0047287510371694

Diep, Q. B., Phan, H. Y., & Truong, T. C. (2024). Crossmixed convolutional neural network for digital speech recognition. *PLoS One*, *19*(4), e0302394. DOI: 10.1371/journal.pone.0302394 PMID: 38669233

Dikmen, I., Birgonul, M. T., & Arikan, A. E. (2006). Application of An Integrated Risk Management System (IRMS) to An International Construction Project. *Management*, *153*, 163.

Dinkel, H., Yan, Z., Wang, Y., Zhang, J., Wang, Y., & Wang, B. (2024). Scaling up masked audio encoder learning for general audio classification. *arXiv preprint arXiv:2406.06992*. DOI: 10.21437/Interspeech.2024-246

Doborjeh, Z., Hemmington, N., Doborjeh, M., & Kasabov, N. (2022). Artificial intelligence: A systematic review of methods and applications in hospitality and tourism. *International Journal of Contemporary Hospitality Management*, *34*(3), 1154–1176. DOI: 10.1108/IJCHM-06-2021-0767

Domingos, P. (2015). *The Master Algorithm: How the Quest for the Ultimate Learning Machine Will Remake Our World*. Basic Books.

Donthu, N., S. Kumar, D Mukherjee, N. Pandey, and N., & Lim, W. M. n.d.

Donthu, N., Kumar, S., Mukherjee, D., Pandey, N., & Lim, W. M. (2021). How to conduct a bibliometric analysis: An overview and guidelines. *Journal of Business Research*, *133*, 285–296. DOI: 10.1016/j.jbusres.2021.04.070

Dos Santos, R. A., Méxas, M. P., Meirino, M. J., Sampaio, M. C., & Costa, H. G. (2020). Criteria for assessing a sustainable hotel business. *Journal of Cleaner Production*, *262*, 121347. DOI: 10.1016/j.jclepro.2020.121347

Duan, L., Yang, L., & Guo, Y. (2024). SIAlex: Species identification and monitoring based on bird sound features. *Ecological Informatics*, *81*, 102637. DOI: 10.1016/j.ecoinf.2024.102637

Dwivedi, A., Sassanelli, C., Agrawal, D., Gonzalez, E. S., & D'Adamo, I. (2023). Technological innovation toward sustainability in manufacturing organizations: A circular economy perspective. *Sustainable Chemistry and Pharmacy*, *35*, 101211. Advance online publication. DOI: 10.1016/j.scp.2023.101211

Dwivedi, Y. K., Hughes, L., Baabdullah, A. M., Ribeiro-Navarrete, S., Giannakis, M., Al-Debei, M. M., Dennehy, D., Metri, B., Buhalis, D., Cheung, C. M. K., Conboy, K., Doyle, R., Dubey, R., Dutot, V., Felix, R., Goyal, D. P., Gustafsson, A., Hinsch, C., Jebabli, I., & Wamba, S. F. (2022). Metaverse beyond the hype: Multidisciplinary perspectives on emerging challenges, opportunities, and agenda for research, practice and policy. *International Journal of Information Management*, *66*, 102542. DOI: 10.1016/j.ijinfomgt.2022.102542

Dwivedi, Y. K., Ismagilova, E., Hughes, D. L., Carlson, J., Filieri, R., Jacobson, J., Jain, V., Karjaluoto, H., Kefi, H., Krishen, A. S., Kumar, V., Rahman, M. M., Raman, R., Rauschnabel, P. A., Rowley, J., Salo, J., Tran, G. A., & Wang, Y. (2021). Setting the future of digital and social media marketing research: Perspectives and research propositions. *International Journal of Information Management*, *59*, 102168. DOI: 10.1016/j.ijinfomgt.2020.102168

Dwivedi, Y. K., Kshetri, N., Hughes, L., Slade, E. L., Jeyaraj, A., Kar, A. K., Baabdullah, A. M., Koohang, A., Raghavan, V., Ahuja, M., Albanna, H., Albashrawi, M. A., Al-Busaidi, A. S., Balakrishnan, J., Barlette, Y., Basu, S., Bose, I., Brooks, L., Buhalis, D., & Wright, R. (2023). Opinion Paper:"So what if ChatGPT wrote it?" Multidisciplinary perspectives on opportunities, challenges and implications of generative conversational AI for research, practice and policy. *International Journal of Information Management*, *71*, 102642. DOI: 10.1016/j.ijinfomgt.2023.102642

Edelman, D., & Singer, M. (2015, November). Competing on Customer Journeys. *Harvard Business Review*.

Eggenschwiler, M., Linzmajer, M., Roggeveen, A. L., & Rudolph, T. (2024). Retailing in the Metaverse: A framework of managerial considerations for success. *Journal of Retailing and Consumer Services*, *79*, 103791. DOI: 10.1016/j.jretconser.2024.103791

Eichmüller, O. L., & Knoblich, J. A. (2022). Human cerebral organoids — A new tool for clinical neurology research. *Nature Reviews. Neurology*, *18*(11), 661–680. DOI: 10.1038/s41582-022-00723-9 PMID: 36253568

el Kaliouby, R. (2017, October). We Need Computers with Empathy. *MIT Technology Review*.

el Kaliouby, R. (2020). *Girl Decoded: A Scientist's Quest to Reclaim Our Humanity by Bringing Emotional Intelligence to Technology*. Currency.

Elias, R. (2024). The influence of family social capital toward the entrepreneurial intention among prospective graduates in Tanzanian universities. *Journal of Applied Research in Higher Education*.

Elliott, C. (2016), https://www.hrotoday.com/news/engaged- workforce/learning/bite-sized-learning-2/

Escamilla-Fajardo, P., Alguacil, M., & López-Carril, S. (2021). Incorporating Tik-Tok in higher education: Pedagogical perspectives from a corporal expression sport sciences course. *Journal of Hospitality, Leisure, Sport and Tourism Education*, *28*, 100302. DOI: 10.1016/j.jhlste.2021.100302

Espasandín-Bustelo, F., Ganaza-Vargas, J., & Diaz-Carrion, R. (2021). Employee happiness and corporate social responsibility: The role of organizational culture. *Employee Relations*, *43*(3), 609–629. DOI: 10.1108/ER-07-2020-0343

Esposito, M., Valente, G., Calaña, Y. P., Dumontier, M., Giordano, B. L., & Formisano, E. (2024). Bridging Auditory Perception and Natural Language Processing with Semantically informed Deep Neural Networks. bioRxiv, 2024-04.

Etori, N. A., Temesgen, E., & Gini, M. (2023, January 1). What We Know So Far: Artificial Intelligence in African Healthcare. Cornell University. https://doi.org/ DOI: 10.48550/arXiv.2305

European Commission. (2018). General Data Protection Regulation (GDPR). European Commission. Available: https://gdpr.eu/

European Commission. (2021). Europe fit for the Digital Age: Commission proposes new rules and actions for excellence and trust in Artificial Intelligence. European Commission Press Release. https://ec.europa.eu/commission/presscorner/detail/en/IP_21_1682

European Commission. (2021). Proposal for a Regulation laying down harmonised rules on artificial intelligence. European Commission. Available: https://digital -strategy.ec.europa.eu/en/library/proposal-regulation-laying-down-harmonised -rules-artificial-intelligence

Evan, W. M., & Freeman, R. E. (1988). A Stakeholder Theory of the Modern Corporation: Kantian Capi talism. In Beauchamp, T., & Bowie, N. (Eds.), *Ethical Theory and Business* (pp. 75–93). Prentice Hall.

Faisal, A., & Ekawanto, I. (2022). The role of Social Media Marketing in increasing Brand Awareness, Brand Image and Purchase Intention. *Indonesian Management and Accounting Research*, *20*(2), 185–208. DOI: 10.25105/imar.v20i2.12554

Felemban, H., Sohail, M., & Ruikar, K. (2024). Exploring the Readiness of Organisations to Adopt Artificial Intelligence. *Buildings (Basel, Switzerland)*, *14*(8), 2460. DOI: 10.3390/buildings14082460

Feng, X., Conrad, M., & Hussein, K. (2022). NHS big data intelligence on Blockchain applications. In *Big Data Intelligence for Smart Applications* (pp. 191–208). Springer International Publishing. DOI: 10.1007/978-3-030-87954-9_8

Fermat's Library | Eliza - A Computer Program For the Study of Natural Language Communication Between Man And Machine. (2023, April 14). https://fermatslibrary.com/s/a-computer-program-for-the-study-of-natural-language-communication-between-man-and-machine

Fernando, E., Sutomo, R., Prabowo, Y. D., Gatc, J., & Winanti, W. (2023). Exploring customer relationship management: Trends, challenges, and innovations. *Journal of Information Systems and Informatics*, *5*(3), 984–1001. DOI: 10.51519/journalisi.v5i3.541

Figueroa-Armijos, M., & Johnson, T. G. (2016). Entrepreneurship policy and economic growth: Solution or delusion? Evidence from a state initiative. *Small Business Economics*, *47*(4), 1033–1047. https://www.jstor.org/stable/26154684. DOI: 10.1007/s11187-016-9750-9

Filho, M. G., & Gonella, J. (2024, September). dos S. L., Latan, H., Ganga, G. M. D. (2024). Awareness as a catalyst for sustainable behaviors: A theoretical exploration of planned behavior and value-belief-norms in the circular economy. *Journal of Environmental Management*, *368*, 122181. Advance online publication. DOI: 10.1016/j.jenvman.2024.122181

Filieri, R., Alguezaui, S., & McLeay, F. (2015). Why do travelers trust TripAdvisor? Antecedents of trust towards consumer-generated media and its influence on recommendation adoption and word of mouth. *Tourism Management*, *51*, 174–185. DOI: 10.1016/j.tourman.2015.05.007

Fischer, J., Orescanin, M., & Eckstrand, E. (2024). VI-PANN: Harnessing Transfer Learning and Uncertainty-Aware Variational Inference for Improved Generalization in Audio Pattern Recognition. *IEEE Access*.

Flavián, C., Ibáñez-Sánchez, S., & Orús, C. (2019). The impact of virtual, augmented and mixed reality technologies on the customer experience. *Journal of Business Research*, *100*, 547–560. DOI: 10.1016/j.jbusres.2018.10.050

Floridi, L., Cowls, J., Beltrametti, M., Chatila, R., Chazerand, P., Dignum, V., Luetge, C., Madelin, R., Pagallo, U., Rossi, F., Schafer, B., Valcke, P., & Vayena, E. (2021). People—an ethical framework for a good AI society: Opportunities, risks, principles, and recommendations. *Minds and Machines*, *28*(4), 689–707. DOI: 10.1007/s11023-018-9482-5 PMID: 30930541

Forman, C., Ghose, A., & Wiesenfeld, B. (2008). Examining the Relationship Between Reviews and Sales: The Role of Reviewer Identity Disclosure in Electronic Markets. *Information Systems Research*, *19*(3), 291–313. DOI: 10.1287/isre.1080.0193

Forrester. (2018). Insights-Driven Businesses Set The Pace For Global Growth. Forrester.

Fotis, J., Buhalis, D., & Rossides, N. (2012). Social Media Use and Impact during the Holiday Travel Planning Process. In *Information and Communication Technologies in Tourism 2012* (pp. 13–24). Springer Vienna., DOI: 10.1007/978-3-7091-1142-0_2

Fountaine, T., McCarthy, B., & Saleh, T. (2019). Building the AI-powered organization. *Harvard Business Review*, *97*(4), 62–73.

Fournier, S., & Avery, J. (2011). The uninvited brand. *Business Horizons*, *54*(3), 193–207. DOI: 10.1016/j.bushor.2011.01.001

Frank, D.-A., Peschel, A. O., Otterbring, T., DiPalma, J., & Steinmann, S. (2024). Does Metaverse fidelity matter? Testing the impact of fidelity on consumer responses in virtual retail stores. *International Review of Retail, Distribution and Consumer Research*, *34*(2), 251–284. DOI: 10.1080/09593969.2024.2304810

Frawley, T., Goh, E., & Law, R. (2019). Quality assurance at hotel management tertiary institutions in Australia: An insight into factors behind domestic and international student satisfaction. Journal of Hospitality and Tourism Education, 31(1), 1–9. .1480961DOI: 10.1080/10963758.2018

Frechette, M. Trudeau, M. Alamdari, H.D. Boily S. "Introductory Remarks on Nano Dielectrics" IEEE Transactions on Dielectrics and Electrical Insulation (Volume: 11, Issue: 5, Oct. 2004)

Frese, M., & Gielnik, M. M. (2023). The psychology of entrepreneurship: Action and process. *Annual Review of Organizational Psychology and Organizational Behavior*, *10*(1), 137–164. DOI: 10.1146/annurev-orgpsych-120920-055646

Friedman, A. (n.d.). *The Power of Lotka's Law Through the Eyes of R Domain-analytical information and knowledge organisation as research infrastructure View project Visual Peer Review View project.* https://www.researchgate.net/publication/280156919

Fry, J. (2024). Revisiting student evaluation of teaching during the pandemic. *Applied Economics Letters*, *31*(14), 1259–1263. DOI: 10.1080/13504851.2023.2178623

Fry, J., & Brint, A. (2025). Customer satisfaction scores: New models to estimate the number of fake reviews. *Tourism Management, 106,* 105030. DOI: 10.1016/j.tourman.2024.105030

Fujiwara, K. (2009). Environmental policies in a differentiated oligopoly revisited. *Resource and Energy Economics, 31*(3), 239–247. DOI: 10.1016/j.reseneeco.2009.03.002

Gabrielli, S., Kimani, S., & Catarci, T. 2006.The Design of MicroLearning Experiences: A Research Agenda.https://www.researchgate.net/publication/253150976_The_Design_of_Micro Learning_Experiences_A_Research_Agenda

Gaines, D., & Balac, N. (2000). Using mobile robots to teach artificial intelligence research skills. In *Proceedings of the 2000 ASEE Annual Conference.*

Garella, P. G. (2021). The effects of taxes and subsidies on environmental qualities in a differentiated duopoly. *Letters in Spatial and Resource Sciences, 14*(2), 197–209. DOI: 10.1007/s12076-021-00272-7

Garg, P. (2016). CSR and corporate performance: Evidence from India. *Decision (Washington, D.C.), 43,* 333–349.

Garriga, E., & Melé, D. (2004). Corporate social responsibility theories: Mapping the territory. *Journal of Business Ethics, 53*(1), 51–71. DOI: 10.1023/B:BUSI.0000039399.90587.34

Gartner. (2019). Gartner Predicts 80% of Marketers Will Abandon Personalization Efforts by 2025. Gartner Press Release. Available: https://www.gartner.com/en/newsroom/press-releases/2019-12-02-gartner-predicts-80--of-marketers-will-abandon-person

Gartner. (2019). Gartner Says the Future of Self-Service Is Customer-Led Automation. www.gartner.com/en/newsroom/press-releases/2019-05-28-gartner-says-the-future-of-self-service-is-customer-l

Gartner. (2020). Gartner Top Strategic Technology Trends for 2021. Available: https://www.gartner.com/smarterwithgartner/gartner-top-strategic-technology-trends-for-2021/

Gartner. (2021). Gartner Predicts Proactive Customer Service to Reduce Costs by 40% for Companies. Gartner Press Release. Available: https://www.gartner.com/en/newsroom/press-releases/2021-02-25-gartner-predicts-proactive-customer-service-to-reduce-costs-by-40-percent-for-companies

Gasbarro, F., Annunziata, E., Rizzi, F., & Frey, M. (2017). The Interplay Between Sustainable Entrepreneurs and Public Authorities: Evidence from Sustainable Energy Transitions. *Organization & Environment, 30*(3), 226–252. https://www.jstor.org/stable/26408339. DOI: 10.1177/1086026616669211

Gates, B. (1999). *Business @ the Speed of Thought: Succeeding in the Digital Economy.* Grand Central Publishing.

Gatti, L., Vishwanath, B., Seele, P., & Cottier, B. (2019). Are we moving beyond voluntary CSR? Exploring theoretical and managerial implications of mandatory CSR resulting from the new Indian companies act. *Journal of Business Ethics, 160*(4), 961–972. DOI: 10.1007/s10551-018-3783-8

Gaubienė, N. (2024). Can Artificial Intelligence Engage in the Practice of Law as the Art of Good and Justice? *Filosofija, Sociologija, 35*(2 Special Issue), 54–63. DOI: 10.6001/fil-soc.2024.35.2Priedas.Special-Issue.6

Geetha, A. (2021). A Study on Artificial Intelligence (Ai) in Banking and Financial Services. *International Journal of Creative Research Journal, 9*(9), 110–114.

GEM (Global Entrepreneurship Monitor). (2023). Global Entrepreneurship Monitor 2023/2024 Global Report: 25 Years and Growing. *London: GEM.* https://www.gemconsortium.org/reports/latest-global-report

Genovese, A., Acquaye, A., Figueroa, A., & Lenny Koh, S. C. (2017). Sustainable supply chain management and the transition towards a circular economy: Evidence and some applications. *Omega, 66-B,* 344–357. DOI: 10.1016/j.omega.2015.05.015

Geraldi, J., Locatelli, G., Dei, G., Söderlund, J., & Clegg, S. (2024). AI for Management and Organization Research: Examples and Reflections from Project Studies. *Project Management Journal, 55*(4), 339–351. DOI: 10.1177/87569728241266938

Gharleghi, B., Shafighi, N., and Nawaser, K. (2024). Green finance and its role in sustainability in the EU. *Journal of Economy and Technology.* DOI: 10.1016/j.ject.2024.07.004

Ghasemaghaei, M., & Calic, G. (2020). Assessing the Impact of Big Data on Firm Innovation Performance: Big Data Is Not Always Better Data. *Journal of Business Research, 108,* 147–162. DOI: 10.1016/j.jbusres.2019.09.062

Giallonardo, L., & Mulino, M. (2024). Green Consumerism and Firms' Environmental Behaviour Under Monopolistic Competition: A Two-Sector Model. *Italian Economic Journal: Springer., 10*(1), 347–376. DOI: 10.1007/s40797-023-00223-9

Giandomenico, S. L., Mierau, S. B., Gibbons, G. M., Wenger, L. M., Masullo, L., Sit, T., Sutcliffe, M., Boulanger, J., Tripodi, M., Derivery, E., Paulsen, O., Lakatos, A., & Lancaster, M. A. (2019). Cerebral organoids at the air–liquid interface generate diverse nerve tracts with functional output. *Nature Neuroscience*, *22*(4), 669–679. DOI: 10.1038/s41593-019-0350-2 PMID: 30886407

Gibbs, D. (2006). Sustainability Entrepreneurs, Ecopreneurs and the Development of a Sustainable Economy. *Greener Management International, 55*, 63–78. https://www.jstor.org/stable/greemanainte.55.63

Gibson, M. C., & Arnott, D. R. (2005). The evaluation of business intelligence: A case study in a major financial institution. In *Australasian Conference on Information Systems 2005* (pp. 1-12). Australasian Chapter of the Association for Information Systems.

Gingras and Yves. 2016. *Bibliometrics and Research Evaluation: Uses and Abuses.* Cambridge: MA: MIT Press.

Gliedt, T., & Parker, P. (2007). Green community entrepreneurship: Creative destruction in the social economy. *International Journal of Social Economics*, *34*(8), 538–553. DOI: 10.1108/03068290710763053

Gocher, H., Taterh, S., & Dadheech, P. (2023). Impact Analysis to Detect and Mitigate Distributed Denial of Service Attacks with Ryu-SDN Controller: A Comparative Analysis of Four Different Machine Learning Classification Algorithms. *SN Computer Science*, *4*(5), 456. DOI: 10.1007/s42979-023-01842-w

Gocher, H., Taterh, S., & Dadheech, P. (2023). *Reinforcing Network Resilience From DDOS: A Review Of Advanced Distributed Denial Of Service (Ddos).* Attacks And Its Mitigation Techniques.

Godes, D., & Mayzlin, D. (2002). Using Online Conversations to Study Word of Mouth Communication. SSRN *Electronic Journal*. DOI: 10.2139/ssrn.327841

Gonçalves, M. J. A., da Silva, A. C. F., & Ferreira, C. G. (2022). The future of accounting: How will digital transformation impact the sector? [). MDPI.]. *Informatics (MDPI)*, *9*(1), 19. DOI: 10.3390/informatics9010019

Goodfellow, I., Bengio, Y., & Courville, A. (2016). *Deep Learning.* MIT Press.

Google. (2021). What-If Tool. Google AI. https://pair-code.github.io/what-if-tool/

Goopio, J., & Cheung, C. (2021). The MOOC dropout phenomenon and retention strategies. *Journal of Teaching in Travel & Tourism*, *21*(2), 177–197. DOI: 10.1080/15313220.2020.1809050

Gouda, S., Khan, A. G., & Hiremath, S. L. (2017). *Corporate social responsibility in India. Trends, issues and strategies.* Anchor Academic Publishing.

Gounaris, S., Dimitriadis, S., & Stathakopoulos, V. (2010). An examination of the effects of service quality and satisfaction on customers' behavioral intentions in e-shopping. *Journal of Services Marketing, 24*(2), 142–156. DOI: 10.1108/08876041011031118

Gourisaria, M. K., Agrawal, R., Sahni, M., & Singh, P. K. (2024). Comparative analysis of audio classification with MFCC and STFT features using machine learning techniques. *Discover Internet of Things, 4*(1), 1. DOI: 10.1007/s43926-023-00049-y

Grand View Research. (2020). Computer Vision Market Size. Grand View Research. Available: https://www.grandviewresearch.com/industry-analysis/computer-vision -market

Greely, H. T. H., & Kreitmair, K. V. (2021). Should Cerebral Organoids be Used for Research if they Have the Capacity for Consciousness?. *Cambridge quarterly of healthcare ethics: CQ: the international journal of healthcare ethics committees, 30*(4), 575–584. https://doi.org/DOI: 10.1017/S0963180121000050

Guden, N., & Safaeimanesh, F. (2024). How has blended learning impacted tourism and hospitality graduates in employability? Feedback from graduates and tourism industry employers. *Worldwide Hospitality and Tourism Themes, 16*(1), 72–81. DOI: 10.1108/WHATT-01-2024-0007

Gümüsay, A. A. (2020). The Potential for Plurality and Prevalence of the Religious Institutional Logic. *Business & Society, 59*(5), 855–880. DOI: 10.1177/0007650317745634

Gupta, R., & Kumar, V. Sound Classification in Indian Cities Using Multi-Label Data and Transfer Learning. In *The Second Tiny Papers Track at ICLR2024.*

Gupta, S. (2021). Impact of artificial intelligence on financial decision 2015: A qualitative study. *Journal of Cardiovascular Disease Research, 12*(6), 2130–2137.

Haenlein, M., & Kaplan, A. (2019). A brief history of artificial intelligence: On the past, present, and future of artificial intelligence. *California Management Review, 61*(4), 5–14. DOI: 10.1177/0008125619864925

Hajihashemi, V., Alavigharahbagh, A., Machado, J. J. M., & Tavares, J. M. R. (2024). Novel sound event and sound activity detection framework based on intrinsic mode functions and deep learning. *Multimedia Tools and Applications, •••*, 1–29. DOI: 10.1007/s11042-024-19557-2

Hamid, M., Zeshan, F., Ahmad, A., Munawar, S., Aimeur, E., Ahmed, S., Abu El-soud, M., & Yousif, M. (2020). An intelligent decision support system for effective handling of IT projects. *Journal of Intelligent & Fuzzy Systems*, *38*(3), 2635–2647. DOI: 10.3233/JIFS-179550

Hamsagayathri, P., & Vigneshwaran, S. (2021). Symptoms based disease prediction using machine learning techniques. 2021 Third international conference on intelligent communication technologies and virtual mobile networks (ICICV). IEEE. DOI: 10.1109/ICICV50876.2021.9388603

Hanai, A., Ishikawa, T., Sugao, S., Fujii, M., Hirai, K., Watanabe, H., ... Kawakami, E. (2024). Explainable Machine Learning Classification to Identify Vulnerable Groups Among Parenting Mothers: Web-Based Cross-Sectional Questionnaire Study. *JMIR Formative Research, 8*(1), e47372.

Haque, A., Chowdhury, M N., & Soliman, H. (2023, June 1). Transforming Chronic Disease Management with Chatbots: Key Use Cases for Personalized and Cost-effective Care. https://doi.org/DOI: 10.1109/IS3C57901.2023.00104

Hartung, T., Morales Pantoja, I. E., & Smirnova, L. (2024). Brain organoids and organoid intelligence from ethical, legal, and social points of view. *Frontiers in Artificial Intelligence*, *6*, 1307613. Advance online publication. DOI: 10.3389/frai.2023.1307613 PMID: 38249793

Hasan, H. E., Jaber, D., Khabour, O. F., & Alzoubi, K. H. (2024). Ethical considerations and concerns in the implementation of AI in pharmacy practice: A cross-sectional study. *BMC Medical Ethics*, *25*(1), 55. Advance online publication. DOI: 10.1186/s12910-024-01062-8 PMID: 38750441

Hazaea, S. A., Zhu, J., Khatib, S. F., Bazhair, A. H., & Elamer, A. A. (2022). Sustainability assurance practices: A systematic review and future research agenda. *Environmental Science and Pollution Research International*, *29*(4), 4843–4864. DOI: 10.1007/s11356-021-17359-9 PMID: 34787810

He, H., Chen, J., Chen, H., Zeng, B., Huang, Y., Zhaopeng, Y., & Chen, X. (2024). Enhancing Insect Sound Classification Using Dual-Tower Network: A Fusion of Temporal and Spectral Feature Perception. *Applied Sciences (Basel, Switzerland)*, *14*(7), 3116. DOI: 10.3390/app14073116

Heji, A. E., Alansari, O. E., & Al-Sartawi, A. (2023). Artificial intelligence and its impact on accounting systems. In *Artificial Intelligence, Internet of Things, and Society 5.0* (pp. 363–376). Springer Nature Switzerland. DOI: 10.1007/978-3-031-43300-9_30

Hendricks, S., & Mwapwele, S. D. (2024). A systematic literature review on the factors influencing e-commerce adoption in developing countries. *Data and Information Management*, *8*(1), 100045. DOI: 10.1016/j.dim.2023.100045

Henke, N., Bughin, J., Chui, M., Manyika, J., Saleh, T., Wiseman, B., & Sethupathy, G. (2016). The age of analytics: Competing in a data-driven world. McKinsey Global Institute. https://www.mckinsey.com/business-functions/mckinsey-analytics/our-insights/the-age-of-analytics-competing-in-a-data-driven-world

Hennig-Thurau, T., Gwinner, K. P., Walsh, G., & Gremler, D. D. (2004). Electronic word-of-mouth via consumer-opinion platforms: What motivates consumers to articulate themselves on the Internet? *Journal of Interactive Marketing*, *18*(1), 38–52. DOI: 10.1002/dir.10073

Henry & Garg. (2024). Predictive Analytics Examples & Benefits on CX. *InMoment*. inmoment.com/blog/predictive-analytics/.

Hentati, H., & Boulila, N. (2023). Digital maturity index for accounting firms. *Journal of Accounting & Organizational Change*, (ahead-of-print).

Hentzen, J. K., Hoffmann, A., Dolan, R., & Pala, E. (2022). Artificial intelligence in customer-facing financial services: A systematic literature review and agenda for future research. *International Journal of Bank Marketing*, *40*(6), 1299–1336. DOI: 10.1108/IJBM-09-2021-0417

Hermann, E. (2022). Artificial intelligence and mass personalization of communication content—An ethical and literacy perspective. *New Media & Society*, *24*(5), 1258–1277. DOI: 10.1177/14614448211022702

Hessels, J., van Gelderen, M., & Thurik, R. (2008). Entrepreneurial aspirations, motivations, and their drivers. *Small Business Economics*, *31*(3), 323–339. https://www.jstor.org/stable/40650947. DOI: 10.1007/s11187-008-9134-x

Hidalgo, G., Monticelli, J. M., & Vargas Bortolaso, I. (2024). Social capital as a driver of social entrepreneurship. *Journal of Social Entrepreneurship*, *15*(1), 182–205. DOI: 10.1080/19420676.2021.1951819

Hirono, Y., Sato, I., Kai, C., Yoshida, A., Kodama, N., Uchida, F., & Kasai, S. (2024). The Approach to Sensing the True Fetal Heart Rate for CTG Monitoring: An Evaluation of Effectiveness of Deep Learning with Doppler Ultrasound Signals. *Bioengineering (Basel, Switzerland)*, *11*(7), 658. DOI: 10.3390/bioengineering11070658 PMID: 39061740

Hokmabadi, H., Rezvani, S. M. H. S., & de Matos, C. A. (2024). Business Resilience for Small and Medium Enterprises and Startups by Digital Transformation and the Role of Marketing Capabilities—A Systematic Review. *Systems*, *12*(6), 220. Advance online publication. DOI: 10.3390/systems12060220

Hollebeek, L. D., Sharma, T. G., Pandey, R., Sanyal, P., & Clark, M. K. (2022). Fifteen years of customer engagement research: A bibliometric and network analysis. *Journal of Product and Brand Management*, *31*(2), 209–309. DOI: 10.1108/JPBM-01-2021-3301

Hoppe, M., Habib, A., Desai, R., Edwards, L., Kodavali, C., Sherry Psy, N. S., & Zinn, P. O. (2023). Human brain organoid code of conduct. *Frontiers in Molecular Medicine*, *3*, 1143298. Advance online publication. DOI: 10.3389/fmmed.2023.1143298 PMID: 39086687

Hoque, Z. (2017). Appreciative inquiry for accounting research. In *The Routledge Companion to Qualitative Accounting Research Methods* (pp. 129–144). Routledge. DOI: 10.4324/9781315674797

Hossain, M. S., Yahya, S. B., & Khan, M. J. (2020). The effect of corporate social responsibility (CSR) health-care services on patients' satisfaction and loyalty–a case of Bangladesh. *Social Responsibility Journal*, *16*(2), 145–158. DOI: 10.1108/SRJ-01-2018-0016

Hsiao, S.-H., Wang, Y.-Y., & Lin, T. L. J. (2024). The impact of low-immersion virtual reality on product sales: Insights from the real estate industry. *Decision Support Systems*, *178*, 114131. DOI: 10.1016/j.dss.2023.114131

Hsu, C. H. (2018). Tourism education on and beyond the horizon. *Tourism Management Perspectives*, *25*, 181–183. DOI: 10.1016/j.tmp.2017.11.022

Hsu, L.-C., Wang, K.-Y., & Chih, W.-H. (2013). Effects of web site characteristics on customer loyalty in B2B e-commerce: Evidence from Taiwan. *Service Industries Journal*, *33*(11), 1026–1050. DOI: 10.1080/02642069.2011.624595

Huang. (2020). Gartner Keynote: The Future of Customer Service. Gartner Customer Service & Support Leader Conference.

Huang, K., Teng, Y., Chen, Y., & Wang, Y. (2024). From Pixels to Principles: A Decade of Progress and Landscape in Trustworthy Computer Vision. *Science and Engineering Ethics*, *30*(3), 26. Advance online publication. DOI: 10.1007/s11948-024-00480-6 PMID: 38856788

Huang, S., & Crotts, J. (2019). Relationships between Hofstede's cultural dimensions and tourist satisfaction: A cross-country cross-sample examination. *Tourism Management*, *72*, 232–241. DOI: 10.1016/j.tourman.2018.12.001

Hug, T. (2005). Microlearning: A New Pedagogical Challenge (Introductory Note). In Microlearning: Emerging Concepts, Practices and Technologies after eLearning. Proceeding of Microlearning 2005, Learning & Working in New Media. Book Editors: Theo Hug, Martin Lindberg, Peter A. Bruck, Innsbruck university press, 7–1

Hu, N., Bose, I., Koh, N. S., & Liu, L. (2012). Manipulation of online reviews: An analysis of ratings, readability, and sentiments. *Decision Support Systems*, *52*(3), 674–684. DOI: 10.1016/j.dss.2011.11.002

Hunt, V., Prince, S., Dixon-Fyle, S., & Yee, L. (2021). *Diversity Wins: How Inclusion Matters*. McKinsey & Company.

Huynh, D., Hille, E. L., & Nasir, A. (2020). Diversification in the age of the 4th industrial revolution: The role of artificial intelligence, greenbonds and crypto currencies. *Technological Forecasting and Social Change*, *15*(9), 120–188. DOI: 10.1016/j.techfore.2020.120188

Hye, Q. M., Ul-Haq, J., Visas, H., & Rehan, R. (2023). The role of eco-innovation, renewable energy consumption, economic risks, globalization, and economic growth in achieving sustainable environment in emerging market economies. *Environmental Science and Pollution Research International*, *30*(40), 92469–92481. DOI: 10.1007/s11356-023-28945-4 PMID: 37491494

Hyken, S. (2023). AI Will Not Eliminate Jobs. *Forbes*. www.forbes.com/sites/shephyken/2023/08/27/ai-will-not-eliminate-jobs/

Hyun, I., Scharf-Deering, J. C., & Lunshof, J. E. (2020). Ethical issues related to brain organoid research. *Brain Research*, *1732*, 146653. DOI: 10.1016/j.brainres.2020.146653 PMID: 32017900

Iannantuono, G. M., Bracken-Clarke, D., Floudas, C. S., Roselli, M., Gulley, J. L., & Karzai, F. (2023, September 4). Applications of large language models in cancer care: Current evidence and future perspectives. *Frontiers of Medicine*, *13*, 1268915. Advance online publication. DOI: 10.3389/fonc.2023.1268915 PMID: 37731643

IBM. (2019). Everyday Ethics for Artificial Intelligence. IBM. https://www.ibm.com/watson/assets/duo/pdf/everydayethics.pdf

IBM. (2021). AI Explainability 360. IBM Research. https://github.com/Trusted-AI/AIX360

Ibrahim, A.-J., & Kahf, M. (2020). Instruments for investment protection when structuring Islamic venture capital. *Journal of Islamic Accounting and Business Research, 11*(9), 1907–1920. DOI: 10.1108/JIABR-01-2019-0025

Ifeanyichukwu, E. E. (2024). Technological Implementation in the Service Sector: A Case Study. In *Artificial Intelligence for Smart Technology in the Hospitality and Tourism Industry* (pp. 305-336). Apple Academic Press. DOI: 10.1201/9781003432951-18

Imai, T., Sawa, F., Yoshimitsu, T., Ozaki, T., & Shimizu, T. "Preparation and insulation properties of epoxy-layered silicate nanocomposite insulating material applications" The 17th Annual Meeting of the IEEE Lasers and Electro-Optics Society, 2004. LEOS 2004 DOI: DOI: 10.1109/CEIDP.2004.1364272

Ince, H., Imamoglu, S. Z., & Karakose, M. A. (2023). Entrepreneurial orientation, social capital, and firm performance: The mediating role of innovation performance. *International Journal of Entrepreneurship and Innovation, 24*(1), 32–43. DOI: 10.1177/14657503211055297

Indrakumari, R., Poongodi, T., & Jena, S. R. (2020). Heart Disease Prediction using Exploratory Data Analysis. *Procedia Computer Science, 173*, 130–139. DOI: 10.1016/j.procs.2020.06.017

Information Commissioner's Office UK. (2023). Guidance on AI and Data Protection. *ICO.* ico.org.uk/for-organisations/uk-gdpr-guidance-and-resources/artificial-intelligence/guidance-on-ai-and-data-protection/

Isaak, R. (1997). Globalisation and Green Entrepreneurship. *Greener Management International, 18*, 80–90. https://www.jstor.org/stable/45259405

Isaak, R. (2002). The Making of the Ecopreneur. *Greener Management International, 38*, 81–91. https://www.jstor.org/stable/greemanainte.38.81

Ishibe, S. Mori, M. Kozako, M. Hikita M. "A New Concept Varistor With Epoxy Microvaristor Composite" IEEE Transactions on Power Delivery (Volume: 29, Issue: 2, April 2014)

Ismagilova, E., Slade, E., Rana, N. P., & Dwivedi, Y. K. (2020). The effect of characteristics of source credibility on consumer behaviour: A meta-analysis. *Journal of Retailing and Consumer Services, 53*, 101736. DOI: 10.1016/j.jretconser.2019.01.005

Ivanov, S., & Soliman, M. (2023). Game of algorithms: ChatGPT implications for the future of tourism education and research. *Journal of Tourism Futures, 9*(2), 214–221. DOI: 10.1108/JTF-02-2023-0038

Iyer, G., & Soberman, D. A. (2016). Social Responsibility and Product Innovation. *Marketing Science, 35*(5), 727–742. https://www.jstor.org/stable/44012185. DOI: 10.1287/mksc.2015.0975

Izogo, E. E., & Jayawardhena, C. (2018). Online shopping experience in an emerging e-retailing market. *Journal of Research in Interactive Marketing, 12*(2), 193–214. DOI: 10.1108/JRIM-02-2017-0015

Javed, A., Rapposelli, A., Khan, F., Javed, A., & Abid, N. (2024). Do green technology innovation, environmental policy, and the transition to renewable energy matter in times of ecological crises? A step towards ecological sustainability. *Technological Forecasting and Social Change, 207*, 123638. Advance online publication. DOI: 10.1016/j.techfore.2024.123638

Javeed, S. A., Teh, B. H., Ong, T. S., Chong, L. L., Abd Rahim, M. F. B., & Latief, R. (2022). How does green innovation strategy influence corporate financing? Corporate social responsibility and gender diversity play a moderating role. *International Journal of Environmental Research and Public Health, 19*(14), 8724. DOI: 10.3390/ijerph19148724 PMID: 35886576

Jayanthi, S., Priyadharshini, V., Kirithiga, V., & Premalatha, S. (2024). Mental health status monitoring for people with autism spectrum disorder using machine learning. *International Journal of Information Technology : an Official Journal of Bharati Vidyapeeth's Institute of Computer Applications and Management, 16*(1), 43–51. DOI: 10.1007/s41870-023-01524-z

Jelili Amuda, Y., & Alabdulrahman, S. (2023). Reinforcing policy and legal framework for Islamic insurance in Islamic finance: Towards achieving Saudi Arabia Vision 2030. *International Journal of Law and Management, 65*(6), 600–613. DOI: 10.1108/IJLMA-03-2023-0045

Ji, Q., Wang, Y., & Sun, L. (2024). Mixer is more than just a model. *arXiv preprint arXiv:2402.18007*.

Ji, Q., Zhang, J., & Wang, Y. (2024). ASM: Audio Spectrogram Mixer. *arXiv preprint arXiv:2401.11102*.

Jiang, H., Mutahira, H., Park, U., & Muhammad, M. S. (2024). Scanning dial: The instantaneous audio classification transformer. *Discover Applied Sciences, 6*(3), 96. DOI: 10.1007/s42452-024-05731-6

Jiang, K., Zheng, J., & Luo, S. (2024). Green power of virtual influencer: The role of virtual influencer image, emotional appeal, and product involvement. *Journal of Retailing and Consumer Services, 77*, 103660. DOI: 10.1016/j.jretconser.2023.103660

Johnson, T. G. (2007). Measuring the benefits of entrepreneur ship development policy. *ICFAI Journal of Entrepreneurship Development, 4*(2), 35–44.

Jones, M. T. (1996). Social Responsibility and the "Green" Business Firm. *Industrial & Environmental Crisis Quarterly, 9*(3), 327–345. https://www.jstor.org/stable/26162491

Joshi, S. (2024). Ai vs. Human Agents: How to Strike the Right Balance in AI Customer Service. *Sendbird.* sendbird.com/blog/ai-vs-human-intelligence.

Jovanovic, M., Báez, M., & Casati, F. (2021, May 1). Chatbots as Conversational Healthcare Services. *IEEE Internet Computing, 25*(3), 44–51. DOI: 10.1109/MIC.2020.3037151

Juniper Research. (2017). Chatbots: A Game Changer for Banking & Healthcare, Saving $8 billion Annually by 2022. Juniper Research. Available: https://www.juniperresearch.com/press/chatbots-a-game-changer-for-banking-healthcare

Kachroo, P., Saiewitz, A., Raschke, R., Agarwal, S., & Huang, A. J. (2020). A New Language and Input–Output Hidden Markov Model for Automated Audit Inquiry. *IEEE Intelligent Systems, 35*(6), 39–49. DOI: 10.1109/MIS.2019.2963653

Kadet, K. (2021). Data from Entrust Reveals Contradictions in Consumer Sentiment Toward Data Privacy and Security. www.entrust.com/newsroom/press-releases/2021/data-from-entrust-reveals-contradictions-in-consumer-sentiment-toward-data-privacy-and-security

Kaesehage, K., Leyshon, M., Ferns, G., & Leyshon, C. (2019). Seriously Personal: The Reasons that Motivate Entrepreneurs to Address Climate Change. *Journal of Business Ethics, 157*(4), 1091–1109. https://www.jstor.org/stable/45106464. DOI: 10.1007/s10551-017-3624-1

Kahn, M. (2018). Co-authorship as a proxy for collaboration: A cautionary tale. *Science & Public Policy, 45*(1), 117–123. DOI: 10.1093/scipol/scx052

Kakeesh, D., Weshah, G. A., Al, , Ma', N., & Al, . (2024). Building e-loyalty through online banking features: Mediating role of e-trust. *International Journal of Management Practice, 17*(5), 577–599. DOI: 10.1504/IJMP.2024.140867

Kanakov, F., & Prokhorov, I. (2022). Analysis and applicability of artificial intelligence technologies in the field of RPA software robots for automating business processes. *Procedia Computer Science, 213*, 296–300. DOI: 10.1016/j.procs.2022.11.070

Kanini, K. S., & Muathe, S. M. (2019). Nexus between social capital and firm performance: A critical literature review and research agenda. *International Journal of Business and Management*, *14*(8), 70–82. DOI: 10.5539/ijbm.v14n8p70

Kaplan, J., & Haenlein, M. (2020). Rulers of the world, unite! The challenges and opportunities of artificial intelligence. *Business Horizons*, *63*(1), 37–50. DOI: 10.1016/j.bushor.2019.09.003

Kasparov, G. (2017). *Deep Thinking: Where Machine Intelligence Ends and Human Creativity Begins*. PublicAffairs.

Kassim, N., & Asiah Abdullah, N. (2010a). The effect of perceived service quality dimensions on customer satisfaction, trust, and loyalty in e-commerce settings. *Asia Pacific Journal of Marketing and Logistics*, *22*(3), 351–371. DOI: 10.1108/13555851011062269

Kautto, P. (2010). New instruments – old practices? The implications of environmental management systems and extended producer responsibility for design for the environment. *Business Strategy and the Environment*, *15*(6), 377–388. DOI: 10.1002/bse.454

Kehinde, S., Oluseun, B., Moses, C., Borishade, T., Kehinde, O., Simon-ilogho, B., & Kehinde, K. (2024). Effective Sales Promotion: A Necessary Catalyst for Elongating the Product Life Cycle in A Developing Economy. *International Journal of Advances in Social Sciences and Humanities*, *3*(1), 1–7. DOI: 10.56225/ijassh.v3i1.231

Kemp, R., & Pearson, P. (2007). *Final Report MEI Project About Measuring Eco-Innovation*. UM Merit.

Khaldy, M. A., Ishtaiwi, A., Al-Qerem, A., Aldweesh, A., Alauthman, M., Almomani, A., & Arya, V. (2023). Redefining E-Commerce Experience: An Exploration of Augmented and Virtual Reality Technologies. *International Journal on Semantic Web and Information Systems*, *19*(1), 1–24. Advance online publication. DOI: 10.4018/IJSWIS.334123

Khanom, M. T. (2023). Using social media marketing in the digital era: A necessity or a choice. *International Journal of Research in Business and Social Science (2147- 4478)*, *12*(3), 88–98. DOI: 10.20525/ijrbs.v12i3.2507

Kim, G., Wu, H. H., Bondi, L., & Liu, B. (2024, April). Multi-Modal Continual Pre-Training For Audio Encoders. In *ICASSP 2024-2024 IEEE International Conference on Acoustics, Speech and Signal Processing (ICASSP)* (pp. 691-695). IEEE. DOI: 10.1109/ICASSP48485.2024.10446424

Kim, P. H., & Aldrich, H. E. (2005). Social capital and entrepreneurship. Foundations and Trends® in Entrepreneurship, 1(2), 55-104.

Kim, D. J., Ferrin, D. L., & Rao, H. R. (2009). Trust and satisfaction, the two wheels for successful e-commerce transactions: A longitudinal exploration. *Information Systems Research*, 20(2), 237–257. DOI: 10.1287/isre.1080.0188

Kim, H. J., & Jeong, M. (2018). Research on hospitality and tourism education: Now and future. *Tourism Management Perspectives*, 25, 119–122. DOI: 10.1016/j.tmp.2017.11.025

Kim, S., & Chang, M. (2023). Application of human brain Organoids—Opportunities and challenges in modeling human brain development and neurodevelopmental diseases. *International Journal of Molecular Sciences*, 24(15), 12528. DOI: 10.3390/ijms241512528 PMID: 37569905

Kim, Y. J., Shin, T. S., Do Choi, H., Kwon, J. H., Chung, Y.-C., & Yoon, H. G. (2005). Electrical conductivity of chemically modified epoxy composites. *Carbon*, 43(1), 23–30. DOI: 10.1016/j.carbon.2004.08.015

Kobbacy, K. A. (2012). Application of artificial intelligence in maintenance modelling and management. *IFAC Proceedings Volumes, 45*(31), 54-59.

Kogut, L., & Komvopoulos, K. (2004). Electrical contact resistance theory for conductive rough surfaces separated by a thin insulating film. *Journal of Applied Physics*, 95(2), 576–585. DOI: 10.1063/1.1629392

Kolesnichenko, A., McVeigh-Schultz, J., & Isbister, K. (2019). Understanding Emerging Design Practices for Avatar Systems in the Commercial Social VR Ecology. *Proceedings of the 2019 on Designing Interactive Systems Conference*, 241–252. DOI: 10.1145/3322276.3322352

Konak, F., & Demir, Y. (2023). Bibliometric Analysis on Islamic Insurance (Takaful). *Hitit Theology Journal*, 22(1), 11–46. DOI: 10.14395/hid.1232415

Konietzko, J., Bocken, N., & Hultink, E. J. (2019). Online platforms and the circular economy. *Innovation for sustainability: Business transformations towards a better world*, 435-450.

Kotowska, B., & Sikorska, M. (2023). Digital transformation of a Polish accounting firm: Tools, impediments, business performance benefits and implications–case study. *Procedia Computer Science*, 225, 327–336. DOI: 10.1016/j.procs.2023.10.017

Kou, I. T., & Liu, T. (2020). Could the adoption of Quick Response (QR) code in lectures enhance University students' satisfaction? A case study of hospitality and tourism programs in Macau. In *Cultural and Tourism Innovation in the Digital Era: Sixth International IACuDiT Conference, Athens 2019* (pp. 161-170). Springer International Publishing.

Kujur, O. (2023). How Corporate Social Responsibility and Sustainable Development Practices Affect the Economic Performance of Indian Businesses: A Critical Analysis.

Kumar. (2021). The Future of Retail: How Technology is Transforming the Shopping Experience. National Retail Federation Big Show Keynote.

Kumari, V., Bala, P. K., & Chakraborty, S. (2024). A text mining approach to explore factors influencing consumer intention to use Metaverse platform services: Insights from online customer reviews. *Journal of Retailing and Consumer Services, 81,* 103967. DOI: 10.1016/j.jretconser.2024.103967

Kunhibava, S. (2011). Reasons on the Similarity of Objections with Regards to Gambling and Speculation in Islamic Finance and Conventional Finance. *Journal of Gambling Studies, 27*(1), 1–13. DOI: 10.1007/s10899-010-9201-5 PMID: 20514512

Kureljusic, M., & Karger, E. (2023). Forecasting in financial accounting with artificial intelligence–A systematic literature review and future research agenda. *Journal of Applied Accounting Research*, (ahead-of-print).

Kusmin, K. L., Normak, P., & Ley, T. (2024). A Methodology for Planning, Implementation and Evaluation of Skills Intelligence Management-Results of a Design Science Project in Technology Organisations. *Frontiers in Artificial Intelligence, 7,* 1424924. DOI: 10.3389/frai.2024.1424924 PMID: 39169913

Kye, B., Han, N., Kim, E., Park, Y., & Jo, S. (2021). Educational applications of Metaverse: Possibilities and limitations. *Journal of Educational Evaluation for Health Professions, 18,* 32. DOI: 10.3352/jeehp.2021.18.32 PMID: 34897242

Kyshakevych, B., Maksyshko, N., Hrytsenko, K., Voronchak, I., & Demediuk, B. (2024). ANALYZING THE EFFICIENCY OF DIGITALIZATION IN SMALL AND MEDIUMSIZED ENTERPRISES ACROSS EU COUNTRIES USING DEA MODELS. *Financial and Credit Activity: Problems of Theory and Practice, 3*(56), 215–229. DOI: 10.55643/fcaptp.3.56.2024.4344

Lai, S., Zhang, S., Hassan, A., & Mushtaq, R. T. (2024). Analyzing Role of Artificial Intelligence in Project Management and Investment Risk: A CiteSpace Insight. In *Proceedings of the 2024 16th International Conference on Machine Learning and Computing* (pp. 713-719). DOI: 10.1145/3651671.3651776

Lal, M., & Neduncheliyan, S. (2024, March 11). Conversational artificial intelligence development in healthcare. Springer Science+Business Media. https://doi.org/DOI: 10.1007/s11042-024-18841-5

Lancaster, M. A., & Knoblich, J. A. (2014). Generation of cerebral organoids from human pluripotent stem cells. *Nature Protocols*, *9*(10), 2329–2340. DOI: 10.1038/nprot.2014.158 PMID: 25188634

Lancaster, M. A., Renner, M., Martin, C., Wenzel, D., Bicknell, L. S., Hurles, M. E., Homfray, T., Penninger, J. M., Jackson, A. P., & Knoblich, J. A. (2013). Cerebral organoids model human brain development and microcephaly. *Nature*, *501*(7467), 373–379. DOI: 10.1038/nature12517 PMID: 23995685

Laszkiewicz, A., & Kalinska-Kula, M. (2023). Virtual influencers as an emerging marketing theory: A systematic literature review. *International Journal of Consumer Studies*, *47*(6), 2479–2494. DOI: 10.1111/ijcs.12956

Lau, K. W., Rehman, Y. A. U., & Po, L. M. (2024). AudioRepInceptionNeXt: A lightweight single-stream architecture for efficient audio recognition. *Neurocomputing*, *578*, 127432. DOI: 10.1016/j.neucom.2024.127432

Lavazza, A., & Pizzetti, F. G. (2020). Human cerebral organoids as a new legal and ethical challenge†. *Journal of Law and the Biosciences*, *7*(1), lsaa005. Advance online publication. DOI: 10.1093/jlb/lsaa005 PMID: 34221418

Lebourdais, M., Mariotte, T., Almudévar, A., Tahon, M., & Ortega, A. (2024). Explainable by-design Audio Segmentation through Non-Negative Matrix Factorization and Probing. *arXiv preprint arXiv:2406.13385*. DOI: 10.21437/Interspeech.2024-791

Lee, K., & Superpowers, A. I. (2018). *China, Silicon Valley, and the New World Order*. Houghton Mifflin Harcourt.

Lee, M., & Lin, Y. (2020). Game-based microlearning for improving university students' physics learning. *Computers & Education*, *154*, 103933.

Lee, T., & Sawai, T. (2023). Global governance of human brain organoid research and applications: A role for the World Health Organization? *Molecular Psychology: Brain. Molecular Psychology*, *2*, 11. DOI: 10.12688/molpsychol.17548.1

Leggett, K. (2019). *The Future of Customer Service: 10 Trends to Watch*. Forrester.

Lehner, O. M., & Knoll, C. (2022). *Artificial Intelligence in Accounting*. Routledge. DOI: 10.4324/9781003198123

Lei, S. I., & So, A. S. I. (2021). Online teaching and learning experiences during the COVID-19 pandemic–A comparison of teacher and student perceptions. *Journal of Hospitality & Tourism Education, 33*(3), 148–162. DOI: 10.1080/10963758.2021.1907196

Lemonade. (2021). Lemonade Insurance Review. Lemonade. https://www.lemonade.com/insuropedia/reviews/lemonade-insurance-review/

Leonidou, L. C., Christodoulides, P., Kyrgidou, L. P., & Palihawadana, D. (2017). Internal Drivers and Performance Consequences of Small Firm Green Business Strategy: The Moderating Role of External Forces. *Journal of Business Ethics, 140*(3), 585–607. https://www.jstor.org/stable/44164312. DOI: 10.1007/s10551-015-2670-9

Li, Y. (2015). Blended learning in English for tourism: A case study. *Exploring Learning & Teaching in Higher Education*, 331-345.

Li. (2018). How to Make A.I. That's Good for People. The New York Times. www.nytimes.com/2018/03/07/opinion/artificial-intelligence-human.html

Liao, C., Palvia, P., & Chen, J.-L. (2009). Information technology adoption behavior life cycle: Toward a Technology Continuance Theory (TCT). *International Journal of Information Management, 29*(4), 309–320. DOI: 10.1016/j.ijinfomgt.2009.03.004

Liao, J., Yi, L., Shi, W., Yang, W., Fang, Y., & Yang, X. (2024). Imperceptible backdoor watermarks for speech recognition model copyright protection. *Visual Intelligence, 2*(1), 1–10. DOI: 10.1007/s44267-024-00055-w

Lillich, M. (2012). Entrepreneurship, Jobs, Sustainable Businesses = Economic Prosperity. *Perspectives on Work, 16*(1/2), 3–7. https://www.jstor.org/stable/41810199

Lim, C. H., Chung, J. J., & Pedersen, P. M. (2012). Effects of Electronic Word - of - Mouth Messages. *Chorigia, 8*(1), 55–76. DOI: 10.4127/ch.2012.0064

Lin, M.-J., & Wang, W.-T. (2015). Examining E-Commerce Customer Satisfaction and Loyalty: An Integrated Quality-Risk-Value Perspective. *Journal of Organizational Computing and Electronic Commerce, 25*(4), 379–401. DOI: 10.1080/10919392.2015.1089681

Little, M. H. (2017). Organoids: A special issue. *Development (Cambridge, England), 144*(6), 935–937. DOI: 10.1242/dev.150292 PMID: 28292836

Litvin, S. W., Goldsmith, R. E., & Pan, B. (2008). Electronic word-of-mouth in hospitality and tourism management. *Tourism Management, 29*(3), 458–468. DOI: 10.1016/j.tourman.2007.05.011

Liu, N., Nikitas, A., & Parkinson, S. (2020). Exploring expert perceptions about the cyber security and privacy of connected and autonomous vehicles: A thematic analysis approach. *Transportation Research Part F: Traffic Psychology and Behaviour*, *75*, 66–86. DOI: 10.1016/j.trf.2020.09.019

Liu, Z., Han, S., Yao, M., Gupta, S., & Laguir, I. (2023). Exploring drivers of eco-innovation in manufacturing firms' circular economy transition: An awareness, motivation, capability perspective. *Annals of Operations Research*. Advance online publication. DOI: 10.1007/s10479-023-05473-5

Long, T., Cummins, J., & Waugh, M. (2019). Investigating the factors that influence higher education instructors' decisions to adopt a flipped classroom instructional model. *British Journal of Educational Technology*, *50*(4), 2028–2039. DOI: 10.1111/bjet.12703

Lumpkin, G. T., Moss, T. W., Gras, D. M., Kato, S., & Amezcua, A. S. (2013). Entrepreneurial processes in social contexts: How are they different, if at all? *Small Business Economics*, *40*(3), 761–783. https://www.jstor.org/stable/23360622. DOI: 10.1007/s11187-011-9399-3

Lu, X., Yang, J., & Xiang, Y. (2022). Modeling human neurodevelopmental diseases with brain organoids. *Cell Regeneration (London, England)*, *11*(1), 1. Advance online publication. DOI: 10.1186/s13619-021-00103-6 PMID: 34982276

Madi, J., Al Khasawneh, M., & Dandis, A. O. (2024). Visiting and revisiting destinations: Impact of augmented reality, content quality, perceived ease of use, perceived value and usefulness on E-WOM. *International Journal of Quality & Reliability Management*, *41*(6), 1550–1571. DOI: 10.1108/IJQRM-10-2023-0314

Mady, K., Anwar, I., & Abdelkareem, R. S. (2023). Nexus between regulatory pressure, eco-friendly product demand and sustainable competitive advantage of manufacturing small and medium-sized enterprises: The mediating role of eco-innovation. *Environment, Development and Sustainability*. Advance online publication. DOI: 10.1007/s10668-024-05096-1

Mahmood, N., Zhao, Y., Lou, Q., & Geng, J. (2022). Role of environmental regulations and eco-innovation in energy structure transition for green growth: Evidence from OECD. *Technological Forecasting and Social Change*, *183*, 121890. Advance online publication. DOI: 10.1016/j.techfore.2022.121890

Maione, G., & Leoni, G. (2021). Artificial intelligence and the public sector: the case of accounting. In *Artificial Intelligence and Its Contexts: Security, Business and Governance* (pp. 131–143). Springer International Publishing. DOI: 10.1007/978-3-030-88972-2_9

Makinde, O. G., Olubiyi, T. O., & Ogundipe, F. (2023). New Insights from Entrepreneurial Characteristics and Business Performance: Empirical Findings from Nigeria. *Asian Journal of Management. Entrepreneurship and Social Sciences*, *3*(04), 72–100.

Mansour, A. A., Gonçalves, J. T., Bloyd, C. W., Li, H., Fernandes, S., Quang, D., Johnston, S., Parylak, S. L., Jin, X., & Gage, F. H. (2018). An in vivo model of functional and vascularized human brain organoids. *Nature Biotechnology*, *36*(5), 432–441. DOI: 10.1038/nbt.4127 PMID: 29658944

MarketsandMarkets. (2021). Emotion Detection and Recognition Market. MarketsandMarkets. Available: https://www.marketsandmarkets.com/Market-Reports/emotion-detection-recognition-market-23376176.html

MarketsandMarkets. (2021). Natural Language Processing Market. MarketsandMarkets. Available: https://www.marketsandmarkets.com/Market-Reports/natural-language-processing-nlp-825.html

Marr, B. (2020). *Artificial Intelligence in Practice: How 50 Successful Companies Used AI and Machine Learning to Solve Problems*. Wiley.

Marshall, R. S. (2011). Conceptualizing the International For-Profit Social Entrepreneur. *Journal of Business Ethics*, *98*(2), 183–198. https://www.jstor.org/stable/41475810. DOI: 10.1007/s10551-010-0545-7

Mayer, R. E., & Johnson, C. I. (2020). Microlearning for engineering education: A review. *Journal of Engineering Education*, *109*(1), 1–11.

McAlone. (2016). Why Netflix Thinks Its Personalized Recommendation Engine Is Worth $1 Billion per Year. *Business Insider*. www.businessinsider.com/netflix-recommendation-engine-worth-1-billion-per-year-2016-6

McCoy, T. H., Castro, V. M., Cagan, A., Roberson, A. M., Kohane, I. S., & Perlis, R. H. (2015, August 24). Sentiment Measured in Hospital Discharge Notes Is Associated with Readmission and Mortality Risk: An Electronic Health Record Study. *PLoS One*, *10*(8), e0136341–e0136341. DOI: 10.1371/journal.pone.0136341 PMID: 26302085

McKercher, B. (2018). What is the state of hospitality and tourism research–2018? *International Journal of Contemporary Hospitality Management*, *30*(3), 1234–1244. DOI: 10.1108/IJCHM-12-2017-0809

McKinsey & Company. (2013). How retailers can keep up with consumers. McKinsey & Company.

McKinsey & Company. (2020). How COVID-19 has pushed companies over the technology tipping point—and transformed business forever. McKinsey & Company. Available: https://www.mckinsey.com/business-functions/strategy-and-corporate -finance/our-insights/how-covid-19-has-pushed-companies-over-the-technology -tipping-point-and-transformed-business-forever

McKinsey & Company. (2021). The value of getting personalization right—or wrong—is multiplying. McKinsey & Company. Available: https://www.mckinsey .com/business-functions/growth-marketing-and-sales/our-insights/the-value-of -getting-personalization-right-or-wrong-is-multiplying

Meenakshy, M., Prasad, K. D. V., Bolar, K., & Shyamsunder, C. (2024). Electronic word-of-mouth intentions in personal and public networks: A domestic tourist perspective. *Humanities & Social Sciences Communications*, *11*(1), 1226. DOI: 10.1057/s41599-024-03753-4

Mehta, M., Pancholi, G., & Saxena, D. A. (2023). Metaverse changing realm of the business world: A bibliometric snapshot. *Journal of Management Development*, *42*(5), 373–387. DOI: 10.1108/JMD-01-2023-0006

Mei, X., Meng, C., Liu, H., Kong, Q., Ko, T., Zhao, C., Plumbley, M. D., Zou, Y., & Wang, W. (2024). Wavcaps: A chatgpt-assisted weakly-labelled audio captioning dataset for audio-language multimodal research. *IEEE/ACM Transactions on Audio, Speech, and Language Processing*, *32*, 3339–3354. DOI: 10.1109/ TASLP.2024.3419446

Menaga, A., & Vasantha, S. Role of Corporate Social Responsibility in Achieving Sustainable Development Goals. In *Interdisciplinary Perspectives on Sustainable Development* (pp. 188–191). CRC Press. DOI: 10.1201/9781003457619-39

Meng, L., Wen, K. H., Brewin, R., & Wu, Q. (2020). Knowledge atlas on the relationship between urban street space and residents' health—A bibliometric analysis based on VOSviewer and CiteSpace. *Sustainability (Basel)*, *12*(6), 2384. DOI: 10.3390/su12062384

Michelli, J. (2021). How Zappos Uses Customer Support as an Opportunity to WOW Customers. *Calix*. www.calix.com/blog/2021/08/how-zappos-uses-customer-support -as-an-opportunity-to-wow-customers.html

Midavaine, N., Go, G. H. T., Canez, D., Simion, I., & Chatterji, S. [Re] On the Reproducibility of Post-Hoc Concept Bottleneck Models. *Transactions on Machine Learning Research*.

Mikalef, P., Boura, M., Lekakos, G., & Krogstie, J. (2019). Big data analytics capabilities and innovation: The mediating role of dynamic capabilities and moderating effect of the environment. *British Journal of Management*, *30*(2), 272–298. DOI: 10.1111/1467-8551.12343

Miller, G. A. (1956). The magical number seven, plus or minus two: Some limits on our capacity for processing information. *Psychological Review*, *63*(2), 81–97. DOI: 10.1037/h0043158 PMID: 13310704

Mishchuk, H., Bilan, Y., Androniceanu, A., & Krol, V. (2023). Social capital: Evaluating its roles in competitiveness and ensuring human development. *Journal of Competitiveness*, *15*(2).

Mishra, R. K., Srivastava, S., Singh, S., & Upadhyay, M. K. (2024). Exploring the Opportunities of AI Integral with DL and ML Models in Financial and Accounting Systems. In *2024 4th International Conference on Advance Computing and Innovative Technologies in Engineering (ICACITE)* (pp. 999-1003). IEEE. DOI: 10.1109/ICACITE60783.2024.10616847

Modafferi, S., Zhong, X., Kleensang, A., Murata, Y., Fagiani, F., Pamies, D., Hogberg, H. T., Calabrese, V., Lachman, H., Hartung, T., & Smirnova, L. (2021). Gene–environment interactions in developmental neurotoxicity: A case study of synergy between Chlorpyrifos and CHD8 knockout in human BrainSpheres. *Environmental Health Perspectives*, *129*(7), 077001. Advance online publication. DOI: 10.1289/EHP8580 PMID: 34259569

Mohaimenuzzaman, M., Bergmeir, C., & Meyer, B. (2024). Deep Active Audio Feature Learning in Resource-Constrained Environments. *IEEE/ACM Transactions on Audio, Speech, and Language Processing*, *32*, 3224–3237. DOI: 10.1109/TASLP.2024.3416697

Mola, F., & Jusoh, J. (2011). Service quality in Penang hotels: A gap score analysis. *World Applied Sciences Journal*, *12*(1), 19–24.

Moliterni, F. (2017). Sustainability-oriented Business Model Innovation: Context and Drivers. *Fondazione Eni Enrico Mattei (FEEM)*. https://www.jstor.org/stable/resrep16412

Mondal, S., Singh, S., & Gupta, H. (2023). Assessing enablers of green entrepreneurship in circular economy: An integrated approach. *Journal of Cleaner Production*, *388*, 135999. Advance online publication. DOI: 10.1016/j.jclepro.2023.135999

Mondal, S., Singh, S., & Gupta, H. (2023). Green entrepreneurship and digitalization enabling the circular economy through sustainable waste management: An exploratory study of emerging economy. *Journal of Cleaner Production, 422*, 422–433. DOI: 10.1016/j.jclepro.2023.138433

Montenegro, J L Z., Costa, C A D., & Righi, R D R. (2019, September 1). Survey of conversational agents in health. Elsevier BV, 129, 56-67. https://doi.org/DOI: 10.1016/j.eswa.2019.03.054

Morgan, B. (2018). 10 Examples of Predictive Customer Experience Outcomes Powered by Ai. *Forbes*. www.forbes.com/sites/blakemorgan/2018/12/20/10-examples -of-predictive-customer-experience-outcomes-powered-by-ai/

Moswete, N., & Lacey, G. (2015). "Women cannot lead": Empowering women through cultural tourism in Botswana. *Journal of Sustainable Tourism, 23*(4), 600–617. DOI: 10.1080/09669582.2014.986488

Mou, A., & Milanova, M. (2024). Performance Analysis of Deep Learning Model-Compression Techniques for Audio Classification on Edge Devices. *Sci, 6*(2), 21. DOI: 10.3390/sci6020021

Moustakas, E., Lamba, N., Mahmoud, D., & Ranganathan, C. (2020). Blurring lines between fiction and reality: Perspectives of experts on marketing effectiveness of virtual influencers. *2020 International Conference on Cyber Security and Protection of Digital Services (Cyber Security)*, 1–6. DOI: 10.1109/CyberSecurity49315.2020.9138861

Mrkajic, B., Murtinu, S., & Scalera, V. G. (2019). Is green the new gold? Venture capital and green entrepreneurship. *Small Business Economics, 52*(4), 929–950. https://www.jstor.org/stable/48701968. DOI: 10.1007/s11187-017-9943-x

Muliadi, M., Muhammadiah, M. U., Amin, K. F., Kaharuddin, K., Junaidi, J., Pratiwi, B. I., & Fitriani, F. (2024). The information sharing among students on social media: The role of social capital and trust. *VINE Journal of Information and Knowledge Management Systems, 54*(4), 823–840. DOI: 10.1108/VJIKMS-12-2021-0285

Müller, R., Locatelli, G., Holzmann, V., Nilsson, M., & Sagay, T. (2024). Artificial intelligence and project management: Empirical overview, state of the art, and guidelines for future research. *Project Management Journal, 55*(1), 9–15. DOI: 10.1177/87569728231225198

Mustak, M., Salminen, J., Plé, L., & Wirtz, J. (2021). Artificial intelligence in marketing: Topic modeling, scientometric analysis, and research agenda. *Journal of Business Research, 124*, 389–404. DOI: 10.1016/j.jbusres.2020.10.044

Mystakidis, S. (2022). Metaverse. *Encyclopedia, 2*(1), 486–497. DOI: 10.3390/ encyclopedia2010031

Nadella, S. (2016). Satya Nadella on Microsoft's New Age of Intelligence. Bloomberg. https://www.bloomberg.com/features/2016-satya-nadella-interview-issue/

Nagy, A. S., Bittner, B., Tuegeh, O. D. M., & Tumiwa, J. R. (2022). Augmented reality improving consumer choice confidence during COVID-19. *Issues in Information Systems, 23*(2), 294–309. DOI: 10.48009/2_iis_2022_126

Nair, A. S., & R., D. K. (2024). *ARise to the Occasion* (pp. 184–203). DOI: 10.4018/979-8-3693-2367-0.ch009

Nandan, S., & Nandan, S. (2020). Microlearning: A Boon for Learning Coding Skills for Beginners. *Journal of Educational Technology & Society, 23*(2), 70–82.

Nanda, W. D., Widianingsih, I., & Miftah, A. Z. (2023). The linkage of digital transformation and tourism development policies in Indonesia from 1879–2022: Trends and implications for the future. *Sustainability (Basel), 15*(13), 10201. DOI: 10.3390/su151310201

Naumov, N. (2019). The impact of robots, artificial intelligence, and service automation on service quality and service experience in hospitality. In *Robots, artificial intelligence, and service automation in travel, tourism and hospitality* (pp. 123-133). Emerald Publishing Limited. DOI: 10.1108/978-1-78756-687-320191007

Naumov, N. (2019). *The Impact of Robots.* Artificial Intelligence, and Service Automation on Service Quality and Service Experience in Hospitality. In Robots, Artificial Intelligence, and Service Automation in Travel, Tourism, and Hospitality Emerald Publishing Limited.

Nebo, C., Nwankwo, P., & Okonkwo, R. (2015). The role of effective communication on organizational performance: A Study of Nnamdi Azikiwe University, Awka. *Review of public administration and management, 4*(8), 132-148.

Nestoroska, I., & Petrovska, I. (2014). Staff training in hospitality sector as benefit for improved service quality. In *Tourism and Hospitality Industry 2014–Congress Proceedings* (pp. 437-448).

Netflix Technology Blog. (2012). Lessons Netflix Learned from the AWS Outage. Netflix. Available: https://netflixtechblog.com/lessons-netflix-learned-from-the -aws-outage-deefe5fd0c04

New Epsilon Research Indicates 80% of Consumers Are More Likely to Make a Purchase When Brands Offer Personalized Experiences. (2018). Epsilon. www.epsilon.com/us/about-us/pressroom/new-epsilon-research-indicates-80-of-consumers-are-more-likely-to-make-a-purchase-when-brands-offer-personalized-experiences

Ng, A. (2017). AI Is the New Electricity. Stanford Graduate School of Business. https://www.gsb.stanford.edu/insights/andrew-ng-ai-new-electricity

Nguyen, A., Ngo, H. N., Hong, Y., Dang, B., & Nguyen, B.-P. T. (2023). Ethical principles for artificial intelligence in education. *Education and Information Technologies*, *28*(4), 4221–4241. DOI: 10.1007/s10639-022-11316-w PMID: 36254344

Nienhaus, V. (2011). Islamic finance ethics and Shari'ah law in the aftermath of the crisis: Concept and practice of Shari'ah compliant finance. *Ethical Perspectives*, *18*(4), 591–623. DOI: 10.2143/EP.18.4.2141849

Noble, S. U. (2018). *Algorithms of Oppression: How Search Engines Reinforce Racism*. NYU Press. DOI: 10.18574/nyu/9781479833641.001.0001

O'Boyle, I. (2014). Mobilising social media in sport management education. *Journal of Hospitality, Leisure, Sport and Tourism Education*, *15*, 58–60. DOI: 10.1016/j.jhlste.2014.05.002

O'Neill, G. D., Jr., Hershauer, J. C., & Golden, J. S. (2006). The Cultural Context of Sustainability Entrepreneurship. *Greener Management International, 55*, 33–46. https://www.jstor.org/stable/greemanainte.55.33

O'Riordan, L., & Fairbrass, J. (2008). Corporate Social Responsibility (CSR): Models and Theories in Stakeholder Dialogue. *Journal of Business Ethics*, *83*(4), 745–758. DOI: 10.1007/s10551-008-9662-y

Obermeyer, Z., Powers, B., Vogeli, C., & Mullainathan, S. (2019). Dissecting racial bias in an algorithm used to manage the health of populations. *Science*, *366*(6464), 447–453. DOI: 10.1126/science.aax2342 PMID: 31649194

Olowoporoku, A. A., & Olubiyi, T. O. (2023). Evaluating Service Quality and Business Outcomes Post-Pandemic: Perspective from Hotels in Emerging Market. *Sawala: Jurnal Administrasi Negara*, *11*(2), 182–201. DOI: 10.30656/sawala.v11i2.6975

Olubiyi T. O. (2023b). Leveraging competitive strategies on business outcomes post-covid-19 pandemic: Empirical investigation from Africa. *Journal of Management and Business: Research and Practice*. DOI: 10.54933/jmbrp-2023-15-2-1

Olubiyi, O., Lawal, A. T., & Adeoye, O. O. (2022). Succession Planning and Family Business Continuity: Perspectives from Lagos State, Nigeria. *Organization and Human Capital Development*, *1*(1), 40–52. DOI: 10.31098/orcadev.v1i1.865

Olubiyi, T. O. (2019). Knowledge management practices and family business profitability: Evidence from Lagos State, Nigeria. *Global Journal of Management and Business Research*, *19*(A11), 21–31. https://journalofbusiness.org/index.php/GJMBR/article/view/2872

Olubiyi, T. O. (2020). Knowledge management practices and family business profitability: Evidence from Lagos State, Nigeria. *Arabian Journal of Business and Management Review*, *6*(1), 23–32.

Olubiyi, T. O. (2022a). Measuring technological capability and business performance post-COVID Era: Evidence from Small and Medium-Sized Enterprises (SMEs) in Nigeria. *Management & Marketing Journal*, *xx*(2), 234–248. DOI: 10.52846/MNMK.20.2.09

Olubiyi, T. O. (2022b). An investigation of sustainable innovative strategy and customer satisfaction in small and medium-sized enterprises (SMEs) in Nigeria. *Covenant Journal of Business and Social Sciences*, *13*(2), 1–24.

Olubiyi, T. O. (2023a). Unveiling the role of workplace environment in achieving the Sustainable Development Goal Eight (SDG8) and employee job satisfaction post-pandemic: Perspective from Africa. *Revista Management & Marketing Craiova*, *XXI*(2), 212–228. DOI: 10.52846/MNMK.21.2.02

Olubiyi, T. O., Adeoye, O. O., Jubril, B., Adeyemi, O. S., & Eyanuku, J. P. (2023). Measuring Inequality in Sub-Saharan Africa Post-Pandemic: Correlation Results for Workplace Inequalities and Implication for Sustainable Development Goal ten. *International Journal of Professional Business Review*, *8*(4), e01405. DOI: 10.26668/businessreview/2023.v8i4.1405

Olubiyi, T. O., & Akpa, V. A. (2023). Does innovative capability and artificial intelligence really matter for the profitability of consumer goods companies in developing economies? *SunText Review of Economics & Business*, *4*(3), 191–204.

Olubiyi, T. O., Jubril, B., Sojinu, O. S., & Ngari, R. (2022). Strengthening gender equality in small business and achieving sustainable development goals (SDGs), Comparative analysis of Kenya and Nigeria. *Sawala Jurnal Administrasi Negara*, *10*(2), 168–186. DOI: 10.30656/sawala.v10i2.5663

Omnisend. (2019). Omnichannel Marketing Automation Statistics Report. Omnisend. Available: https://www.omnisend.com/resources/reports/omnichannel-marketing -automation-statistics-2019/

Omoyele, O. S., Olubiyi, T. O., Lanre-Babalola, F. O., Obadare, G. O., & Onikoyi, I. A. (2023). Business Model Innovation as a Catalyst for Sustainable Entrepreneur-ship: Empirical Findings from Small and Medium Enterprises in Nigeria. *Skyline Business Journal*, *19*(2), 55–64. DOI: 10.37383/SBJ190205

Oyekunle, D., & Boohene, D. (2024). Digital transformation potential: The role of artificial intelligence in business. *International Journal of Professional Business Review: Int.J. Prof. Bus. Rev.*, *9*(3), 1.

P. Tewari. 2023. Rise of digital entrepreneurship during COVID-19 in India. *In Industry 4.0 and the Digital Transformation of International Business* 135–141.

Paissan, F., Della Libera, L., Ravanelli, M., & Subakan, C. (2024). Listenable Maps for Zero-Shot Audio Classifiers. *arXiv preprint arXiv:2405.17615*.

Pakkala, P. G. R., Akhila Thejaswi, R., Rai, B. S., & Nagesh, H. R. (2024). Road safety analysis framework based on vehicle vibrations and sounds using deep learning techniques. *International Journal of System Assurance Engineering and Management*, *15*(3), 1086–1097. DOI: 10.1007/s13198-023-02191-w

Pal, A. K., Rawal, P., Ruwala, R., & Patel, V. (2019). Generic disease prediction using symptoms with supervised machine learning. *Int. J Sci. Res. Comput. Sci. Eng. Inf. Technol*, *5*(2), 1082–1086. DOI: 10.32628/CSEIT1952297

Palanica, A., Flaschner, P., Thommandram, A., Li, M. H., & Fossat, Y. (2019, April 5). Physicians' Perceptions of Chatbots in Health Care: Cross-Sectional Web-Based Survey. *Journal of Medical Internet Research*, *21*(4), e12887–e12887. DOI: 10.2196/12887 PMID: 30950796

Pan, C., Abbas, J., Álvarez-Otero, S., Khan, H., & Cai, C. (2022). Interplay between corporate social responsibility and organizational green culture and their role in employees' responsible behaviour towards the environment and society. *Journal of Cleaner Production*, *366*, 132878. DOI: 10.1016/j.jclepro.2022.132878

Pantelidis, I. S. (2010). Electronic Meal Experience: A Content Analysis of On-line Restaurant Comments. *Cornell Hospitality Quarterly*, *51*(4), 483–491. DOI: 10.1177/1938965510378574

Papaoikonomou, E., Valverde, M., & Ryan, G. (2012). Articulating the Meanings of Collective Experiences of Ethical Consumption. *Journal of Business Ethics*, *110*(1), 15–32. https://www.jstor.org/stable/41684010. DOI: 10.1007/s10551-011-1144-y

Paranayapa, T., Ranasinghe, P., Ranmal, D., Meedeniya, D., & Perera, C. (2024). A comparative study of preprocessing and model compression techniques in deep learning for forest sound classification. *Sensors (Basel), 24*(4), 1149. DOI: 10.3390/s24041149 PMID: 38400306

Parasuraman, A., Zeithaml, V. A., & Berry, L. L. (1985). A conceptual model of service quality and its implications for future research. *Journal of Marketing, 49*(4), 41–50. DOI: 10.1177/002224298504900403

Parrish, B. D., & Foxon, T. J. (2006). Sustainability Entrepreneurship and Equitable Transitions to a Low-Carbon Economy. *Greener Management International, 55,* 47–62. https://www.jstor.org/stable/greemanainte.55.47

Parthymos, A., & Daskalopoulou, I. (2024). Entrepreneurship and social capital: Some evidence on micro-spatial interactions. *Journal of Small Business and Entrepreneurship, 36*(1), 108–129. DOI: 10.1080/08276331.2020.1868839

Paşca, A. M., Sloan, S. A., Clarke, L. E., Tian, Y., Makinson, C. D., Huber, N., Kim, C. H., Park, J., O'Rourke, N. A., Nguyen, K. D., Smith, S. J., Huguenard, J. R., Geschwind, D. H., Barres, B. A., & Paşca, S. P. (2015). Functional cortical neurons and astrocytes from human pluripotent stem cells in 3D culture. *Nature Methods, 12*(7), 671–678. DOI: 10.1038/nmeth.3415 PMID: 26005811

Pathak, K., & Prakash, G. (2023). Exploring the role of augmented reality in purchase intention: Through flow and immersive experience. *Technological Forecasting and Social Change, 196,* 122833. DOI: 10.1016/j.techfore.2023.122833

Pegasystems. (2017). What Consumers Really Think About AI: A Global Study. Pegasystems. Available: https://www.pega.com/ai-survey

Pham, M. T., Pollock, K. M., Rose, M. D., Cary, W. A., Stewart, H. R., Zhou, P., Nolta, J. A., & Waldau, B. (2018). Generation of human vascularized brain organoids. *Neuroreport, 29*(7), 588–593. DOI: 10.1097/WNR.0000000000001014 PMID: 29570159

Phillips, R. A. (1997). Stakeholder Theory and a Principle of Fairness. *Business Ethics Quarterly, 7*(1), 51–66. DOI: 10.2307/3857232

Pishdad, P., & Onungwa, I. O. (2024). ANALYSIS OF 5D BIM FOR COST ESTIMATION, COST CONTROL, AND PAYMENTS. [ITcon]. *Journal of Information Technology in Construction, 29*(24), 525–548. DOI: 10.36680/j.itcon.2024.024

Plummer, S., Wallace, S., Ball, G., Lloyd, R., Schiapparelli, P., Quiñones-Hinojosa, A., Hartung, T., & Pamies, D. (2019). A human iPSC-derived 3D platform using primary brain cancer cells to study drug development and personalized medicine. *Scientific Reports*, *9*(1), 1407. Advance online publication. DOI: 10.1038/s41598-018-38130-0 PMID: 30723234

Prados-Castillo, J. F., Torrecilla-García, J. A., Guaita-Fernandez, P., & De Castro-Pardo, M. (2024). The impact of the Metaverse on consumer behaviour and marketing strategies in tourism. *ESIC Market*, *55*(1), e327. DOI: 10.7200/esicm.55.327

Prasad, K. R., Karanam, S. R., Ganesh, D., Liyakat, K. K. S., Talasila, V., & Purushotham, P. (2024). AI in public-private partnership for IT infrastructure development. *The Journal of High Technology Management Research*, *35*(1), 100496. DOI: 10.1016/j.hitech.2024.100496

Prentice, C., Dominique Lopes, S., & Wang, X. (2020). The impact of artificial intelligence and employee service quality on customer satisfaction and loyalty. *Journal of Hospitality Marketing & Management*, *29*(7), 739–756. DOI: 10.1080/19368623.2020.1722304

Press, G. (2024). Ai Stats News: 86% of Consumers Prefer Humans to Chatbots. *Forbes*. www.forbes.com/sites/gilpress/2019/10/02/ai-stats-news-86-of-consumers-prefer-to-interact-with-a-human-agent-rather-than-a-chatbot/

Pritchard, A. (1969). Statistical Bibliography or Bibliometrics? *The Journal of Documentation*, *25*, 348–349.

Purwati, A., Budiyanto, B., Suhermin, S., & Hamzah, M. (2021). The effect of innovation capability on business performance: The role of social capital and entrepreneurial leadership on SMEs in Indonesia. *Accounting*, *7*(2), 323–330. DOI: 10.5267/j.ac.2020.11.021

PwC. (2018). Experience is everything: Here's how to get it right. PwC. Available: https://www.pwc.com/us/en/advisory-services/publications/consumer-intelligence-series/pwc-consumer-intelligence-series-customer-experience.pdf

Qasim, A., El Refae, G. A., Issa, H., & Eletter, S. (2021). The impact of drone technology on the accounting profession: the case of revenue recognition in long-term construction contracts. In *2021 22nd International Arab Conference on Information Technology (ACIT)* (pp. 1-4). IEEE. DOI: 10.1109/ACIT53391.2021.9677226

Qian, X., Song, H., & Ming, G. L. (2019). Brain organoids: Advances, applications and challenges. *Development (Cambridge, England)*, *146*(8), dev166074. Advance online publication. DOI: 10.1242/dev.166074 PMID: 30992274

Qiao, C., & Wang, H. (2012). The Further Research on the Application of ABC to the Optimization and Control of Project. *Engineering Management Research*, *1*(2), 96. DOI: 10.5539/emr.v1n2p96

Quadrato, G., Nguyen, T., Macosko, E. Z., Sherwood, J. L., Min Yang, S., Berger, D. R., Maria, N., Scholvin, J., Goldman, M., Kinney, J. P., Boyden, E. S., Lichtman, J. W., Williams, Z. M., McCarroll, S. A., & Arlotta, P. (2017). Cell diversity and network dynamics in photosensitive human brain organoids. *Nature*, *545*(7652), 48–53. DOI: 10.1038/nature22047 PMID: 28445462

Quer, G., Muse, E D., Nikzad, N., Topol, E J., & Steinhubl, S R. (2017, July 1). Augmenting diagnostic vision with AI. Elsevier BV, 390(10091), 221-221. https://doi.org/DOI: 10.1016/S0140-6736(17)31764-6

Quinn, K., & Buzzetto-Hollywood, N. (2019). Faculty and student perceptions of the importance of management skills in the hospitality industry.

Radhi, W. A., Hamdan, A., & Binsaddig, R. (2024). Assessing the Role of Artificial Intelligence (AI) on Tax Fraud Detection. In *Business Development via AI and Digitalization* (Vol. 1, pp. 359–364). Springer Nature Switzerland. DOI: 10.1007/978-3-031-62102-4_30

Rahim, R., & Chishti, M. A. (2024). Artificial Intelligence Applications in Accounting and Finance. In *2024 ASU International Conference in Emerging Technologies for Sustainability and Intelligent Systems (ICETSIS)* (pp. 1782-1786). IEEE. DOI: 10.1109/ICETSIS61505.2024.10459526

Rai, A., Constantinides, P., & Sarker, S. (2019). Next-generation digital platforms: Toward human–AI hybrids. *Management Information Systems Quarterly*, *43*(1), iii–ix.

Ramakrishnan, D. (2017). Contribution of CSR towards Development-the Indian perspective. *Available atSSRN* 3059833. DOI: 10.2139/ssrn.3059833

Ramprakash, P. (2020). Heart disease prediction using deep neural network. 2020 international conference on inventive computation technologies (ICICT). IEEE.

Ranmal, D., Ranasinghe, P., Paranayapa, T., Meedeniya, D., & Perera, C. (2024). ESC-NAS: Environment Sound Classification Using Hardware-Aware Neural Architecture Search for the Edge. *Sensors (Basel)*, *24*(12), 3749. DOI: 10.3390/s24123749 PMID: 38931532

Rathod, K., & Sonawane, A. (2022). Application of Artificial Intelligence in Project Planning to Solve Late and Over-Budgeted Construction Projects. In *2022 International Conference on Sustainable Computing and Data Communication Systems (ICSCDS)* (pp. 424-431). IEEE. DOI: 10.1109/ICSCDS53736.2022.9761027

Ravn, T., Sørensen, M. P., Capulli, E., Kavouras, P., Pegoraro, R., Picozzi, M., Saugstrup, L. I., Spyrakou, E., & Stavridi, V. (2023). Public perceptions and expectations: Disentangling the hope and hype of organoid research. *Stem Cell Reports*, *18*(4), 841–852. DOI: 10.1016/j.stemcr.2023.03.003 PMID: 37001517

Rejeba, H. B., Monnierb, E., Rioa, M., Evrarda, D., Tardifb, F., & Zwolinski, P. (2022). From Innovation to Eco-Innovation: Co-Created Training Materials as a Change Driver for Research and Technology Organisations. *29th CIRP Life Cycle Engineering Conference. Science Direct Procedia CIRP, 105*. 98-103. DOI: 10.1016/j.procir.2022.02.017

Replika. (2021). Your AI Friend. Replika. Available: https://replika.ai/

Ridho Kismawadi, E., Irfan, M., & Shah, S. M. A. R. (2023). Revolutionizing islamic finance: Artificial intelligence's role in the future of industry. In *The Impact of AI Innovation on Financial Sectors in the Era of Industry 5.0* (pp. 184–207). DOI: 10.4018/979-8-3693-0082-4.ch011

Roberts, D. L., & Candi, M. (2024). Artificial intelligence and innovation management: Charting the evolving landscape. *Technovation*, *136*, 103081. DOI: 10.1016/j.technovation.2024.103081

Robertson, A. (2022). Predicting Project Outcomes with the Association of Project Management. In *Abu Dhabi International Petroleum Exhibition and Conference* (p. D031S081R003). SPE. DOI: 10.2118/210795-MS

Rodgers, W., Degbey, W. Y., Söderbom, A., & Leijon, S. (2022). Leveraging international R&D teams of portfolio entrepreneurs and management controllers to innovate: Implications of algorithmic decision-making. *Journal of Business Research*, *140*, 232–244. DOI: 10.1016/j.jbusres.2021.10.053

Rodriguez-Aflecht, G., & Fuentealba, R. (2021). The Evolution of Microlearning: A Review. *Sustainability*, *13*(2), 669.

Romito, A., & Cobellis, G. (2015). Pluripotent stem cells: Current understanding and future directions. *Stem Cells International*, *2016*(1), 9451492. Advance online publication. DOI: 10.1155/2016/9451492 PMID: 26798367

Ronchini, F., Comanducci, L., & Antonacci, F. (2024). Synthesizing Soundscapes: Leveraging Text-to-Audio Models for Environmental Sound Classification. *arXiv preprint arXiv:2403.17864.*

Rosário, A. T., & Dias, J. C. (2024). *Innovative Digital Marketing in Business.*, DOI: 10.4018/979-8-3693-1231-5.ch001

Rosati, F., & Faria, L. G. (2019). Addressing the SDGs in sustainability reports: The relationship with institutional factors. *Journal of Cleaner Production, 215*, 1312–1326. DOI: 10.1016/j.jclepro.2018.12.107

Rosca, M., Vatra, A.-D., & Avadanei, M. (2023). The digital transformation of garment product development. *Industria Textila (Bucuresti), 74*(1), 98–106. DOI: 10.35530/IT.074.01.2022148

Rothstein, M. M. *Data Driven Mel Filter Bank Design for Environmental Sound Analysis* (Doctoral dissertation, Worcester Polytechnic Institute).

Royce, R. (2022). How Ai Scales up IOT Capability in Turbofan Jet Engines. www .rolls-royce.com/media/our-stories/discover/2020/intelligentengine-how-ai-scales -up-iot-capability-in-turbofan-jet-engines.aspx

Ruiz-Alba, J. L., Abou-Foul, M., Nazarian, A., & Foroudi, P. (2022). Digital platforms: Customer satisfaction, eWOM and the moderating role of perceived technological innovativeness. *Information Technology & People, 35*(7), 2470–2499. DOI: 10.1108/ITP-07-2021-0572

Russell, S., & Norvig, P. (2020). *Artificial Intelligence: A Modern Approach* (4th ed.). Pearson.

Rust, N. A., Ptak, E. N., Graversgaard, M., Iversen, S., Reed, M. S., de Vries, J. R., Ingram, J., Mills, J., Neumann, R. K., Kjeldsen, C., Muro, M., & Dalgaard, T. (2023). Social capital factors affecting uptake of sustainable soil management practices: A literature review. *Emerald Open Research, 1*(10). Advance online publication. DOI: 10.1108/EOR-10-2023-0002

Sabet, N. S., & Khaksar, S. (2024). The performance of local government, social capital and participation of villagers in sustainable rural development. *The Social Science Journal, 61*(1), 1–29. DOI: 10.1080/03623319.2020.1782649

Sabha, A., & Selwal, A. (2024). A novel Approach for Audio-based Video Analysis via MFCC Features. *Procedia Computer Science, 235*, 1512–1521. DOI: 10.1016/j. procs.2024.04.142

Sabokro, M., Masud, M. M., & Kayedian, A. (2021). The effect of green human resources management on corporate social responsibility, green psychological climate and employees' green behavior. *Journal of Cleaner Production, 313*, 127963. DOI: 10.1016/j.jclepro.2021.127963

Safri, S. N., Mohi, Z., & Hanafiah, M. H. M. (2020). Conceptualization of MOOC E-Learning Service Quality Dimensions in Faculty of Hotel and Tourism Management, UITM, Malaysia. In *3rd Asia Pacific International Conference of Management and Business Science (AICMBS 2019)* (pp. 268-271). Atlantis Press. DOI: 10.2991/aebmr.k.200410.041

Sağsan, M., & Özden, Y. (2021). An Overview of Microlearning: A Historical Development Perspective. [iJET]. *International Journal of Emerging Technologies in Learning*, *16*(7), 188–202.

Sahli, A., Pei, E., & Evans, R. (2023). A Conceptual Framework for Applying Artificial Intelligence to Manufacturing Projects. In *IFIP International Conference on Advances in Production Management Systems* (pp. 650-661). Cham: Springer Nature Switzerland. DOI: 10.1007/978-3-031-43666-6_44

Sakaguchi, H., Ozaki, Y., Ashida, T., Matsubara, T., Oishi, N., Kihara, S., & Takahashi, J. (2019). Self-organized synchronous calcium transients in a cultured human neural network derived from cerebral Organoids. *Stem Cell Reports*, *13*(3), 458–478. DOI: 10.1016/j.stemcr.2019.05.029 PMID: 31257131

Sakinç, İ. (2021). Analysis of the Working Capital Management Efficiency of the Manufacturing Companies in the Islamic Index. *Hitit Theology Journal*, *20*(3), 107–128. DOI: 10.14395/hid.930402

Salehi-Esfahani, S., & Ozturk, A. B. (2018). Negative reviews: Formation, spread, and halt of opportunistic behavior. *International Journal of Hospitality Management*, *74*, 138–146. DOI: 10.1016/j.ijhm.2018.06.022

Salleh, M. H. B., & Aziz, K. A. (2024). Enhancing project management with artificial intelligence: A framework for use case development. In *AIP Conference Proceedings* (Vol. 3153, No. 1). AIP Publishing. DOI: 10.1063/5.0218846

Sancaktar, E., & Bai, L. (2011). Electrically Conductive Epoxy Composites. *Polymers*, *3*(1), 427–466. www.mdpi.com/journal/polymers. DOI: 10.3390/polym3010427

Sancaktar, E., & Bai, L. "Modeling Filler Volume Fraction & Film Thickness Effects on Conductive adhesive Resistivity" *IEEE International Conference on Polymers and Adhesives in Microelectronics* DOI: DOI: 10.1109/POLYTR.2004.1402737

Sánchez, Á. P., & Deza, X. V. (2015). Environmental Policy Instruments and Eco-innovation: An Overview of Recent Studies. *Innovar: Revista de Ciencias Administrativas y Sociales, 25*(58), 65–80. https://www.jstor.org/stable/innrevcieadmsoc.25.58.65

Sánchez, O., Castañeda, K., Vidal-Méndez, S., Carrasco-Beltrán, D., & Lozano-Ramírez, N. E. (2024). Exploring the influence of linear infrastructure projects 4.0 technologies to promote sustainable development in smart cities. *Results in Engineering, 23*, 102824. DOI: 10.1016/j.rineng.2024.102824

Sanyal, P., Singh, R., & Singh, R. (2024). Making of a social buyer: The role of knowledge capital authenticity and inter-firm communication in B2B sales situations. *Journal of Marketing Theory and Practice*, •••, 1–20. DOI: 10.1080/10696679.2023.2291713

Sato, T., Vries, R. G., Snippert, H. J., Van de Wetering, M., Barker, N., Stange, D. E., Van Es, J. H., Abo, A., Kujala, P., Peters, P. J., & Clevers, H. (2009). Single Lgr5 stem cells build crypt-villus structures in vitro without a mesenchymal niche. *Nature, 459*(7244), 262–265. DOI: 10.1038/nature07935 PMID: 19329995

Savari, M., & Khaleghi, B. (2023). The role of social capital in forest conservation: An approach to deal with deforestation. *The Science of the Total Environment, 896*, 165216. DOI: 10.1016/j.scitotenv.2023.165216 PMID: 37392871

Schaefer, K., Corner, P. D., & Kearins, K. (2015). Social, Environmental and Sustainable Entrepreneurship Research: What Is Needed for Sustainability-as-Flourishing? *Organization & Environment, 28*(4), 394–413. https://www.jstor.org/stable/26164745. DOI: 10.1177/1086026615621111

Schaltegger, S. (2002). A Framework for Ecopreneurship: Leading Bioneers and Environmental Managers to Ecopreneurship. *Greener Management International, 38*, 45–58. https://www.jstor.org/stable/greemanainte.38.45

Schaltegger, S., & Burritt, R. (2018). Business Cases and Corporate Engagement with Sustainability: Differentiating Ethical Motivations. *Journal of Business Ethics, 147*(2), 241–259. https://www.jstor.org/stable/45022375. DOI: 10.1007/s10551-015-2938-0

Schaltegger, S., Lüdeke-Freund, F., & Hansen, E. G. (2016). Business Models for Sustainability: A Co-Evolutionary Analysis of Sustainable Entrepreneurship, Innovation, and Transformation. *Organization & Environment, 29*(3), 264–289. https://www.jstor.org/stable/26164769. DOI: 10.1177/1086026616633272

Schaper, M. (2002). *Introduction:* The Essence of Ecopreneurship. *Greener Management International, 38*, 26–30. https://www.jstor.org/stable/greemanainte.38.26

Schick, H., Marxen, S., & Freimann, J. (2002). Sustainability Issues for Start-up Entrepreneurs. *Greener Management International, 38*, 59–70. https://www.jstor.org/stable/greemanainte.38.59

Schmidt, A. (2007). Microlearning and the Knowledge Maturing Process:Micromedia and Corporate Learning. *Proceedings of the 3rd International Microlearning 2007*, Innsbruck, Austria, June 2007, Innsbruck University Press, 99-105.

Schmidt, G., & Botelho, A. (2020). The future of work in the digital era: The rise of the human-machine partnership. *Journal of Business Research*, *115*, 360–367.

Schmitt, L., Epler, R., Casenave, E., & Pallud, J. (2024). An Inquiry into Effective Salesperson Social Media Use in Multinational Versus Local Firms. *Journal of International Marketing*, *32*(1), 72–91. DOI: 10.1177/1069031X231207050

Schreiber, D. (2023). Lemonade Sets a New World Record. *Lemonade Blog*.www .lemonade.com/blog/lemonade-sets-new-world-record/

Schumpeter, J. (1976). The Process of Creative Destruction. In *Capitalism, Socialism and Democracy* (5th ed., pp. 81–86). George Allen & Unwin., DOI: 10.4324/9780203202050

Secinaro, S., Calandra, D., Lanzalonga, F., & Biancone, P. (2024). The Role of Artificial Intelligence in Management Accounting: An Exploratory Case Study. In *Digital Transformation in Accounting and Auditing: Navigating Technological Advances for the Future* (pp. 207–236). Springer International Publishing. DOI: 10.1007/978-3-031-46209-2_8

Sen, A., Rajakumaran, G., Mahdal, M., Usharani, S., Rajasekharan, V., Vincent, R., & Sugavanan, K. (2024). Live event detection for people's safety using NLP and deep learning. *IEEE Access: Practical Innovations, Open Solutions*, *12*, 6455–6472. DOI: 10.1109/ACCESS.2023.3349097

Senecal, S., & Nantel, J. (2004). The influence of online product recommendations on consumers' online choices. *Journal of Retailing*, *80*(2), 159–169. DOI: 10.1016/j. jretai.2004.04.001

Serra-Cantallops, A., Ramón Cardona, J., & Salvi, F. (2020). Antecedents of positive eWOM in hotels. Exploring the relative role of satisfaction, quality and positive emotional experiences. *International Journal of Contemporary Hospitality Management*, *32*(11), 3457–3477. DOI: 10.1108/IJCHM-02-2020-0113

Serrano, W. (2022). Verification and Validation for data marketplaces via a blockchain and smart contracts. *Blockchain: Research and Applications*, *3*(4), 100100.

Shahriar, H. (2024). Into the Metaverse: Technological Advances Shaping the Future of Consumer and Retail Marketing. In *The Future of Consumption* (pp. 55–75). Springer International Publishing. DOI: 10.1007/978-3-031-33246-3_4

Shahriari, S., Mohammadreza, S., & Gheiji, S. (2015). E-commerce and its impacts on global trend and market. *International Journal of Research -GRANTHAALAYAH*, *3*(4), 49–55. DOI: 10.29121/granthaalayah.v3.i4.2015.3022

Shahzad, M., Qu, Y., Zafar, A. U., Ding, X., & Rehman, S. U. (2020). Translating stakeholders' pressure into environmental practices–The mediating role of knowledge management. *Journal of Cleaner Production*, *275*, 124163. DOI: 10.1016/j.jclepro.2020.124163

Shamsudheen, S. V., Mohamad, S., Muneeza, A., & Mahomed, Z. (2024). Ethical discourse of ethical (Islamic) finance: A systematic literature review (1988–2022) and the way forward. *Journal of Islamic Accounting and Business Research*. Advance online publication. DOI: 10.1108/JIABR-11-2022-0315

Sharma, D., & Vishwanath, V. (2024, February). Heart disease prediction using machine learning ensemble. In *AIP Conference Proceedings* (Vol. 2742, No. 1). AIP Publishing. DOI: 10.1063/5.0191651

Sharples, M., Taylor, J., & Vavoula, G. (2010). A theory of learning for the mobile age. In *Medienbildung in neuen Kulturräumen* (pp. 87–99). VS Verlag für Sozialwissenschaften. DOI: 10.1007/978-3-531-92133-4_6

Shen, D., & Chen, W. (2019). Challenges and opportunities of microlearning: A review of the literature. *Journal of Educational Technology & Society*, *22*(3), 62–76.

Sheng, Y., Wang, S., & Wang, Y. (2024). Doing good in times of need: Green finance policy and strategic corporate social responsibility. *Economic Analysis and Policy*, *84*, 1029–1045. Advance online publication. DOI: 10.1016/j.eap.2024.10.008

Shete, N. L., Maddel, M., & Shaikh, Z. (2024). A Comparative Analysis of Cybersecurity Scams: Unveiling the Evolution from Past to Present. In *2024 IEEE 9th International Conference for Convergence in Technology (I2CT)* (pp. 1-8). IEEE.

Shih, R. C., Hsin, H. T., Huang, H. C., & Cheng, Y. M. (2015). The Development of a mobile App for a Hospitality and Tourism course at technological universities. In *2015 Third International Conference on Robot, Vision and Signal Processing (RVSP)* (pp. 252-255). IEEE. DOI: 10.1109/RVSP.2015.67

Shin, D. (2020). How VR Can Bridge Gender Gaps in STEM. *Journal of Educational Technology & Society*, *23*(4), 23–35.

Shin, H., Perdue, R. R., & Kang, J. (2019). Front desk technology innovation in hotels: A managerial perspective. *Tourism Management*, *74*, 310–318. DOI: 10.1016/j.tourman.2019.04.004

Shishkin, S., Hollosi, D., Goetze, S., & Doclo, S. (2024, April). Active Learning for Sound Event Classification Using Bayesian Neural Networks with Gaussian Variational Posterior. In *ICASSP 2024-2024 IEEE International Conference on Acoustics, Speech and Signal Processing (ICASSP)* (pp. 896-900). IEEE. DOI: 10.1109/ICASSP48485.2024.10446970

Shrivastav, P. (2022). Hubspot Annual State of Service in 2022. www.hubspot .com/hubfs/assets/flywheel%20campaigns/HubSpot%20Annual%20State%20of %20Service%20Report%20-%202022.pdf

Shum, H., He, X., & Li, D. (2018). From Eliza to XiaoIce: Challenges and Opportunities with Social Chatbots. *Frontiers of Information Technology & Electronic Engineering*, *19*(1), 10–26.

Silva, D. A., Whitehead, S., Lengerich, C., & Leather, H. (2024). CoLLAT: On adding fine-grained audio understanding to language models using token-level locked-language tuning. *Advances in Neural Information Processing Systems*, ●●●, 36.

Singh, M. P., & Rashmi, P. (2024). Convolution Neural Networks of Dynamically Sized Filters with Modified Stochastic Gradient Descent Optimizer for Sound Classification.

Singh, A. K., & Taterh, S. (2023). Exploring the Significance and Obstacles of Adopting Futuristic Technology Perspectives for Entrepreneurship and Sustainable Innovation. In *Futuristic Technology Perspectives on Entrepreneurship and Sustainable Innovation* (pp. 1–10). IGI Global. DOI: 10.4018/978-1-6684-5871-6.ch001

Singh, K., Han, A., & Wang, W. (2021). AI-Powered Chatbots for Delivering Telemedicine in Remote Areas. *Journal of Medical Internet Research*, *23*(5), e24732.

Sinnott, R. (2024). Predicting and Avoiding Dog Barking Behaviour through Deep Learning. In *Proceedings of the 2024 Australasian Computer Science Week* (pp. 26-35). DOI: 10.1145/3641142.3641176

Skripak, S. J. (2016). *Fundamentals of business*. Virginia Tech.

Slavickova, L. (2022). Hyper-Personalization Can Increase Your Conversion Rates by up to 60%. *Trask*. www.thetrask.com/blog/hyper-personalization-can-increase -your-conversion-rates-by-up-to-60

Small, H. (1973). Co-citation in the scientific literature: A new measure of the relationship between two documents. *Journal of the American Society for Information Science*, *24*(4), 265–269. DOI: 10.1002/asi.4630240406

Smirnova, L., Caffo, B. S., Gracias, D. H., Huang, Q., Morales Pantoja, I. E., Tang, B., Zack, D. J., Berlinicke, C. A., Boyd, J. L., Harris, T. D., Johnson, E. C., Kagan, B. J., Kahn, J., Muotri, A. R., Paulhamus, B. L., Schwamborn, J. C., Plotkin, J., Szalay, A. S., Vogelstein, J. T., & Hartung, T. (2023). Organoid intelligence (OI): The new frontier in biocomputing and intelligence-in-a-dish. *Frontiers in Science*, *1*, 1017235. Advance online publication. DOI: 10.3389/fsci.2023.1017235

Smith, A. V., & Garza-Rubalcava, U. (2019). "Ecological Modernization Theory: Developing a Consensus with the Addition of Green and Sustainable Remediation". Springer Nature (Switzerland W. Leal Filho et al. (eds.)). *Industry, Innovation and Infrastructure, Encyclopedia of the UN Sustainable Development Goals*. DOI: 10.1007/978-3-319-71059-4_39-1

Smith, S. (2019). Digital Voice Assistants in Use to Triple to 8 Billion by 2023, Driven by Smart Home Devices. www.juniperresearch.com/press/digital-voice-assistants-in-use-to-8-million-2023/

Smith, T. W. (1999). Aristotle on the Conditions for and Limits of the Common Good. *The American Political Science Review*, *93*(3), 625–637. DOI: 10.2307/2585578

Solomon, D. D., Sonia, , Kumar, K., Kanwar, K., Iyer, S., & Kumar, M. (2024). Extensive review on the role of machine learning for multifactorial genetic disorders prediction. *Archives of Computational Methods in Engineering*, *31*(2), 623–640. DOI: 10.1007/s11831-023-09996-9

Song, X., Xiong, J., Wang, M., Mei, Q., & Lin, X. (2024). Combined Data Augmentation on EANN to Identify Indoor Anomalous Sound Event. *Applied Sciences (Basel, Switzerland)*, *14*(4), 1327. DOI: 10.3390/app14041327

Spitzer, K. L., & Lippoldt, D. (2019). Microlearning: A Review of the Literature. *Journal of Educational Technology Systems*, *47*(2), 151–172.

Spotify. (2022). Discovery Mode – Spotify for Artists. Discovery Mode – Spotify for Artists. artists.spotify.com/en/discovery-mode

Srivastava, M., & Sivaramakrishnan, S. (2021). The impact of eWOM on consumer brand engagement. *Marketing Intelligence & Planning*, *39*(3), 469–484. DOI: 10.1108/MIP-06-2020-0263

Stancheva-Todorova, E., & Bogdanova, B. (2021). Enhancing investors' decision-making–An interdisciplinary AI-based case study for accounting students. In *AIP Conference Proceedings* (Vol. 2333, No. 1). AIP Publishing.

Stankov, I. (2024). *Natural Audio Data Augmentation Techniques* (Master's thesis, Humboldt-Universität zu Berlin).

Stellefson, M., Dipnarine, K., & Stopka, C. (2013, February 21). The Chronic Care Model and Diabetes Management in US Primary Care Settings: A Systematic Review. *Preventing Chronic Disease*, *10*, 120180. Advance online publication. DOI: 10.5888/pcd10.120180 PMID: 23428085

Stockholm Environment Institute. (2019). *Transformational change through a circular economy*. Stockholm Environment Institute. https://www.jstor.org/stable/resrep22978

Stone, P., Brooks, R., Brynjolfsson, E., Calo, R., Etzioni, O., Hager, G., & Shoham, Y. (2016). Artificial Intelligence and Life in 2030: One Hundred Year Study on Artificial Intelligence: Report of the 2015–2016 Study Panel. Stanford University. https://ai100.stanford.edu/2016-report

Subodh, G., Deepu, V., Mohanan, P., & Sebestian, M. T. (2009). Dielectric response of high permittivity polymer ceramic composite with low loss tangent. *Applied Physics Letters*, *95*(6), 062903. DOI: 10.1063/1.3200244

Suki, N. M., Suki, N. M., & Azman, N. S. (2016). Impacts of Corporate Social Responsibility on the Links Between Green Marketing Awareness and Consumer Purchase Intentions. *Procedia Economics and Finance*, *37*, 262–268. DOI: 10.1016/S2212-5671(16)30123-X

Sun, G., & Zhou, Y. (2023, December 19). AI in healthcare: Navigating opportunities and challenges in digital communication. *Frontiers of Medicine*, *5*, 1291132. Advance online publication. DOI: 10.3389/fdgth.2023.1291132 PMID: 38173911

Sun, X., & Xie, X. (2024). How does digital finance promote entrepreneurship? The roles of traditional financial institutions and BigTech firms. *Pacific-Basin Finance Journal*, *85*, 102316. Advance online publication. DOI: 10.1016/j.pacfin.2024.102316

Sun, Y., Li, J., Wang, L., Xv, J., & Liu, Y. (2024). Deep Learning-based drone acoustic event detection system for microphone arrays. *Multimedia Tools and Applications*, *83*(16), 47865–47887. DOI: 10.1007/s11042-023-17477-1

Sun, Z., Yin, H., Chen, H., Chen, T., Cui, L., & Yang, F. (2020). Disease prediction via graph neural networks. *IEEE Journal of Biomedical and Health Informatics*, *25*(3), 818–826. DOI: 10.1109/JBHI.2020.3004143 PMID: 32749976

Swain, S. (2014). From philanthropy to social entrepreneurship. In Damousi, J., Rubenstein, K., & Tomsic, M. (Eds.), *Diversity in Leadership: Australian women, past and present* (pp. 189–206). ANU Press., https://www.jstor.org/stable/j.ctt13wwvj5.13 DOI: 10.22459/DL.11.2014.10

Swinscoe. (2023). The New Rules of Customer Engagement. www.freshworks.com/assets/resources/freshworks/The-New-Rules-of-Customer-Engagement_US.pdf

Taeusch, C. F. (1935). The Relation between Legal Ethics and Business Ethics. *California Law Review*, *24*(1), 79–95. DOI: 10.2307/3476485

Talwar, S., Kaur, P., Escobar, O., & Lan, S. (2022). Virtual reality tourism to satisfy wanderlust without wandering: An unconventional innovation to promote sustainability. *Journal of Business Research*, *152*, 128–143. DOI: 10.1016/j.jbusres.2022.07.032

Taneja, S. S., Taneja, P. K., & Gupta, R. K. (2011). Researches in Corporate Social Responsibility: A Review of Shifting Focus, Paradigms, and Methodologies. *Journal of Business Ethics*, *101*(3), 343–364. https://www.jstor.org/stable/41475906. DOI: 10.1007/s10551-010-0732-6

Tarımer, İ., & Karadağ, B. C. Genres Classification of Popular Songs Listening by Using Keras. *Gazi University Journal of Science Part A: Engineering and Innovation, 11*(1), 123-136.

Teacher Education International Conference (pp. 2457-2463). Association for the Advancement of Computing in Education (AACE).

Technology Review Insights, M. I. T. (2020). The Global AI Agenda: Promise, Reality, and a Future of Data Sharing. *MIT Technology Review*.

Teece, D. J., Pisano, G., & Shuen, A. (1997). 'Dynamic Capabilities and Strategic Management. *Strategic Management Journal*, *18*(7), 509–533. DOI: 10.1002/(SICI)1097-0266(199708)18:7<509::AID-SMJ882>3.0.CO;2-Z

The Power of Hyper-Personalization: How Ai Elevates Customer Experience. (2024). *Comarch*. www.comarch.com/trade-and-services/loyalty-marketing/blog/the-power-of-hyper-personalization/

Thirumagal, P. G., Vaddepalli, S., Das, T., Das, S., Madem, S., & Immaculate, P. S. (2024). AI-Enhanced IoT Data Analytics for Risk Management in Banking Operations. In *2024 5th International Conference on Recent Trends in Computer Science and Technology (ICRTCST)* (pp. 177-181). IEEE. DOI: 10.1109/ICRTCST61793.2024.10578533

Thirumalesh Madanaguli, A., Dhir, A., Kaur, P., Mishra, S., & Srivastava, S. (2023). A systematic literature review on corporate social responsibility (CSR) and hotels: Past achievements and future promises in the hospitality sector. *Scandinavian Journal of Hospitality and Tourism*, *23*(2-3), 141–175. DOI: 10.1080/15022250.2023.2221214

Thorne, S. L. (2017). The Evolution of Mobile Microlearning: From Personalization to Adaptive Learning. *Journal of Learning Analytics*, *4*(3), 114–122.

Tigges, M., Mestwerdt, S., Tschirner, S., & Mauer, R. (2024). Who gets the money? A qualitative analysis of fintech lending and credit scoring through the adoption of AI and alternative data. *Technological Forecasting and Social Change*, *205*, 123491. Advance online publication. DOI: 10.1016/j.techfore.2024.123491

Tilley, F., & Young, W. (2006). Sustainability Entrepreneurs: Could They Be the True Wealth Generators of the Future? *Greener Management International, 55*, 79–92. https://www.jstor.org/stable/greemanainte.55.79

Tober, M. (2011). PubMed, ScienceDirect, Scopus or Google Scholar – Which is the best search engine for an effective literature research in laser medicine? *Medical Laser Application*, *26*(3), 139–144. DOI: 10.1016/j.mla.2011.05.006

Toledano, D. S., Toledano, J. I. S. S., & Jiménez, I. M. Á. (2024). Implementation of Analytical Accounting Models and Management Indicators and Development of Business Intelligence and Business Analytics Tools in Urban and Metropolitan Collective Transport Operators: A Case Study. *Revista de Gestão Social e Ambiental*, *18*(9), e06070–e06070. DOI: 10.24857/rgsa.v18n9-023

Torgerson, C. (2016). *The Microlearning Guide to Microlearning*. Torgerson Consulting.

Tran, K. T., Vu, X. S., Nguyen, K., & Nguyen, H. D. (2024). NeuProNet: Neural profiling networks for sound classification. *Neural Computing & Applications*, *36*(11), 5873–5887. DOI: 10.1007/s00521-023-09361-8

Trawnih, A., Al-Masaeed, S., Alsoud, M., & Alkufahy, A. (2022). Understanding artificial intelligence experience: A customer perspective. *International Journal of Data and Network Science*, *6*(4), 1471–1484. DOI: 10.5267/j.ijdns.2022.5.004

Trujillo, C. A., Gao, R., Negraes, P. D., Gu, J., Buchanan, J., Preissl, S., Wang, A., Wu, W., Haddad, G. G., Chaim, I. A., Domissy, A., Vandenberghe, M., Devor, A., Yeo, G. W., Voytek, B., & Muotri, A. R. (2019). Complex oscillatory waves emerging from cortical Organoids model early human brain network development. *Cell Stem Cell*, *25*(4), 558–569.e7. DOI: 10.1016/j.stem.2019.08.002 PMID: 31474560

Tursunbayeva, A., & Gal, H. C. B. (2024). Adoption of artificial intelligence: A TOP framework-based checklist for digital leaders. *Business Horizons*, *67*(4), 357–368. DOI: 10.1016/j.bushor.2024.04.006

U.S. Congress. (2019). Algorithmic Accountability Act of 2019. 116th Congress. Available: https://www.congress.gov/bill/116th-congress/house-bill/2231

Ukabi, O. B., Uba, U. J., Ewum, C. O., & Olubiyi, T. O. (2023). Measuring Entrepreneurial Skills and Sustainability in Small Business Enterprises Post-Pandemic: Empirical Study from Cross River State, Nigeria. *International Journal of Business. Management and Economics, 4*(2), 132–149. DOI: 10.47747/ijbme.v4i2.1140

Uma, R., & Uma, K. (2021). Corporate social responsibility in india-an overview. *ACADEMICIA: An International Multidisciplinary Research Journal, 11*(5), 906–912.

Umesh Dwivedi, S. A. R. (2009). Hashmi "SiC dispersed polysulphide epoxy resin based functionally graded material". *Polymer Composites, 30*(2), 162–168. DOI: 10.1002/pc.20546

Unger, J. (2012, March 1). Uncovering undetected hypoglycemic events. Dove Medical Press, 57-57. https://doi.org/DOI: 10.2147/DMSO.S29367

United States Patent and Trademark Office. (2013). Method and System for Anticipatory Package Shipping. Patent No. US 8,615,473 B2.

Uwem, E. I., Oyedele, O. O., & Olubiyi, O. T. (2021). Workplace green behavior for sustainable competitive advantage. In *Human Resource Management Practices for PromotingSustainability.* IGI Global.

van der Ree, K. (2019). Promoting Green Jobs: Decent Work in the Transition to Low-carbon, Green Economies. In C. Gironde & G. Carbonnier (Eds.), *The ILO @ 100: Addressing the Past and Future of Work and Social Protection* (Vol. 11, pp. 248–272). Brill. https://www.jstor.org/stable/10.1163/j.ctvrxk4c6.19

Van Eck, N., & Waltman, L. (2010). Software survey: VOSviewer, a computer program for bibliometric mapping. *Scientometrics, 84*(2), 523-538.

van Eck, N. J., & Waltman, L. (2014). Visualizing Bibliometric Networks. In *Measuring Scholarly Impact* (pp. 285–320). Springer International Publishing., DOI: 10.1007/978-3-319-10377-8_13

van Ewijk, S., & Stegemann, J. (2023). THE CIRCULAR ECONOMY. In *An Introduction to Waste Management and Circular Economy* (pp. 306–348). UCL Press., DOI: 10.2307/jj.4350575.17

Van Marrewijk, M. (2003). Concept and Definitions of CSR and Corporate Sustainability: Between Agency and Corporate Sustainability: Between Agency nd Communion. *Journal of Business Ethics, 44*(2), 95–105. DOI: 10.1023/A:1023331212247

Vargas-Hernandez, J. G., González-Àvila, F. J., Vargas-Gonzàlez, O. C., Castañeda-Burciaga, S., & Guirette-Barbosa, O. A. (2025). Green Technology Innovation and Its Implications in the Sustainable Organizational Environment. In Ullah, A., Pandey, J., & Masengu, R. (Eds.), *Impacts of Technology on Operations Management: Adoption, Adaptation, and Optimization* (pp. 205–234). IGI Global., DOI: 10.4018/979-8-3693-6205-1.ch008

Vargas-Hernandez, J. G., González, F. J., Orozco-Qijano, E. P., & Vargas-Gonzàlez, O. C. (2024). Green Organizational Management and Technological Innovation on Green Sustainable Organizational Performance. In Cepni, E. (Ed.), *Chaos, Complexity, and Sustainability in Management* (pp. 115–140). IGI Global., DOI: 10.4018/979-8-3693-2125-6.ch007

Vegar, B., & Mijač, T. (2024). Artificial Intelligence in Project Management: Insights from Croatia. In *2024 47th MIPRO ICT and Electronics Convention (MIPRO)* (pp. 1766-1771). IEEE.

Venkatesh, V., & Bala, H. (2008). Technology acceptance model 3 and a research agenda on interventions. *Decision Sciences*, *39*(2), 273–315. DOI: 10.1111/j.1540-5915.2008.00192.x

Venkatesh, V., & Davis, F. D. (2000). A theoretical extension of the technology acceptance model: Four longitudinal field studies. *Management Science*, *46*(2), 186–204. DOI: 10.1287/mnsc.46.2.186.11926

Venkatraman. (2021, April). Part 3: The Five Traps That Stall Digitalization. *Strategic*, *CFO360*, 26. strategiccfo360.com/the-five-traps-that-stall-digitalization/

Verhulst, S., & Young, A. (2017). *Open Data in Developing Economies: Toward Building an Evidence Base on What Works and How*. African Minds. DOI: 10.47622/9781928331599

Verint. (2019). Engagement in the Always-on Era: How Humans and Technology Work Hand-in-Hand to Meet Rising Expectations. Verint.

Verma, S. (2011). Why Indian companies indulge in CSR? *Journal of Management and Public Policy*, *2*(2), 52–69.

Verma, S., & Yadav, N. (2021). Past, Present, and Future of Electronic Word of Mouth (EWOM). *Journal of Interactive Marketing*, *53*(1), 111–128. DOI: 10.1016/j.intmar.2020.07.001

Vidmar, M., Fleck, J., & Williams, R. (2023). AI and Data in Engineering and Innovation: Towards a Sustainable Future? In *2023 IEEE International Conference on Engineering, Technology and Innovation (ICE/ITMC)* (pp. 1-2). IEEE. DOI: 10.1109/ICE/ITMC58018.2023.10332336

Viglia, G., Pelloia, M., & Buhalis, D. (2018). Information technology in hospitality education. *Innovation in Hospitality Education: Anticipating the Educational Needs of a Changing Profession*, 87-100.

Volery, T. (2002). An Entrepreneur Commercialises Conservation: The Case of Earth Sanctuaries Ltd. *Greener Management International, 38*, 109–116. https://www.jstor.org/stable/greemanainte.38.109

Vosoughi, A., Bondi, L., Wu, H. H., & Xu, C. (2024, April). Learning Audio Concepts from Counterfactual Natural Language. In *ICASSP 2024-2024 IEEE International Conference on Acoustics, Speech and Signal Processing (ICASSP)* (pp. 366-370). IEEE. DOI: 10.1109/ICASSP48485.2024.10446736

Wallace, C., & Chen, G. (2006). A multilevel integration of personality, climate, self-regulation, and performance. *Personnel Psychology, 59*(3), 529–557. DOI: 10.1111/j.1744-6570.2006.00046.x

Walley, E. E. (Liz), & Taylor, D. W. (David). (2002). Opportunists, Champions, Mavericks …? A Typology of Green Entrepreneurs. *Greener Management International, 38*, 31–43. https://www.jstor.org/stable/greemanainte.38.31

Wang, J. (2023). Intelligent Decision Support System for Building Project Management Based on Artificial Intelligence. [). IOP Publishing.]. *Journal of Physics: Conference Series, 2665*(1), 012022. DOI: 10.1088/1742-6596/2665/1/012022

Wangmo, T., Lipps, M., Kressig, R. W., & Ienca, M. (2019). Ethical concerns with the use of intelligent assistive technology: Findings from a qualitative study with professional stakeholders. *BMC Medical Ethics, 20*(1), 98. Advance online publication. DOI: 10.1186/s12910-019-0437-z PMID: 31856798

Wang, Z., Shahid, M. S., Binh An, N., Shahzad, M., & Abdul-Samad, Z. (2022). Does green finance facilitate firms in achieving corporate social responsibility goals? *Economic research-. Ekonomska Istrazivanja, 35*(1), 5400–5419. DOI: 10.1080/1331677X.2022.2027259

Watson, F., & Wu, Y. (2022). The Impact of Online Reviews on the Information Flows and Outcomes of Marketing Systems. *Journal of Macromarketing, 42*(1), 146–164. DOI: 10.1177/02761467211042552

West, S. M., Whittaker, M., & Crawford, K. (2019). *Discriminating Systems: Gender, Race, and Power in AI.* AI Now Institute.

Williamson, D., Lynch-Wood, G., & Ramsay, J. (2006). Drivers of Environmental Behaviour in Manufacturing SMEs and the Implications for CSR. *Journal of Business Ethics, 67*(3), 317–330. https://www.jstor.org/stable/25123876. DOI: 10.1007/s10551-006-9187-1

Wilson, C., & van der Velden, M. (2022). Sustainable AI: An integrated model to guide public sector decision-making. *Technology in Society, 68,* 101926. Advance online publication. DOI: 10.1016/j.techsoc.2022.101926

Wilson, F., & Post, J. E. (2013). Business models for people, planet (& profits): Exploring the phenomena of social business, a market-based approach to social value creation. *Small Business Economics, 40*(3), 715–737. https://www.jstor.org/stable/23360620. DOI: 10.1007/s11187-011-9401-0

Wimalasena, R. A. L. B., & Ranasinghe, D. D. M. (2024). Audio-Based Vehicle Detection System: Enhancing Safety for Bicycle Riders. *ENGINEER, 57*(01), 85–94. DOI: 10.4038/engineer.v57i1.7613

Windsor, D. (2001). The Future of Corporate Social Responsibility. *The International Journal of Organizational Analysis, 9*(3), 225–256. DOI: 10.1108/eb028934

Wirba, A. V. (2023). Corporate social responsibility (CSR): The role of government in promoting CSR. *Journal of the Knowledge Economy, 15*(2), 1–27. DOI: 10.1007/s13132-023-01185-0

Woebot Health. (2021). Woebot: Your Self-Care Expert. Woebot Health. Available: https://woebothealth.com/

Wood, D. J. (1991). Corporate Social Performance Revisited. *Academy of Management Review, 16*(4), 691–718. DOI: 10.2307/258977

Wu, L., Wang, L., Philipsen, N. J., & Fang, X. (2023). The impact of eco-innovation on environmental performance in different regional settings: New evidence from Chinese cities. *Environment, Development and Sustainability.* Advance online publication. DOI: 10.1007/s10668-023-04280-z

Wu, Y., & Li, Z. (2024). Digital transformation, entrepreneurship, and disruptive innovation: Evidence of corporate digitalization in China from 2010 to 2021. *Humanities & Social Sciences Communications, 11*(1), 163. Advance online publication. DOI: 10.1057/s41599-023-02378-3

Xue, F., Sanderson, A. C., & Graves, R. J. (2009). Multiobjective evolutionary decision support for design–supplier–manufacturing planning. *IEEE Transactions on Systems, Man, and Cybernetics. Part A, Systems and Humans*, *39*(2), 309–320. DOI: 10.1109/TSMCA.2008.2010791

Xue, X., Sun, Y., Resto-Irizarry, A. M., Yuan, Y., Aw Yong, K. M., Zheng, Y., Weng, S., Shao, Y., Chai, Y., Studer, L., & Fu, J. (2018). Mechanics-guided embryonic patterning of neuroectoderm tissue from human pluripotent stem cells. *Nature Materials*, *17*(7), 633–641. DOI: 10.1038/s41563-018-0082-9 PMID: 29784997

Xu, X., & Li, Y. (2016). The antecedents of customer satisfaction and dissatisfaction toward various types of hotels: A text mining approach. *International Journal of Hospitality Management*, *55*, 57–69. DOI: 10.1016/j.ijhm.2016.03.003

Yadav, U. S., Sood, K., Tripathi, R., Grima, S., & Yadav, N. (2023). "Entrepreneurship in India's handicraft industry with the support of digital technology and innovation during natural calamities." *International Journal of Sustainable Development and Planning. International Journal of Sustainable Development and Planning*, *18*(6), 1777–1791. DOI: 10.18280/ijsdp.180613

Yala, A., Lehman, C., Schuster, T., Portnoi, T., & Barzilay, R. (2020). A Deep Learning Mammography-based Model for Improved Breast Cancer Risk Prediction. *Radiology*, *296*(1), 90–98. PMID: 31063083

Yamane, Y. (1967). Mathematical Formulae for Sample Size Determination.

Yang, L., Henthorne, T. L., & George, B. (2020). Artificial intelligence and robotics technology in the hospitality industry: Current applications and future trends. Digital transformation in business and society: Theory and cases, 211-228.

Yang, D., Zhang, J., Sun, Y., & Huang, Z. (2024). Showing usage behavior or not? The effect of virtual influencers' product usage behavior on consumers. *Journal of Retailing and Consumer Services*, *79*, 103859. DOI: 10.1016/j.jretconser.2024.103859

Yang, L., Zhaob, Q., Hou, Y., Sun, R., Chengb, M., Shena, M., Zenga, S., Jib, H., & Qiub, J. (2018). High breakdown strength and outstanding piezoelectric performance in flexible PVDF based percolative nanocomposites through the synergistic effect of topological-structure and composition modulations. *Composites. Part A, Applied Science and Manufacturing*, *114*(November), 13–20. DOI: 10.1016/j.compositesa.2018.07.039

Yang, S. J. H., & Lin, Y. C. (2019). A Personalized Microlearning Approach for EFL Learners: A Study of Learning Effectiveness and Learner Satisfaction. *Journal of Educational Technology & Society*, *22*(1), 103–115.

Yang, S. N., Chang, L. C., & Chang, F. J. (2019). AI-based design of urban storm-water detention facilities accounting for carryover storage. *Journal of Hydrology (Amsterdam)*, *575*, 1111–1122. DOI: 10.1016/j.jhydrol.2019.06.009

Yang, X. (2023). The effects of AI service quality and AI function-customer ability fit on customer's overall co-creation experience. *Industrial Management & Data Systems*, *123*(6), 1717–1735. DOI: 10.1108/IMDS-08-2022-0500

Yang, X., Hu, J., & He, J. "Adjusting Nonlinear Characteristics of ZnO-Silicone Rubber Composites by Controlling Filler's Shape and Size" *2016 IEEE International Conference on Dielectrics (ICD)* 23 August 2016. DOI: 10.1109/ICD.2016.7547607

Yani, A., Eliyana, A., Hamidah, I., & Buchdadi, A. D. (2020). The impact of social capital, entrepreneurial competence on business performance: An empirical study of SMEs. *Systematic Reviews in Pharmacy*, *11*(9), 779–787.

Ye, Z., Ciccarelli, G., & Kulis, B. (2024, April). Maximum-Entropy Adversarial Audio Augmentation for Keyword Spotting. In *ICASSP 2024-2024 IEEE International Conference on Acoustics, Speech and Signal Processing (ICASSP)* (pp. 10826-10830). IEEE. DOI: 10.1109/ICASSP48485.2024.10446557

Yeo, S. F., Tan, C. L., Kumar, A., Tan, K. H., & Wong, J. K. (2022). Investigating the impact of AI-powered technologies on Instagrammers' purchase decisions in digitalization era–A study of the fashion and apparel industry. *Technological Forecasting and Social Change*, *177*, 121551–121567. DOI: 10.1016/j.techfore.2022.121551

Yi, Z., & Luo, X. (2024). Construction cost estimation model and dynamic management control analysis based on artificial intelligence. *Civil Engineering (Shiraz)*, *48*(1), 577–588. DOI: 10.1007/s40996-023-01173-z

Yoo, K., Welden, R., Hewett, K., & Haenlein, M. (2023). The merchants of meta: A research agenda to understand the future of retailing in the Metaverse. *Journal of Retailing*, *99*(2), 173–192. DOI: 10.1016/j.jretai.2023.02.002

Yu, R., Yu, W., & Wang, X. (2024). Kan or mlp: A fairer comparison. *arXiv preprint arXiv:2407.16674*.

Yuan, Y., Chen, Z., Liu, X., Liu, H., Xu, X., Jia, D., . . . Wang, W. (2024). T-CLAP: Temporal-Enhanced Contrastive Language-Audio Pretraining. *arXiv preprint arXiv:2404.17806*. DOI: 10.1109/MLSP58920.2024.10734763

Yudha, P. (2018). Exploring the impact of social capital on entrepreneurial orientation and business performance (Study on members of MSMEs communities in Malang). *Profit: Jurnal Adminsitrasi Bisnis*, *12*(1), 20–31. DOI: 10.21776/ub.profit.2018.012.01.3

Yu, Y., Li, Y., Zhang, Z., Gu, Z., Zhong, H., Zha, Q., Yang, L., Zhu, C., & Chen, E. (2020). A bibliometric analysis using VOSviewer of publications on COVID-19. *Annals of Translational Medicine, 8*(13), 816. DOI: 10.21037/atm-20-4235 PMID: 32793661

Yu, Z., & Gibbs, D. (2020). Unravelling the role of green entrepreneurs in urban sustainability transitions: A case study of China's Solar City. *Urban Studies (Edinburgh, Scotland), 57*(14), 2901–2917. https://www.jstor.org/stable/26959607. DOI: 10.1177/0042098019888144

Zaman, A. (2023). Zero-Waste: A New Sustainability Paradigm for Addressing the Global Waste Problem, in *The Vision Zero Handbook: Theory, Technology and Management for a Zero Casualty Policy*, (Edvardsson, K.et al., (Eds.), *Springer*.) DOI: 10.1007/978-3-030-76505-7_46

Zema, T., Kozina, A., Sulich, A., Römer, I., & Schieck, M. (2022). Deep learning and forecasting in practice: An alternative costs case. *Procedia Computer Science, 207*, 2958–2967. DOI: 10.1016/j.procs.2022.09.354

Zgraggen, M. (2021). Blended learning model in a vocational educational training hospitality setting: From teachers' perspectives. *International Journal of Training Research, 19*(3), 202–228. DOI: 10.1080/14480220.2021.1933568

Zhang, D., Chen, J., Bai, J., & Wang, M. (2024). Sound event localization and classification using WASN in Outdoor Environment. *arXiv preprint arXiv:2403.20130*.

Zhang, Y., Cai, J., Shan, X., Li, S., & Shou, Y. Sales promotion and supply chain finance for shopping days: Strategies of e-commerce platform and seller. Managerial and Decision Economics.

Zhang, H., Zhao, L., & Gupta, S. (2018). The role of online product recommendations on customer decision making and loyalty in social shopping communities. *International Journal of Information Management, 38*(1), 150–166. DOI: 10.1016/j.ijinfomgt.2017.07.006

Zhao, W., Wang, H., Chen, Y., Pan, X., Zhang, K., & Bai, Z. An Environmental Sound Classification Algorithm Based on Multiscale Channel Feature Fusion.

Zhao, Y., Xu, X., & Wang, M. (2019a). Predicting overall customer satisfaction: Big data evidence from hotel online textual reviews. *International Journal of Hospitality Management, 76*, 111–121. DOI: 10.1016/j.ijhm.2018.03.017

Zheng, X. (2023). Application and Prospects of Digital Technology in Teaching Tourism Landscape. *Contemporary Education and Teaching Research, 4*(09), 467–471. DOI: 10.61360/BoniCETR232014900910

Zhong, X., Harris, G., Smirnova, L., Zufferey, V., Sá, R. D., Baldino Russo, F., Baleeiro Beltrao Braga, P. C., Chesnut, M., Zurich, M., Hogberg, H. T., Hartung, T., & Pamies, D. (2020). Antidepressant Paroxetine exerts developmental neurotoxicity in an ipsc-derived 3D human brain model. *Frontiers in Cellular Neuroscience*, *14*, 14. DOI: 10.3389/fncel.2020.00025 PMID: 32153365

Zhou, L., Gao, J., Li, D., & Shum, H. (2020). The Design and Implementation of XiaoIce, an Empathetic Social Chatbot. *Computational Linguistics*, *46*(1), 53–93.

Zhu, F., & Zhang, X. (2010). Impact of Online Consumer Reviews on Sales: The Moderating Role of Product and Consumer Characteristics. *Journal of Marketing*, *74*(2), 133–148. DOI: 10.1509/jm.74.2.133

Zohaib, M., Asim, M., & ELAffendi, M. (2024). Enhancing Emergency Vehicle Detection: A Deep Learning Approach with Multimodal Fusion. *Mathematics*, *12*(10), 1514. DOI: 10.3390/math12101514

About the Contributors

Sanjay Misra is a highly accomplished Senior Scientist at the Institute for Energy Technology (IFE) in Halden, Norway. With a distinguished academic background, including a Ph.D. in Information and Knowledge Engineering from the University of Alcala, Spain, and an M.Tech. in Software Engineering from MLN National Institute of Technology, India, Dr. Misra specializes in Applied Informatics, focusing on Software Engineering, Cyber Security, Health Informatics, and AI-based Intelligent Systems. His expertise and contributions have earned him recognition as one of the top 2% of scientists globally, as ranked by Stanford University for five consecutive years. He is also ranked 6th in Norway in all disciplines (2nd in Science and Technology) in 2024. Dr Misra has received multiple accolades, including the 2014 IET Software Premium Award, underscoring his significant contributions to the field. As a Senior Member of IEEE and an ACM Distinguished Lecturer (2021–2024), Dr. Misra has chaired numerous influential conferences and edited special issues for prestigious journals. He is the Editor-in-Chief of the International Journal of Human Capital & Information Technology Professionals and serves as an editor for journals like Nature: Scientific Reports and Elsevier's Alexandria Engineering Journal. Dr. Misra has authored over 100 books and proceedings and delivered more than 100 keynote speeches in over 60 countries, making him a prominent global speaker in the fields of informatics, cybersecurity, and AI.

Manju Kaushik is a Professor at Amity Institute of Information Technology (AIIT), Deputy Director-Amity Innovation Incubator, President of the Institution's Innovation Council (IIC), Amity University Rajasthan, Branch Counselor IEEE & ACM student Branches and Coordinator of Technical Clubs –Amity University Rajasthan. She has more than 19 years of experience in the field of teaching and research. She was awarded Ph.D. from the Mohan Lal Sukhadia University, Udaipur, India. Her research papers have been published in various journals and conferences of National and International repute like IEEE, Springer, Elsevier, and other SCI, and

Scopus indexed. She has completed her Post Doc Associate Researcher in CENTER OF ICT/ICE (SOFTWARE & INTELLIGENT SYSTEMS) RESEARCH, CURID, Covenant University, Nigeria. She has organized 07 International conferences and more than 350 events (workshops, webinars, competitions, FDPs, awareness programs, Boot Camps, Regional meets, Entrepreneurship fest, etc.). 04 research scholars have successfully been awarded Ph.D. under her supervision. Currently, she is guiding 04 Ph.D. scholars & Guided more than 300 PG Level students (MCA, M.Sc. & M.Tech.). She has been appointed as a paper setter and an external examiner for the viva-voce exams of postgraduate students, Ph.D. Students of various Universities. Presently, she is the executive member of Rajasthan sub-section of IEEE, a member of ACM, and a life member of ISTE, CSI, Global Academy of Doctorates (GAD) & Global Forum for Sustainable Rural Development (GFSRD). She is on the Editorial board of Scopus and reputed journals. She has published four patents and is an editor of eight books.

Amit Jain is currently President of Amity University Rajasthan. Prior to this Prof. Amit has served the university as Pro-Vice Chancellor, Dean Faculty of Management, and Director, Amity Business School. Prof. Jain has held senior positions like Director-International Collaborations, Director-School of Hotel Management, Professor-In Charge Corporate Training and Consultancy with several organizations including Manipal University Jaipur, JK Lakshmipat University, etc. Prof. Amit has been a Visiting Professor with University of Technology (UTS), Sydney, Australia, Szechenyi Istvan University, Hungary & Selye Janos University, Slovakia. Prof. Amit holds a Ph.D. from Sardar Patel University and has completed FDP from IIM Ahmedabad. Prof. Amit is actively involved in administration, teaching, training, research and consultancy. He has been associated with various management and training institutes as a guest faculty such as Bank of Baroda Staff College, Staff College Union Bank, Sardar Patel University, NCHMCT, IICD, etc to name a few. Prof. Amit has Designed and conducted training programmes/ sessions for executives of various organizations including IDBI Bank, Bank of Baroda, BKT Tyres, JK Tyre, JK Paper, JK Lakshmi Cement, Claris Life Sciences, Veeda Clinical Research, Natraj Foods, Trimurti FoodTech, E- mall Infotech, Euro India Foods, Fenner India, Gravita India, ECGC, GAIL, AU Bank, Wonder Home Finance etc., on topics related to Marketing Orientation, Effective Selling Skills, Customer Relationship Management, Managerial Effectiveness, Leadership and Communication Skills. He has several publications to his credit and has presented research papers at National and International conferences organized by institutes like Singapore Management University (SMU), University of Technology, Sydney, IIM, and IIT. He has also received awards for three research papers at reputed Conferences. His professional associations include Life Membership of Indian Society for Training and Development (ISTD). He is a Member of All India Management Association and IIM Ahmedabad Alumni Association. Dr Amit is Reviewer and Committee Member with several Journals and Conferences. He has served as an Editorial Board Member of Indian Journal of Training and Development and Chief Editor of Amity Management Review. His area of interest for Research, Training and Consultancy includes Marketing Strategy, Sales Management, Marketing Communications, Digital Marketing and Entrepreneurship.

Chitresh Banerjee is a Science Graduate (B.Sc.), he holds a Master's Degree in Computer Science M.Tech. (CS), another Master's Degree in Computer Applications (MCA) with GNIIT (3 years) program from NIIT Ltd. He is Ph.D. CS from Pacific University, Udaipur, Rajasthan, India. He has rich academic experience of about 27+ years in the field of Computer Applications / Science / IT. He is currently working as Associate Professor, Amity Institute of Information Technology, Amity University, Jaipur. He has also worked as Executive Officer in the Board of Studies, The Institute of Chartered Accountants of India (Set up by an Act of Parliament), New Delhi. He is presently Vice President, Institution's Innovation Council, an initiative of MHRD, Govt. of India, Amity University Rajasthan. He was also a member of management committee of Jaipur Chapter of CSI (Computer Society of India) for the year 2011-12 & 2012-13. In addition, he is also member in 23 International Societies/ Associations. He has been awarded Best Researcher in Online Award Ceremony organized by Institute of Technical and Scientific Research, Jaipur, India on 13th Sep., 2020. He has an excellent academic background with very sound academic and research experience. Under the Institute-Industry linkage programme, he delivers expert lectures on varied themes pertaining to IT. As a prolific writer in the arena of Computer Sciences and Information Technology, he penned down number of books/learning material on Multimedia Systems, Information Technology, Software Engineering, E-banking Security Transactions, System Analysis and Design, Web Technologies, etc. He has contributed 78 research papers/chapters/articles in the conferences / journals / seminar / publications of international and national repute. He also provides consultancy in the area of software and project management to a no. of IT companies. He has acted as Editor in five International Journals and in two International Convention of Climate Change and Water (during 2012 & 2013). He has acted as organizing member in 18 National and International Conference. He is also Reviewer of 7 International Journals. He is Guest Editor in 15 International Journal and Books of Inderscience Publisher, IGI Global, Taylor and Francis (CRC Press), and Springer. His area of interest includes software security, software engineering, IoT, smart systems and e-systems.

***

Arpita Agarwal serves as an Assistant Professor in the Faculty of Management at the Centre for Distance and Online Education, Manipal University Jaipur. An accomplished academician, she holds a Ph.D. in Management and is a qualified member of Institute of Company Secretaries of India (ICSI). With over five years of experience in both industry and academia, Dr. Arpita has made significant contributions to the field of management studies. Her work includes the publication of multiple research papers and articles in esteemed national and international journals. Additionally, she has actively participated in and conducted numerous workshops and conferences at the international level.

Rachit Agarwal is working as an Assistant Professor at University School of Business, Chandigarh University, Mohali. He is having more than 8 years of work experience in teaching and research. He is UGC-NET qualified and also completed executive financial data analytics from IIM Kashipur. His areas of interest are financial analysis, investment behavior analysis and behavioral study. He has published various papers in national and international journals including Scopus and ABDC listed journals. He has also organized many workshops and Faculty Development Programme on Research Methodology and on different statistical tools.

Ramneek Ahluwalia is presently working as an Assistant Professor in University School of Business, Chandigarh University, Mohali. She has more than 5 years of teaching and industrial experience and has worked as a Human Resource Manager in one of the Multinational Companies. She holds a doctorate in Management in the domain of Human Resources and has published many papers in UGC Care as well as Scopus. She has presented various papers in international as well as national conferences. She is a keen researcher and has attended various workshops on Data Analysis.

Mukta Arora is a seasoned expert in management and gender studies, with over 36 years of professional and academic experience. She holds a Ph.D. in Management and an MBA specializing in Human Resource Development and Financial Management. Dr. Mukta Arora has held pivotal roles with international organizations like UN Women, the Asian Development Bank, and UNICEF, driving gender mainstreaming, policy formation, and capacity building initiatives. She has served as the Deputy General Manager of the Rajasthan Skill & Livelihoods Development Corporation and contributed to key government programs. In academia, Dr. Mukta Arora has over two decades of teaching experience, having served as an Associate Professor and Lecturer in Management. She is also an accomplished author, having co-authored the book Functional Management. Her areas of expertise include skill development, gender-based violence, human resource management, and strategic management.

Pranshuta Arora has graduated from Delhi University with a B.A. in Economics & Entrepreneurship. She has done her M.B.A. (Marketing) as a major and finance as minor specialisation from West Bengal University of Technology, Kolkata. She is pursuing her PhD from Bhartiya Skill Development University-BSDU, Jaipur. She has 10 years of experience in providing consultancy services in E- governance, bid process management, project planning, ICT along with an extensive experience in Documentation, Preparing reports, etc. in government sector. She has expertise in coordination, research, and has worked with various internal and external stakeholders, prepared frameworks and templates for monitoring and tracking progress of schemes and programs.

Meenakshi Bajpai, an Assistant Professor (stage 3) at the Department of Psychology, Arya Mahila P.G. College, admitted to the privileges of Banaras Hindu University, Varanasi, India, obtained her Master's Degree from Chhatrapati Shahuji Maharaj University, Kanpur, qualified UGC-NET followed by a doctorate for her work on "Role stress and Physical ailments among employed women" in 2009. She has been teaching psychology since last 20 years. Dr. Meenakshi has authored various chapters in books of psychology as well as social concern. She also published more than 35 research papers at various National and International repute journals. She is life member of various journals including National Academy of Psychology (NOAP) & Indian Academy of Health Psychology (IAHP). Her research interest includes Positive Psychology, Health Psychology, Developmental psychology, Cross Cultural Psychology and Applied Social Psychology. She is Co-ordinator & Member in various College committees. She served as Programme Officer in National Service Scheme (NSS), guided various students in research and dissertations. She is also serving as counselor at IGNOU. She is co-ordinator of various Diploma courses run by UGC, New Delhi. Besides Banaras Hindu University (BHU) she is examiner of various other Universities. Dr. Meenakshi has been selected to be part and parcel of COVID-19 help group and has been providing free counseling and psychological assistance to the public during COVID pandemic. She was called as resource person in various institutions and other social bodies. She was awarded by Uttar Pradesh government for the same. She was also awarded by various agencies as Inner wheel Club, Indian academy of Health Psychology, etc.

Durgesh Batra is currently professor in CDOE, Manipal University Jaipur with 20+ years of working in academics, Industry and government with 20+ published paper with 4 patents. Currently also associated with VIT Australia as adjunct

Professor. His research areas include analytics,, statistics, psychology, marketing and employment branding

Mahesh Deshpande is a Senior Principal Consultant with over 15 years of Program Management experience in leading cross-functional transformation initiatives across geographies to drive innovation and change. He specializes in Data Product Program Management and is known for his passion for data storytelling and design thinking. Mahesh is also the Vice President of PMI San Francisco Bay Area Chapter's Mentorship Program. He introduced novel innovative ideas to mentorship and fostered a strong mentor community. Under his leadership, the mentorship program grew to support 200 participants annually, became self-sustainable and transformed into a platform that supports any project professional seeking mentorship.

Vertika Goswami is an Assistant Professor at the Faculty of Management, Centre for Distance and Online Education, Manipal University Jaipur. With a career that began as a Financial Analyst at Zucol Services Pvt. Ltd., she later transitioned into academia, bringing over 7 years of combined corporate and academic experience. Her research areas include Financial Management, Behavioural Finance, Financial Analytics, Investor Sentiment Analysis, and Operations Management.

Monika Gwalani is an assistant professor at the Amity Institute of Behavioural and Allied Sciences, Amity University Rajasthan, with five years of teaching experience. She is passionate about research, having published 16 articles and book chapters in Scopus-indexed, UGC Care-listed, and peer-reviewed journals. Dr. Gwalani actively participates in national and international conferences, presenting her work. Notably, she supervised her first PhD candidate, awarded in 2024.

Saad Ullah Khan has more than a decade long extensive experience at university level teaching and research. Dr. Khan has authored various research papers in journals of national as well as international repute. He was awarded 'Rajat Jayanti Vigyan Sancharak Fellowship-2009' by DST, Ministry of Science and Technology (Government of India) for his contribution to the field of communication and its social implications. Further he was bestowed with MANF Fellowship 2011 by UGC to carry out his research work. A PhD from prestigious Aligarh Muslim University, Dr. Saad Ullah Khan has presented 90 Research Papers in various National and International Conferences. He was awarded 'Best Researcher Award' twice (2020 & 2023) and 'Best Journalism Teacher Award' in 2021.

Early Ridho Kismawadi, S.E.I, MA, is a lecturer at the Department of Islamic Banking, Faculty of Islamic Economics and Business IAIN Langsa, Aceh, Indonesia, he has been a lecturer since 2013, he has completed a doctoral program in 2018 majoring in Sharia economics at the State Islamic University of North Sumatra. He was appointed head of the Islamic economics Law study program (2023) Islamic banking study program (2020) and Islamic financial management study program (2019) at Langsa State Islamic Institute (IAIN Langsa), Aceh, Indonesia. His research interests include financial economics, applied econometrics, Islamic economics, banking, and finance. He has published articles in national and international journals. In addition, he is also a reviewer of several reputable international journals such as Finance Research Letters, Financial Innovation, Cogent Business &; Management, Journal of Islamic Accounting and Business Research. He has also presented his papers at various local and international seminars.

Ajay Kumar is a Professor at Chanakya National Law University. He also serves as the Dean, Academic Affairs & Law; and Head, Centre for Post-Graduate Legal Studies at Chanakya National Law University, Patna.

Tanya Kumar is currently working at Chandigarh Group of Colleges, Landran Mohali. She has experience in teaching and research. She has completed her Ph.D. in commerce. Her areas of interest are finance and marketing. She has also presented various papers in national and international conferences. She provides supervision to students in financial analysis, blockchain technology, artificial intelligence and cybersecurity and fintech. She has guided more than 90 master's students to complete their dissertation. She has published papers in national and international journals, including Scopus, Web of Sciences and ABDC-listed journals.

Monica Kunte is an Assistant Professor at SCMHRD. She specializes in qualitative techniques and her areas of interest are Talent Management and Talent Acquisition.

Pallavi Mishra is an accomplished academic and media professional, currently serving as an Associate Professor at the Amity School of Communication, Amity University Rajasthan with a passion for communication and education. A seasoned freelance writer, Dr. Mishra has been contributing articles and opinion pieces to various newspapers and magazines since 2013, covering a wide array of topics from current affairs to social issues. Dr. Mishra has been recognized for her exceptional contributions to journalism, literature, and academia with numerous awards. She has received numerous accolades including Third Prize in the prestigious International Reportage Writing Competition by the World Hindi Secretariat, Mauritius. Additionally, she represented India at the International Hindi Literature Meet in Dubai, presenting a self-composed poem and being honored by Shabd Yatra in Abu Dhabi on the 75th Republic Day. In education she received the ASIAN EDUCATION

AWARD 2021 for excellence in student development and innovation in teaching. Her research prowess earned them the Young Achiever Award 2021, along with two Best Paper Awards at international conferences. Additionally, she is a NET/ JRF holder. Dr. Mishra has previously held positions as an Assistant Professor at Banasthali Vidyapith and University of Lucknow, where she contributed to the academic and professional development of students in media and communication studies. Additionally, she has served as a Sub-editor at Hindustan Times, gaining editorial experience that complements her academic expertise. Dr. Mishra holds a Ph.D. from the University of Lucknow, with a research focus on the impact of social media on mainstream journalism. In addition to her academic and research roles, Dr. Mishra serves as the Editor-in-Chief of Amity University Rajasthan's newsletter and magazine, and as Media Coordinator for the university's media team. Her multifaceted expertise and contributions to education, media, and society make her a distinguished professional in her field.

Shweta Misra is a Senior Lecturer at IHM Pusa, New Delhi since 2010. In her present role, she is taking classes for Accommodation Operations (Rooms Division), Food Costing and Management Subjects. She has also been working as an Editor for the Pusa Journal of Hospitality and Applied Sciences since the inception of the journal, which is going to publish its 11th Volume this year. Dr. Misra has guided the students of Ph. D., M.B.B.S., M. Sc., and B. Sc. in their research work. She has more than 20 publications and has organised various conferences and seminars. Dr. Misra has also contributed in the development of National Guidelines for Engagement of Housekeeping Services in Government Hospitals through Outsourcing.

Tanushri Mukherjee is a top-performing, self-motivated and result-oriented teaching professional with a rich teaching experience of 19.2 years. Highly experienced in teaching Undergraduate as well as Post Graduate Students of Journalism and Mass Communication and students of Business School (BBA & Executive MBA) specialized subjects like Public Relations, Corporate Communication and Event Management, Design Thinking & Innovation. Has done Ph.D in Public Relations from Amity University Rajasthan in the field of Mass Communication in the Faculty of Humanities and Social Sciences in 2016. She is also UGC NET Qualified. As a mark of academic excellence, she has bagged many awards of high repute like featured in the July, 2021 edition of the renowned " IMPACT" Magazine under " Top 50 Women Leaders in PR and Corporate Communication, "Communication Excellence Honor" by Public Relations Society of India, "Excellence in Education Award" by the Hon'ble Education Minister of Rajasthan, Shree B.D.Kalla ji in the 12th Principals & Teachers Award-23 organized by Simply Jaipur Magazine group, Thar Sarvoday Sansthan, Raghu Sinha Mala Mathur Charity Trust and Technology

Partner, Imagine on 2nd Sept, 2023, Distinguished Associate Professor Award, Educator of the Year Award by Exchange4media Group, Distinguished Teacher for Holistic Development of Students Award by International Centre for Excellence in Education (INCEED) and SPACE (India) {Society for Perpetuation of Art, Culture & Education), Teaching Excellence Award in "PR and Communication" from Public Relations Society of India, Jaipur Chapter, Media Innovation Award on the occasion of All India Media Conference, Women's Excellence Award, Empowered Women Award, Certificate of Excellence in EET CRS 2ND Faculty Branding Awards" and has also bagged Medal, Citation & Certificate of Merit in the "Gold Medallion National Award for the Best PR Case Study by PRSI, Hyderabad Chapter. Presently she is Deputy Director Outcome, Amity University Rajasthan and Former Presiding Officer of Amity Website-Standing Review Committee. She is responsible for handling various key academic and administrative tasks of the department like, NAAC Coordinator, IQAC Coordinator, Placement Coordinator, Alumni Affairs Coordinator, Ph.D Coordinator, Exam Coordinator etc. As a mark of Academic Excellence, she has been the Chief Coordinator of Curriculum and Syllabus making/revision of all Professional Programmes of Amity School of Communication. Has presented quality Research Papers at various National & International Conferences. Has been Session Chair, Guest of Honor & Panelist in various National and International Conferences and has a good number of Research Papers published in various reputed National and International Journals along with Publications in Book of Conference Proceedings, Book Chapters, Souvenir, Newsletters etc to her credit. As a mark of excellence in the field of research, she is also a Member of the Editorial Board as well as Reviewer of Eminent Journals. She is also in the panel of External Examiner for judging Practical Projects of leading Jaipur based Universities and is also an eminent Paper Setter & Examiner for various courses of her specialization area in reputed Universities. She, besides being convener of Board of Studies (Film & Animation) of Amity School of Communication, has also been an eminent member of Board of Studies in Public Relations of reputed Universities. She is an active member of various Professional Bodies like Lifetime Member of Public Relations Society of India, Life-Time Member of Indian Society for Training and Development (ISTD), Jaipur.

Preeti Nagar is an Assistant Professor at the Centre for Distance and Online Education, She has experience working in industry and academia for 6 years. Her academic interests include teaching courses on human resource practices, organisational behaviour, and Business Analytics, Artificial Intelligence & Fintech.

Timi Olubiyi is an Entrepreneurship and Small Business Management expert with a Ph.D in Business Administration. A Babcock University Business School

Affiliated Scholar and senior government official. He is a prolific investment coach, business engineer, Chartered Member of the Chartered Institute for Securities & Investment (CISI), and a financial literacy specialist.

Garima Pancholi is currently pursuing a PhD from Amity Business School, Amity University Rajasthan. She has also qualified the UGC-NET exam twice. She has completed her Graduation and Post-Graduation from Mohanlal Sukhadia University, specializing in Accountancy and Business Statistics as well as Business Administration. Her areas of focus in her research involve exploring topics such as Organizational Behavior, Organizational Resilience, Artificial Intelligence, Metaverse and Sustainability.

Revti Rani Roy, former assistant professor at Manipal Law School, (MAHE) Bengaluru and currently a Ph.D. scholar at CNLU, completed her B.A. LL.B. (Hons.) in the year 2021 from Chanakya National Law University, Patna. In the year 2022, she completed her LL.M. from Chanakya National Law University, Patna itself. She has secured a Gold Medal in her LL.M. Programme (CNLU). Ms. Revti has also qualified for UGC- NET in the year 2022 and is a JRF Holder.

Mani Sachdev's research interests include theories of epistemic investigations, philosophical study and review of music and other performing arts and their role in social reforms, the debate between monism and pluralism in the traditional philosophical systems, the relationship between performing arts & human behavior and the nature of metaphors. Designing and developing curricula for students taking up innovative approaches, delivering a range of programs of teaching for students, ensuring courses that fulfil the quality assurance framework of the University, assessing research proposals & dissertations. Besides developing, implementing and coordinating University research strategy, supervising students' projects and field trips wherever necessary, she is interested in developing the ability of students to engage in critical discourses and rational thinking, promoting and developing team spirit and team coherence, ensuring teaching design and methods in humanities. She has also been involved with teaching ethics to medical fraternity in a recognised medical university in Jaipur and is a distinguished member of their ethics committee. She has presented many research papers in International and National Conferences & Seminars and written Book Chapters, Book Reviews in several Journals and Magazines. She has also produced, directed and written the corporate films for Hospitals and other Organizations and has written Scripts for various documentaries of Doordarshan for national telecast. She has interviewed over 50 globally renowned celebrities including the Governors, Philosophers, academicians and important film celebrities & personalities.

Abhineet Saxena is an incisive professional with 12 years of experience spanning diversified areas such as Corporate Training, Research, and Teaching. He received an MBA degree in Finance and Marketing from Rajasthan technical university, in 2010 and a Ph.D. degree in Microfinance from the University of Rajasthan in 2016. He is NET qualified and currently working as an Assistant professor (AP-3) at the Amity Business School, Amity University Rajasthan. Currently, he is also in charge of Yunus social business Centre – AUR Jaipur. His current research interests include Banking and Finance, corporate finance, Impact assessment of social security schemes, and Holistic rural development. He also received a post-Doctorate fellowship and award from ICSSR New Delhi in the year 2017. He is an experienced professional and has been guiding individuals and Research scholars in eminent educational Institutions. He has worked in various capacities such as Research Project Principal investigator, Educational Consultant, Data analytics Trainer, Examiner, etc., and has honed his professional training and management skills working with globally reputed organizations.

Antima Sharma is currently working as an Assistant Professor, Senior Scale at the Centre for Distance and Online Education, Manipal University, Jaipur. She holds a PhD, a bachelor's degree in commerce, and a master's degree in finance and human resource management. Dr. Sharma, having over ten years of combined industry and academic experience, worked with multinational corporations, prestigious institutions, and universities. Her areas of expertise are Human Resource Management, Organizational Behaviour, Basic Trades, Commerce, and Accounting. Dr. Sharma is actively engaged in conducting research and publishing high-quality papers in esteemed journals. Her research encompasses employee learning, employee well-being, the impact of AI on the workforce, and employer branding. Moreover, she has successfully managed international conferences, workshops, and seminars.

Monika Sharma, a Ph.D. scholar at Amity Business School, Amity University Rajasthan, holds degrees in Economic and financial management and Business Administration from Rajasthan University. Her research area is Robotic Process Automation, Artificial intelligence.

Suyesha Singh is presently working as an Assistant Professor (Senior) at Manipal University Jaipur, Rajasthan, India. She has experience of 14 years in teaching, research, and life skills training. Her research interests include Clinical Psychology and Applied Psychology. She has published around 30 research papers in national and international journals of repute (SCOPUS, WOS, UGC-CARE, ABDC indexed), including 3 edited books. She is a Certified Counsellor and an Art Analyst.

Jai Sonker is having 21+ years of multi-dimensional experience. He has faced many practical, technical, and professional challenges. He was able to deal with and overcome them with zeal, confidence, and assurance of positive results. From being an Executive Housekeeper in hotels to a program co-coordinator in academics, has given him a wide range of opportunities, responsibilities & duties and with it learning and experience. His range of knowledge varies from skill development to management and academic administration. The variety of professional knowledge which includes Facility Management, Hostel Management, Vendor Management, Negotiation skills and Academic Administration gives a complete strength to his professional curriculum. He has done several publications which are inclusive of Scopus journal, ABDC Listed and UGC index journal. A part of that he is author of 2 books on sustainable hotel operation.

Neharshi Srivastava is presently working as an Assistant Professor, Program & Ph.D. Coordinator Amity University Rajasthan. She is a Gold Medalist of Lucknow University and RCI licensed Rehabilitation Psychologist. Dr. Srivastava has 12 years of Counseling, research and Industry experience. She is trained CBT therapist from London. Her area of expertise are Mindfulness and Positive & Yogic Psychology based Intervention. She has more than 40 publications in international journals of Scopus/ Web of Science/ UGC recognized/UGC Care Listed/Peer review and presented more than 45 papers in National and International Conferences such as South Korea, Malaysia, Prague Czech Republic, contributed 10 Chapters in edited books and authored 8 books on different domains of Psychology. She Received "Best Young Faculty Award", "Dhruv Ratan Award" & "Jan Shakti Sema Samman" for outstanding contribution in the field of Psychology. She has been invited as a Guest Speaker and Resource Person in various reputed national/International Universities News Channels and Organizations like Vigyan Prasar, Department of Science and Technology, NISD, Narcotics Control Bureau, Government of India. She supervised more than 25 Masters and bachelor's Psychology Students 2 Ph.D. awards under her supervision and presently 4 Ph.D. Scholars doing their Ph.D. under her guidance. She is also a Core Committee Member of Indian Academy of Health Psychology. Apart from her academics she has around 14 years' experience in the field of electronic Media. She is Radio Jockey of AIR FM Rainbow 100.7 Mhz. Lucknow.

Swapnesh Taterh is currently serving as Professor & Head, Amity Institute of Information Technology, Amity University Rajasthan, Jaipur, India. Experienced Researcher with 24 years of rich experience in teaching and research. In an academic career spanning over 24 years, he had served with various responsibilities Departmental Coordinator, Program Coordinator, Library Faculty In-charge, Mentor and Member of various committees like – Research Committee, Board of Studies,

Academic Council, Examination Committee, Discipline Committee, Technical Program Committee, and others. He has served as Expert Panel Member in RPSC. Rajasthan. As an ingenious researcher, had presented several research papers at national and international conferences and published more than 65 research papers in international journals. His doctoral thesis work focuses on software security. He holds two master's degrees in computer science and a bachelor's degree in science. His current and prospective research & academic interests include Cryptography, Information Security, Data Analytics, and Open-Source software. Received Grant from SERB as a Principal Investigator. As a research supervisor, Seven research scholar has been awarded with Ph.D. degree under his supervision. He has published Patents and research paper in reputed Indexed journals with Impact Factor 13.473, ranking it 2 out of 162 in Computer Science Journal, SCIE, 2021. Besides these he also participated in UGC sponsored Orientation & Refresher Course, Awarded with A+ Grade and associates as a Chief Editor, Guest Editor, and Reviewer of International and National Journal of repute including IGI and InderScience. Dedicated to showing keen interest, determination, and confidence to execute all responsibilities rested upon me in professional teaching, research and administrative. Respond positively and effectively in demanding situations too. He is the pioneer to take initiative and started the post graduate program in Cyber Security and Data Science at the university level.

Anadi Trikha is currently working as an Assistant Professor and Course Coordinator at the Centre for Distance and Online Education, Manipal University Jaipur. Her area of specialization is Marketing. She has over 8 years of experience in academics and teaches subjects like Consumer Behaviour, Digital Marketing, Sales, and Brand Management.

Index